RUTLEY'S ELEMENTS OF
MINERALOGY

D1381723

CBS GEOLOGICAL SCIENCE TEXTS

RUTLEY'S ELEMENTS OF MINERALOGY

27th Edition

Revised by

C. D. GRIBBLE

Department of Geology, University of Glasgow

CBS

CBS Publishers & Distributors Pvt. Ltd.

New Delhi • Bengaluru • Chennai • Kochi • Kolkata • Mumbai
Hyderabad • Nagpur • Patna • Pune • Vijayawada

ISBN: 81-239-0916-0

First Indian Reprint: 1991
Reprint: 2001, 2002, 2004, 2005

Published by **Satish Kumar Jain** and produced by **Varun Jain** for
CBS Publishers & Distributors Pvt. Ltd.,
4819/XI Prahlad Street, 24 Ansari Road, Daryaganj, New Delhi - 110002
delhi@cbspd.com, cbspubs@airtelmail.in • www.cbspd.com
Ph.: 23289259, 23266861, 23266867 • Fax: 011-23243014

Corporate Office: 204 FIE, Industrial Area, Patparganj, Delhi - 110 092
Ph: 49344934 • Fax: 011-49344935
E-mail: publishing@cbspd.com • publicity@cbspd.com

Branches:
• *Bengaluru:* 2975, 17th Cross, K.R. Road, Bansankari 2nd Stage,
  Bengaluru - 70 • Ph: +91-80-26771678/79 • Fax: +91-80-26771680
  E-mail: cbsbng@gmail.com, bangalore@cbspd.com
• *Chennai:* No. 7, Subbaraya Street, Shenoy Nagar, Chennai - 600030
  Ph: +91-44-26681266, 26680620 • Fax: +91-44-42032115
  E-mail: chennai@cbspd.com
• *Kochi:* Ashana House, 39/1904, A.M. Thomas Road, Valanjambalam,
  Ernakulum, Kochi • Ph: +91-484-4059061-65
  Fax: +91-484-4059065 • E-mail: cochin@cbspd.com
• *Kolkata:* 6-B, Ground Floor, Rameshwar Shaw Road, Kolkata - 700014
  Ph: +91-33-22891126/7/8 • E-mail: kolkata@cbspd.com
• *Mumbai:* 83-C, Dr. E. Moses Road, Worli, Mumbai - 400018
  Ph: +91-9833017933, 022-24902340/41 • E-mail: mumbai@cbspd.com

Representatives:

• Hyderabad: 9885175004 • Nagpur: 9021734563 • Patna: 9334159340
• Pune: 9623451994   • Vijayawada: 9000660880

Printed at:
Neekunj Print Process, Delhi

Preface

Rutley's elements of mineralogy has been around for a long time, certainly throughout my own lifetime; and if my great grandfather had read geology, it would have been prescribed reading for him too! It has been rewritten and revised frequently since first conceived by Frank Rutley in the late 19th century. Major revisions occurred in 1902, and then in 1914, when H. H. Read first took over the authorship, and thereafter in 1936 and in 1965 when the last major changes occurred.

It was with some trepidation that I agreed to attempt this revision. I had been asked to do it by Janet Watson in 1979, but various commitments delayed my start on it until 1984.

This 27th edition encompasses a number of changes. Chapters 1–5 have the same headings as before, but considerable changes have been made in all of them, particularly 1, 3, 4 and 5. Comments sought prior to the revision revealed considerable disagreement about the role of blowpipe analyses in the book. I have only once had blowpipe analyses demonstrated to me, and have never used them; but there is no doubt that they *are* employed in many countries, and many of the tests (flame colour, bead, etc.) are still useful as rapid indicators of which element is present in a mineral. I have therefore kept blowpipe analysis information in *Rutley*, but have relegated it to an appendix. None the less, I have retained the small paragraph on [blowpipe] tests for each mineral where it is appropriate.

After Chapter 6, which deals with the layout of Chapters 7, 8 and 9, Chapter 7 discusses the periodic table and elemental groupings, and lists the various minerals which contain a particular element; but detailed information on each mineral is now found either in Chapter 8, where the non-silicates are described, or in Chapter 9, where the silicate minerals are described. In Chapter 8, the non-silicate minerals have been subdivided on the basis of the anion groups present, using the scheme adopted by the American mineralogist, J. D. Dana. The silicate minerals, in Chapter 9, have been subdivided on the basis of their crystal chemistry, using the scheme employed by the British mineralogists W. A. Deer, R. A. Howie and J. Zussman. In Chapters 8 and 9, the physical properties and optical properties (where appropriate) are given for each mineral, with details on their occurrence, as well as details on their composition, [blowpipe] tests and varieties.

The referees were most helpful, and their comments have unquestionably improved this new edition, and I would wish to thank all those involved. The guidance of Roger Jones of Unwin Hyman has been invaluable, and I am most grateful to Dorothy Rae, of the Geology Department of the University of Glasgow, for typing Chapters 6, 7 and 8. The university and the publisher provided me with a word-processor, thus ensuring that the text would be prepared by myself at least one year sooner than I had thought possible. I would like finally to thank Professor Bernard Leake and my colleagues at Glasgow for their support over this long period of time, and also members of the technical staff at Glasgow for the help I received from them at various times.

I have been encouraged by comments from many friends and colleagues that I have managed to retain much of the 'flavour' of *Rutley* and, if so, I am extremely pleased that my *Rutley* will remain one of the family, as it were. The previous editions of *Rutley's elements of mineralogy* have certainly guided me in producing this one but, of course, any mistakes and inaccuracies present in the text are mine.

Colin D. Gribble
Glasgow

Contents

List of Tables

1

The chemistry of minerals

1.1 States of matter

Matter may exist in three states: **solid, liquid** and **gaseous**. A **mineral** is a naturally occurring inorganic substance which has a definite chemical composition, and which commonly, but not always, is crystalline. Most minerals are solid, but some 'minerals' such as native mercury are normally liquid in their natural state, and others, such as natural gas, are gaseous. Gases and liquids are termed **fluids**; that is, they flow, unlike solids, under the action of gravity at atmospheric temperature (t) and pressure (p); solids *may flow* under the influence of gravity but at *higher t* and *p*. A gas will entirely fill the space containing it, whereas a liquid may not, but may be bounded by an upper, horizontal surface. Most pure substances can exist in all three states depending upon the combination of temperature *and* pressure acting on the mineral. At specific temperatures, called **melting points**, many minerals melt to form liquids, although some may actually be *decomposed* by the heat before reaching their melting points. A **sublimate** is formed by the direct condensation of a gas into a solid.

1.2 Elements, compounds and mixtures

A **pure substance** is one that possesses characteristic and invariable properties. Pure substances may be of two types; **elements** and **compounds**.

Elements are substances which have not as yet been split up into simpler substances by any ordinary *chemical* means. Over 100 elements are known at the present time, but many of these are either rare or unstable, and of little importance to the mineralogist. The estimated composition of each of the three layers of the Earth, for the most important elements (in weight %) is as shown in Table 1.1.

From Table 1.1 it can be seen that just over 99% of the Earth's crust is composed of just eight elements which, in descending order, are oxygen (O), silicon (Si), aluminium (Al), iron (Fe), calcium (Ca), magnesium (Mg), sodium (Na) and potassium (K). Most of the economic elements are missing from the above list.

Table 1.1 The composition of the three main layers of the Earth.

Elements	Crust	Mantle	Core
oxygen	45.6	36.4	
silicon	27.3	28.8	
aluminium	8.36	1.9	
iron	6.22	6.57	86.3
calcium	4.66	2.2	
magnesium	2.76	22.6	
sodium	2.27	0.42	
potassium	1.84	0.1	
titanium	0.63	0.4	

The amounts of important trace elements are given below in parts per million (10 000 ppm = 1%)

hydrogen	1520		
phosphorus	1120	300	
manganese	1060	1000	
fluorine	544		
barium	390		
strontium	384		
sulphur	340		5.96%

All others are below 200 ppm, but nickel and cobalt are given:

nickel	99	1600	7.28%
cobalt	29	100	0.40%

Compounds are pure substances made up of two or more elements. They are formed as a result of *chemical change*, and are different from mere **mixtures** in the following ways:

(1) The elements constituting a compound are combined in definite proportions by weight.
(2) A compound cannot easily be split up, whereas the components of a mixture can usually be separated by mechanical means. Such components may themselves be either elements or compounds.
(3) The properties of a compound are often very different from those of the elements it contains, whereas a mixture usually possesses the properties of its constituents.
(4) Heat is either given out or absorbed when a compound is formed. This may not occur when substances are mixed.

Minerals are compounds of their constituent elements, whereas **rocks**

are mixtures of their component minerals. Thus, the mineral quartz, SiO_2, is a compound of the elements silicon (Si) and oxygen (O), whereas the rock granite is a mixture of several minerals, one of which is quartz.

1.3 Atoms and molecules

An **atom** is the smallest part of an element that can enter into chemical combination with another element. The atom of a particular element is represented by an abbreviation, or **symbol**, which is usually the first letter, or the first two letters, of the *English* or *Latin* name of the element. The elements, their symbols, and their atomic weights and atomic numbers are given in Table 1.2. Thus O is the symbol for an atom of oxygen, but O_2 is the **formula** for a **molecule** of oxygen. Similarly, CO_2 represents the formula for one molecule of the gas carbon dioxide. In the gaseous state, the molecules of a substance are widely separated from each other, and are in a state of rapid, random motion. These molecules may consist of single atom, as with the gas helium (He), or of two or more atoms of the same element, as a hydrogen (H_2), or oxygen (O_2 – as given above) or, in the case of compounds, of two or more atoms of different elements; for example, steam (H_2O) and carbon monoxide (CO).

When a gas condenses to a liquid, the molecules are no longer separated in space but come together. When the liquid solidifies, the atoms are arranged in a fairly rigid pattern, and it is no longer possible to segregate any one group of atoms from the rest. The term 'molecule' can be described as the smallest possible particle of a compound; that is, it still contains the basic combinations of atoms. However, solid or liquid compounds are also represented by formulae. Thus the mineral calcite has the formula $CaCO_3$; which means that calcite is composed of the elements calcium (Ca), carbon (C) and oxygen (O), in the proportions of one atom of calcium, one atom of carbon and three atoms of oxygen. A molecule of calcite represents the smallest possible particle of the mineral which still contains the basic combinations of atoms.

Although a large number of **sub-atomic particles** are known to exist, the atom can be regarded as consisting of three sub-atomic particles, namely **protons, neutrons** and **electrons**. The proton carries a positive charge, the neutron has no charge, and the electron carries a negative charge. The proton and neutron have the same mass (or weight), which is taken as the unit of mass. Protons and neutrons together form the **nucleus** of an atom, except for the lightest element, hydrogen, in which the nucleus consists of a single proton. The nucleus is surrounded by electrons, which are equal in

Table 1.2 Atomic weights of the elements.

Element	Symbol	Atomic no.	Atomic weight	Element	Symbol	Atomic no.	Atomic weight
aluminium	Al	13	26.98154	neodymium	Nd	60	144.24
antimony	Sb	51	121.7	neon	Ne	10	20.179
argon	Ar	18	39.948	neptunium	Np	93	237.0482
arsenic	As	33	74.9216	nickel	Ni	28	58.70
barium	Ba	56	137.33	niobium	Nb	41	92.9064
beryllium	Be	4	9.01218	nitrogen	N	7	14.0067
bismuth	Bi	83	208.9804	osmium	Os	76	190.2
boron	B	5	10.81	oxygen	O	8	15.9994
bromine	Br	35	79.904	palladium	Pd	46	106.4
cadmium	Cd	48	112.41	phosphorus	P	15	30.97376
calcium	Ca	20	40.08	platinum	Pt	78	195.09
carbon	C	6	12.011	potassium	K	19	30.0983
cerium	Ce	58	140.12	praseodymium	Pr	59	140.0977
caesium	Cs	55	132.9054	protactinium	Pa	91	231.0395
chlorine	Cl	17	35.453	radium	Ra	88	226.0254
chromium	Cr	24	51.996	rhenium	Re	75	186.207
cobalt	Co	27	58.9332	rhodium	Rh	45	102.9055
copper	Cu	29	63.546	rubidium	Rb	37	85.4678
dysprosium	Dy	66	162.50	ruthenium	Ru	44	101.07
erbium	Er	68	167.26	samarium	Sm	62	150.4
europium	Eu	63	151.96	scandium	Sc	21	44.9559
fluorine	F	9	18.9984	selenium	Se	34	78.96
gadolinium	Gd	64	157.25	silicon	Si	14	28.0855
gallium	Ga	31	69.72	silver	Ag	47	107.868
germanium	Ge	32	72.59	sodium	Na	11	22.98977
gold	Au	79	196.9665	strontium	Sr	38	87.62
hafnium	Hf	72	178.49	sulphur	S	16	32.06

Table 1.2 *(Continued)*

helium	He	2	4.00260		tantalum	Ta	73	180.9479
holmium	Ho	67	164.9304		tellurium	Te	52	127.60
hydrogen	H	1	1.0079		terbium	Tb	65	158.9254
indium	In	49	114.82		thallium	Tl	81	204.37
iodine	I	53	126.9045		thorium	Th	90	232.0381
iridium	Ir	77	192.22		thulium	Tm	69	168.9342
iron	Fe	26	55.847		tin	Sn	50	118.69
krypton	Kr	36	83.80		titanium	Ti	22	47.90
lanthanum	La	57	138.906		tungsten	W	74	183.85
lead	Pb	82	207.2		uranium	U	92	238.029
lithium	Li	3	6.941		vanadium	V	23	50.9414
lutetium	Lu	71	174.97		xenon	Xe	54	131.30
magnesium	Mg	12	24.305		ytterbium	Yb	70	173.04
manganese	Mn	25	54.9380		yttrium	Y	39	88.9059
mercury	Hg	80	200.59		zinc	Zn	30	65.38
molybdenum	Mo	42	95.94		zirconium	Zr	40	91.22

number to the total number of protons in the nucleus; thus the total negative charge on the electrons equals the total positive charge on the protons, so that the whole atom is electrically balanced. The mass of an electron is about 1/1850 of the mass of a proton (or neutron), so that almost the whole mass of an atom is concentrated in the nucleus. In summary therefore:

nucleus	{ proton	1 positive charge
	neutron	no charge
electron		1 negative charge

In the Rutherford–Bohr theory of atomic structure, the electrons are pictured as revolving in fixed orbits around the nucleus, rather in the way that the planets revolve around the Sun. Although this picture is nowadays far too simple and inaccurate (with mathematical functions that express the probability of finding electrons in specific places), the Rutherford–Bohr picture is, nevertheless, a suitable model for dealing with the majority of elements dealt with in this elementary book.

1.4 Atomic number, valency and atomic weight

The number of electrons present in the atoms of elements varies from 1 in hydrogen, to 103 in lawrentium (Lr), one of the trans-uranic elements. Uranium, the heaviest naturally occurring element, contains 92 protons. This number is known as the **atomic number**, and is denoted by the letter Z. The number of electrons around a nucleus in a neutral atom also equals the atomic number, since the number of electrons equals the number of protons in an element. Table 1.2 lists the elements and gives their atomic numbers.

There are seven shells which may contain electrons, concentrically disposed about the nucleus, and these are shown below, with the maximum number of electrons in each shell also given:

shells	K	L	M	N	O	P	Q
electrons	2	8	18	32	50	72	98

Shells O, P and Q are not fully occupied in any known element. The simplest atomic structure is that of hydrogen where $Z = 1$, with a single proton occupying the nucleus and a single electron occupying the K shell. In silicon ($Z = 14$), the nucleus contains 14 protons (and 14 neutrons), and is surrounded by 14 electrons, with the **K** and **L** shells fully occupied, thus leaving four electrons in the **M** shell. The **M** shell is not fully occupied here, and this is the case in the outer shell of all elements except for the inert gases helium ($Z = 2$), neon ($Z = 10$), krypton ($Z = 36$; made up of $2 + 8 + 18 + 8$), xenon ($Z = 54$; made up of $2 + 8 + 18 + 18 + 8$) and radon ($Z = 86$; made up of $2 + 8 + 18 + 32 + 18 + 8$); in which the electron configurations are extremely stable, with all the inert gases having their outermost shells *filled* with eight electrons (except for helium, which has its **K** shell completely filled).

This stability of the inert gases and their atomic configuration underlies the theory of **valency**. The atoms of other elements try to attain the configuration of these inert gases and, to do this, must add electrons to, or lose electrons from, the outer shell. These electrons are the **valency electrons**; if there are few electrons, the outer shell can lose them, whereas if the outer shell is nearly complete, electrons can be gained to fill the gaps and attain a stable configuration. For example, an atom of sodium has an atomic number of 11 ($Z = 11$), and its electronic configuration must be $2 + 8 + 1$; that is, with one electron in the outermost shell. Thus sodium has a **valency** of *one* (it needs to lose the outermost electron to attain the configuration of neon, $Z = 10$). Sodium, and other similar elements, are **metals**, and it can be stated that *metals are elements which can be ionized by electron loss*.

There are also a number of elements which have a few electrons *less* than the stable configuration. For example fluorine, with $Z = 9$, requires only one electron to attain stability. Fluorine, oxygen and other similar elements are **non-metals**, and it can be stated that *non-metals are elements which can be ionized by electron gain*.

An inert gas possesses *no* valency. The nucleus of an atom does not change with gain or loss of valency electrons, so that the elements retain their essential properties. The hydrogen atom is unique in that it can gain or lose one electron to achieve a stable configuration, that is, it behaves both like a metal and a non-metal. From this it can be stated that *the valency of an element is measured by the number of hydrogen atoms that can combine with, or can replace, one atom of the element*.

An atom which has lost or gained electrons in its outer shell is no longer neutral; that is, its protons are no longer equal in number to its electrons, and such an atom is called an **ion**. If the atom loses electrons, the ion is *positively* charged, since the number of positive protons in the nucleus exceeds the number of negative electrons; thus **sodium**, which loses an electron, is a positive ion and is denoted by **Na$^+$**. Positively charged ions are those of metals, and are called **cations**. The number of charges that the cation carries is called the **electrovalency**, and this can be regarded as the valency that the metal possesses. The common cations in minerals include Si^{4+}, Al^{3+}, Fe^{2+} and Fe^{3+} (originally called ferrous and ferric iron, but now more simply iron (II) and iron (III)), Mg^{2+}, Ca^{2+}, K^+, etc. Non-metals, on the other hand, *gain* electrons, and thus acquire additional *negative* charges since there are more negative electrons than positive protons; these negatively charged ions are called **anions**, and include O^{2-}, F^- and Cl^-.

The **atomic weight** of an element is the weight of an atom of the element compared with the weight of an atom of hydrogen. The weight of an atom of hydrogen is not significantly different from that of the single proton making up its nucleus, since the weight of a single electron is only in the order of 1/1850 of the weight of a proton (or neutron). Therefore the atomic weight of helium (2 protons, 2 neutrons and 2 electrons) will be just about four times that of hydrogen (since protons and neutrons are of equal weight). In fact, the atomic weight of hydrogen is 1.0079 and that of helium is 4.00260.

The atomic weights of all elements are given in Table 1.2: it is obvious from inspection of this table that none of the atomic weights are whole numbers. The reason for this is that a particular element may be represented by several atoms, each having a different atomic weight, due to the nuclei of each atom having different numbers of neutrons. If iron ($Z = 26$) is taken as an example, it is represented by four atoms, of atomic weights 54, 56, 57 and 58. Each atom has an atomic number of 26; that is, each contains 26 electrons and a similar number of protons, and thus each atom possesses all the properties of iron. Such atoms of an element are called **isotopes**, and the atomic weight of an element is determined by the proportions of each isotope that are present in a sample. For most elements the relative abundances of the different isotopes remains virtually constant in all samples. The standard atomic weight in Table 1.2 is taken as that of one of the isotopes of carbon; namely ^{12}C, which is equal to 12.000.

The **molecular weight** of a substance is the sum of the atomic weights of the atoms composing a molecule of the substance. In the case of solids, the **formula-weight** is a convenient quantity, and is the sum of the weights of the atoms making up the formula of the compound. Nowadays this just tends to be called the atomic weight of the compound. Thus the formula-weight of calcite $CaCO_3$) is $40.08 + 12.011 + (3 \times 15.9994) = 100.0892$. The elements which comprise minerals are conventionally given as oxides, and the atomic weights (formula-weights) of the most common of these are given in Table 1.3.

Table 1.3 Atomic weights of common elemental oxides in rocks.

Oxides	Atomic weights (formula-weights)	To two decimal places
SiO_2	60.0843	60.08
TiO_2	79.8988	79.90
Al_2O_3	101.96128	101.96
Fe_2O_3	159.6922	159.69
MgO	40.3044	40.30
FeO	71.8464	71.85
MnO	70.9374	70.94
CaO	56.0794	56.08
Na_2O	61.97894	61.98
K_2O	94.196	94.20
P_2O_5	141.94452	141.94
H_2O	18.0152	18.02
CO_2	44.0098	44.01

1.5 Atomic bonding

There are four main types of bonding in atomic lattices, as follows:

(1) **Ionic** or heteropolar bonding between ions of opposite electrical charge.
(2) **Covalent** or homopolar bonding, in which atoms share electrons.
(3) **Metallic** bonding, which is found in metals and is responsible for their cohesion. This is given here for the purposes of comparison.
(4) **Van der Waals** (or **residual**) bonding due to weak forces present in all crystals.

Although certain groups of substances are characterized by certain types of bonding, two or more types may operate between atoms, or groups of atoms, in a single substance.

Ionic or heteropolar bonding

Atoms held together by this type of bond are in the ionized state, each atom having gained or lost one or more electron, so that they have acquired a positive or negative charge. The forces holding the ions together are those of electrical attraction between oppositely charged bodies. Each ion is surrounded by ions of opposite charge, and the whole structure is neutral, that is with no overall charge.

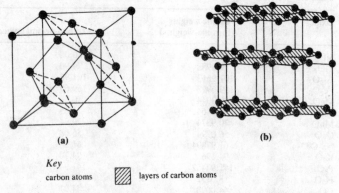

Key

carbon atoms   ▨ layers of carbon atoms

Figure 1.1 The structure of (a) diamond and (b) graphite

The *halides*, which include the two common minerals halite (NaCl) and fluorite (CaF_2), possess ionic bonding. In halite, one cation of sodium (Na^+) is united with one anion of chlorine (Cl^-), and in fluorite, one cation of calcium (Ca^{2+}) is united with two anions of fluorine ($2F^-$). In both these minerals a stable configuration is achieved, with each ion obtaining a complete outer shell. The ionic bond is strong and results in crystals with a high degree of symmetry, a high melting point and a low coefficient of expansion.

Covalent or homopolar bonding

In this type of bonding, electrons are shared between two atoms, with the outermost shells of the atoms overlapping. In hydrogen, for example, the single electrons of two atoms are shared to make the hydrogen molecule and, in this way, the stable configuration of the nearest inert gas is achieved. An atom of oxygen ($Z = 8$) has six electrons in its outer shell; when a molecule is formed, two of the electrons in the outer shell are shared by both atoms, thus achieving the stable configuration of eight electrons in each outer shell.

Diamond has a covalent structure composed entirely of carbon atoms. The carbon atom ($Z = 6$, i.e. $2 + 4$) has four electrons in its outer shell, and thus can form four covalent bonds with other carbon atoms. This is the basis of the diamond structure, in which each carbon atom is surrounded by four others, arranged as seen in Figure 1.1a. Atoms are located at the corners and the face centres of the cube, and also at points 1/4 or 3/4 along a diagonal. Such a structure has great strength, which is reflected in the extreme hardness of diamond.

The other crystalline form of carbon, graphite, has a totally different structure, with the carbon atoms arranged in layers at the corners of regular, plane hexagons, each carbon atom being linked to three others. There is a covalent bond between each pair of atoms, and the planes or layers of atoms, which are mutually displaced, are held together by weak van der Waals bonding (see above, and p. 14). The atomic structure accounts for the well developed cleavage (p. 223) which graphite possesses, parallel to the sheets of atoms (see Fig. 1.1b).

Metallic bonding

A piece of metal is composed of a mosaic of crystals, each of which is itself composed of closely packed atoms of the particular metallic element. Such a crystal may be regarded as an aggregate of positive cations surrounded by a 'cloud' of free electrons. All the atoms contribute electrons to this 'cloud', which serves to bind the metallic ions together. The **metallic bond** is, therefore, an attraction between the positive metal cations and the 'cloud' of negative electrons. Crystals with this type of bonding are called **metallic crystals**.

Many properties which are characteristic of metals, such as their opacity and their thermal and electrical conductivity, are due to the presence of free electrons – the electron 'cloud' of the last paragraph. When an electrical field is applied to a metal, the electrons are able to move under the influence of this field; that is, the metal conducts by electron transport. This method of conduction applies to *true metals*, such as iron, copper, nickel, etc.; but other metallic elements, such as zinc and lead, possess this property to a lesser degree, since their atomic structure is more complex, with more than one type of bond occurring between the atoms.

In general, there are three ways in which the ions of most metals are packed together to form crystals, corresponding to three methods of stacking spheres of equal size. These are **cubic close packing** (such as occurs in copper), **body-centred cubic packing** (such as occurs in iron at room temperatures) and **hexagonal close packing** (such as occurs in magnesium). Some metals have more complex structures, and detailed information may be obtained from specialist texts.

When identical spheres are arranged in a single layer, the closest packing possible is obtained when the centres of the spheres lie at the corners of equilateral triangles, each sphere touching six others. Layers of this type occur in both cubic and hexagonal close packing.

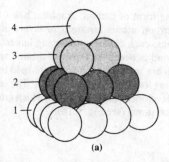

(a)

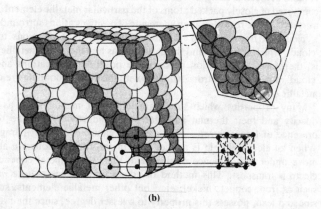

(b)

Figure 1.2 Cubic close packing. (a) The tetrahedron produced by cubic close packing. (b) The cube and tetrahedron produced by cubic close packing, and the relationship to the face-centred cubic space lattice.

Cubic close packing

The structure is as shown in Figure 1.2, where the spheres represent identical metallic ions. The layers with closest packing lie at right angles to the *triad axes* of a cube; that is at right angles to a line joining opposite corners of the cube (through its centre). Since there are four triad axes, there must be four directions in which layers of this type occur. In the complete structure, any one such layer fits against the adjoining layers, such that each sphere touches three in the layer above *or* below. A succession of three layers can be placed one on top of the other in different

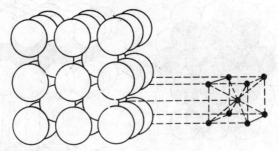

Figure 1.3 Body-centred cubic packing, and the relationship to the body-centred space lattice shown in Figure 3.3.

positions, but the fourth layer repeats the position of the first layer; that is, the pattern repeats after every third layer.

In a crystal of copper, movement or **gliding** may take place along these layers of closest packing when the crystal is subject to stress, and it is this property that gives a crystal its ductility and malleability (p. 38). Even when large numbers of crystals are present, as in a piece of copper wire, gliding still takes place when the wire is stretched.

Ductile metals in this group include *copper, gold, platinum, silver* and *iron* (at high temperatures).

Body-centred cubic packing
This structure is shown in Figure 1.3, where the spheres again represent identical metallic ions. In this structure the spheres lie at the corners of cubes and with one sphere at the centre of each cube. Each sphere touches eight others, and the packing is not as close as in the other two types. Since close-packed layers are absent, gliding is difficult to induce, and a metal with this type of structure is harder and more brittle. Iron has this structure at normal temperatures, but at high temperatures it possesses the cubic close-packed structure, and in consequence, becomes malleable. The importance of iron in metallurgy lies in these two different structures, since the metal can be made to assume different properties according to the heat treatment it receives.

Other brittle metals with this structure include *tungsten, barium, molybdenum, vanadium* and, as already mentioned, *iron* (at room temperatures).

Hexagonal close packing
This structure contains close-packed layers, with the centres of identical spheres lying at the corners of equilateral triangles. Adjoining layers are

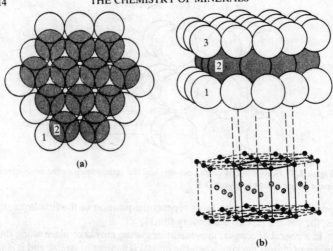

Figure 1.4 Hexagonal close packing. (a) The hexagonal close packing of spheres in plan. (b) An oblique view of hexagonal close packing, and the relationship to the hexagonal space lattice.

displaced so that any one sphere lies above or below three spheres in the layer next to it, with every alternate layer identical; that is, the structure repeats after every second layer, instead of after every third layer as in cubic close packing. The structure is shown in Figure 1.4. Planes of gliding in this structure are fewer since they occur in one direction only, and metals with this structure are correspondingly less soft and ductile than those in the first group (with cubic close packing).

Metals with this structure include *magnesium*, *titanium*, one form of *nickel*, and *calcium* (above 450°C).

Van der Waals (or residual) bonding

This bond differs from all the others in that it is rarely responsible for the coherence of any common substances, although napthalene is a notable exception. It is a weak force of attraction between the ions or atoms of *all* solids, but its effect is often masked in structures where ionic, covalent or metallic bonding already occurs The only *solids* entirely with van der Waals bonding are inert gases (neon, argon, etc.) in the solid state.

However, this type of bonding is of importance in organic substances. For example, organic carbon compounds are formed of molecules containing carbon, hydrogen and oxygen atoms arranged in different ways. The bonding *within* these molecules is usually covalent, but *between* the molecules it is, in many cases, of the van der Waals type. Thus paraffin wax is composed of molecules of a 'long-chain' hydrocarbon, and the molecules are packed together, like a series of rods, to form crystals ('molecular crystals'). The soft nature of such a substance is due to the weakness of the van der Waals bonding between these molecular crystals.

1.6 Ion size and ionic radii

The relative sizes of ions play an important part in the construction of crystals. It is convenient to think of atoms and ions as spheres having a definite radius. When two ions are brought close together, a force of repulsion between them occurs when they are a set distance apart, and any closer approach is resisted. The distance between them is taken as the sum of the radii of the two ions. In this way the ions are treated as spheres in contact, and their radii can be measured by various methods. Furthermore, ionic radius *and* charge were recognized by Goldschmidt (1937) as of great importance in determining element distribution in crystal lattices. His ideas are known as **Goldschmidt's rules**, and are as follows:

(1) Ions of similar radii ($\pm 15\%$) and the same charge will enter into a crystal in amounts proportional to their concentration in the liquid.
(2) When two ions with the same charge compete for a lattice site, the ion of smaller radius will be preferentially incorporated into the growing crystal.
(3) When two ions of the same radius ($\pm 15\%$) compete for a lattice site, the ion with the higher charge will be preferentially incorporated into a growing crystal.

Later work by Ahrens, Ringwood, Pauling and others showed that bonding characteristics are also of importance in determining the order in which elements are incorporated into mineral lattices, with those elements whose cations have ionic bonding being preferentially incorporated before more covalently bonded ions. Properties such as electronegativity, ionization potential and bonding energy were used to determine bonding characteristics, and the reader is referred to more specialized geochemical texts such as Henderson (1982) for a fuller explanation of element substitution.

Ionic radii are given in Table 1.4, and it should be noted that the values of these change depending upon both the valency of the ion and its coordination state in the lattice.

The size of the $(OH)^-$ ion is the same as that of oxygen, O^{2-}. Oxygen is

Table 1.4 Ionic radii, valency and coordination numbers.

Group	Ion				
Ia	Li^+ 0.76(VI)	Na^+ 1.02(VI) 1.18(VIII) 1.39(XII)	K^+ 1.38(VI) 1.51(VIII) 1.64(XII)	Rb^+ 1.52(VI) 1.61(VIII) 1.72(XII)	Cs^+ 1.67(VI) 1.74(VIII) 1.88(XII)
IIa	Be^{2+} 0.27(IV)	Mg^{2+} 0.72(VI)	Ca^{2+} 1.00(VI) 1.12(VIII)	Sr^{2+} 1.18(VI) 1.26(VIII)	Ba^{2+} 1.35(VI) 1.42(VIII) 1.61(XII)
IIIa			Sc^{3+} 0.745(VI)	Y^{3+} 0.90(VI)	La^{3+} 1.032(VI)
IVa			Ti^{4+} 0.42(IV) 0.605(VI)	Zr^{4+} 0.72(VI)	Hf^{4+} 0.71(VI)
Va			V^{3+} 0.64(VI)	Nb^{5+} 0.64(VI)	Ta^{5+} 0.64(VI)
VIa			Cr^{3+} 0.615(VI)	Mo^{4+} 0.65(VI)	W^{6+} 0.66(VI)
VIIa			Mn^{2+} 0.67(VI)		
VIII			Fe^{2+} 0.78(VI) Ni^{2+} 0.69(VI) Co^{2+} 0.745(VI)	Fe^{3+} 0.645(VI)	
Ib			Cu^+ 0.77(VI)		
IIb			Zn^{2+} 0.74(VI)	Cd^{2+} 0.95(VI)	
IIIb	B^{3+} 0.11(IV)	Al^{3+} 0.39(IV) 0.535(VI)	Ga^{3+} 0.47(IV) 0.62(VI)		
IVb		Si^{4+} 0.26(IV)	Ge^{4+} 0.39(IV)	Sn^{4+} 0.69(VI)	Pb^+ 1.19(VI)
Vb		P^{5+} 0.38(IV)			
VIb	O^{2-} 1.40(VI)	S^{2-} 1.72(VI)			
VIIb	F^- 1.33(VI)	Cl^- 1.72(VI)	Br^- 1.96(VI)		

Radius measurements are in Angstrom (Å) = 10^{-10} m.

Roman numbers following radius sizes are the coordination number for the ion; thus Fe^{2+} = 0.78(VI) means that the iron (II) ion has a radius size of 0.78Å when in six-fold coordination.

Table 1.5 Radius ratios and predicted ion coordination numbers.

Radius ratio range	Arrangement of anions around cation	Predicted coordination number
0.15–0.22	corners of equilateral triangle	3
0.22–0.41	corners of tetrahedron	4
0.41–0.73	corners of octahedron	6
0.73–1.00	corners of cube	8
>1.00	closest packing	12

the most common element in the Earth's crust (see Table 1.1), and most minerals are oxygen compounds. The size of the O^{2-} ion means that the crust is composed of >90% by volume of this one element. The large oxygen ions are closely packed together in most minerals and the much smaller cations are situated in the interstices between the oxygens. The most important cation in silicate minerals, silicon (Si^{4+}), forms the silicate oxyanion $[SiO_4]^{4-}$, which resembles a tetrahedron in outline with a silicon ion at the centre and four large oxygen ions at the apices. Other anions such as F^-, S^{2-}, etc., are also very large.

The coordination number of an ion can be predicted by determining a ratio, called the **radius ratio**, which represents the radius of the cation divided by that of the radius of the oxygen anion; that is, $r_x/r_{O^{2-}}$, where x is any cation (K^+, Fe^{2+}, Al^{3+}, etc.). From Table 1.5 the predicted coordination numbers for the most common cations in rock-forming minerals can be determined, as shown in Table 1.6.

Table 1.6 Coordination numbers of common cations.

Coordination number Predicted	Observed	Cations
12	12	Cs^+
12	8–12	Rb^+
8	8–12	K^+, Ba^{2+}
8	8	Sr^{2+}
6	6–8	Ca^{2+}, Na^+
6	6	Fe^{2+}, Mg^{2+}, Fe^{3+}, Mn^{2+}, Li^+ Cr^{3+}, Ti^{4+}
4	4–6	Al^{3+}
4	4	Si^{4+}, Be^{2+}, P^{5+}
3	3–4	B^{3+}

Table 1.7 Ionic potential and ion behaviour in magmas.

Cation	φ	Behaviour in magma
Cs^+	0.60	these occur as free ions
Rb^+	0.68	
K^+	0.75	
Na^+	1.03	
Li^+	1.47	
Ba^{2+}	1.49	
Sr^{2+}	1.79	
Ca^{2+}	2.02	
Mn^{2+}	2.50	
REE[a]	2.64–3.48	
Fe^{2+}	2.70	if the magma has a high (Si + Al): O
Ni^{2+}	3.00	ratio these cations will enter lattice
Mg^{2+}	3.04	vacant sites
Th^{4+}	3.92	these occur as free ions
U^{4+}	4.15	
Fe^{3+}	4.68	
Cr^{3+}	4.77	
Zr^{4+}	5.08	these ions form complexes which may
Sn^{4+}	5.64	combine into lattices; the stability of the
Al^{3+}	5.88	complexes decreases with increasing φ
Ti^{4+}	5.88	
Ge^{4+}	7.56	
Si^{4+}	9.52	
B^{3+}	13.05	
P^{5+}	14.30	
C^{4+}	25.00	

[a] REE = rare earth elements.

1.7 Ionic potential and behaviour of ions in magma

The hydration of an ion is proportional to its change and inversely proportional to its radius; this factor, namely charge of cation/radius of cation is called the **ionic potential** and is denoted by the letter φ. Ionic potentials of the main cations are given in Table 1.7.

In Table 1.7 the values of φ generally increase with increasing radius size. For those ions forming tetrahedral complexes, the higher the charge on the central cation, the less likely is the incorporation of the complex into the silicate structure. Thus complexes with a central cation whose valency is greater than four are concentrated in the residual phase. Furthermore, those tetrahedal complexes with a central cation larger than that of Si^{4+} (>0·42Å) are less likely to be incorporated into the silicate structure, and will preferentially be incorporated into the residual phase.

Table 1.8 The periodic table of the elements.

Period	Group Ia	Group IIa	Group IIIa	Group IVa	Group Va	Group VIa	Group VIIa	Group VIII			Group Ib	Group IIb	Group IIIb	Group IVb	Group Vb	Group VIb	Group VIIb	Group O
1 1s	1 H																	2 He
2 2s2p	3 Li	4 Be											5 B	6 C	7 N	8 O	9 F	10 Ne
3 3s3p	11 Na	12 Mg											13 Al	14 Si	15 P	16 S	17 Cl	18 Ar
4 4s3d 4p	19 K	20 Ca	21 Sc	22 Ti	23 V	24 Cr	25 Mn	26 Fe	27 Co	28 Ni	29 Cu	30 Zn	31 Ga	32 Ge	33 As	34 Se	35 Br	36 Kr
5 5s4d 5p	37 Rb	38 Sr	39 Y	40 Zr	41 Nb	42 Mo	43 Tc	44 Ru	45 Rh	46 Pd	47 Ag	48 Cd	49 In	50 Sn	51 Sb	52 Te	53 I	54 Xe
6 6s (4f) 5d 6p	55 Cs	56 Ba	57☆ La	72 Hf	73 Ta	74 W	75 Re	76 Os	77 Ir	78 Pt	79 Au	80 Hg	81 Tl	82 Pb	83 Bi	84 Po	85 At	86 Rn
7 7s (5f) 6d	87 Fr	88 Ra	89☆☆ Ac															

☆Lanthanide series 4f	58 Ce	59 Pr	60 Nd	61 Pm	62 Sm	63 Eu	64 Gd	65 Tb	66 Dy	67 Ho	68 Er	69 Tm	70 Yb	71 Lu
☆☆Actinide series 5f	90 Th	91 Pa	92 U	93 Np	94 Pu	95 Am	96 Cm	97 Bk	98 Cf	99 Es	100 Fm	101 Md	102 No	103 Lr

1.8 Periodic classification of the elements

When the elements are listed in order of their atomic numbers, they can be divided into **groups** so that elements of similar chemical properties are brought together. In Table 1.8 each element is shown by its symbol and its atomic number. The rows across correspond to the original **periods** of Mendeleev, with the elements running from top to bottom corresponding to his **groups**. The groups contain elements with similar valencies, and are split into 'a' and 'b' subdivisions on the basis of their chemical similarities.

The periodic table is used as a basis for the groupings of elements in Chapter 7 of this book, where the various minerals associated with a particular element are also listed, *but not described*. Detailed mineral descriptions are given in Chapters 8 (non-silicates) and 9 (silicates), where they are dealt with in a more conventional manner.

1.9 Oxides, acids and bases, and salts

Compounds of oxygen with another element are called **oxides**, and these constitute an important group of minerals which are dealt with in Chapter 8. It is important to note that *all* silicates, whether oxides or not, are dealt with together in Chapter 9. Thus quartz (silicon dioxide, silica or SiO_2), although an oxide, is discussed as a member of the *silicate* group of minerals.

Most **oxides** of *non-metals* react with water (itself an oxide – H_2O) to give a solution that turns blue litmus paper red. Such a solution is called **acidic**, and is said to contain an **acid**. The essential feature of all acids is that they contain positive hydrogen ions (H^+), formed by the loss of the single hydrogen electron from the hydrogen atom. Compared with the few H^+ ions and compensatory $(OH)^-$ ions that are present in water, acids contain countless numbers of H^+ ions only, and acid properties result. In a particular acid the H^+ ions combine with simple anions (Cl^-, F^-, etc.) or complexes ($[CO_3]^{2-}$, $[SO_4]^{2-}$, etc.). Examples of acids that are important in mineralogy are given in Table 1.9

Most **oxides** of *metals* do not dissolve in water, although a few react to form **hydroxides**, such as, for example, $NaOH$, KOH and $Ca(OH)_2$, which are **alkaline** in character, turning red litmus paper blue. These hydroxides are called **bases**, and their alkaline properties are related to the abundant $(OH)^-$ ions that are present in solution. Many metal hydroxides, some of which occur as minerals (for example brucite – $Mg(OH)_2$), cannot form in the way just described since their oxides are insoluble, but they lose their

Table 1.9 Acids and their corresponding salts.

Acid	Formula	Salt	Example
carbonic	H_2CO_3	carbonate	calcite ($CaCO_3$)
hydrobromic	HBr	bromide	bromyrite (AgBr)
hydrochloric	HCl	chloride	halite (NaCl)
hydrofluoric	HF	fluoride	fluorite (CaF_2)
nitric	HNO_3	nitrate	nitre (KNO_3)
phosphoric	H_3PO_4	phosphate	apatite ($Ca_5[PO_4]_3$)
sulphuric	H_2SO_4	sulphate	barite ($BaSO_4$)
sulphuretted hydrogen	H_2S	sulphide	galena (PbS)
tetraboric	$H_2B_4O_7$	borate	borax ($Na_2B_4O_7.10H_2O$)

hydroxyl 'water' on heating and form *basic oxides*. The term 'base' has been extended to include metal hydroxides.

When an acid reacts with a base, part or whole of the hydrogen of the acid is replaced by the metal of the base, and the result is the formation of a **salt**. Thus the reaction of hydrochloric acid (HCl) with the base sodium hydroxide (NaOH) gives the salt sodium chloride (NaCl), with water (H_2O) being released during the reaction, as follows:

$$HCl + NaOH \rightarrow NaCl + H_2O$$

If this reaction is written in terms of ions, it becomes:

$$[H^+ + Cl^-] + [Na^+ + (OH)^-] \rightarrow [Na^+ + Cl^-] + H_2O$$

When all the H^+ ions from the acid have been 'partnered' with the $(OH)^-$ ions from the base, **neutralization** will have taken place. Insoluble metal hydroxides, when they are neutralized by acids, act like true soluble bases and form salts, and on this account are often included among the bases.

Many minerals are salts and examples of these are included in Table 1.9. Where all the hydrogen of the acid has been replaced by metallic elements, the resulting salts are called **normal salts**. When only a part of the hydrogen has been replaced **acid salts** are produced. Thus potassium sulphate (K_2SO_4) is a normal salt, but potassium hydrogen sulphate ($KHSO_4$) is an acid salt. In a **basic salt** the whole of the base has not been neutralized by the acid portion; thus the mineral malachite is a basic salt with a composition $CuCO_3.Cu(OH)_2$.

Some salts are deposited from saturated solutions, and contain a definite and constant number of water molecules incorporated into the crystal structure. Such salts are called **hydrates**, and the attached water is called **water of crystallization**. Common hydrated minerals include gypsum ($CaSO_4$. $2H_2O$) and limonite ($FeO(OH)$. nH_2O). The water of crystallization can be driven off at a moderate temperature, the **anhydrous salt** being left behind.

1.10 Oxidation and reduction

A chemical change by which oxygen is added to an element or compound is called **oxidation**. The term **reduction** is applied to a change in which oxygen or other non-metal is taken away from a compound. When oxidation occurs, reduction occurs as well, and the reaction is called a **redox reaction**. For example, if metallic copper is heated in air it is changed to black copper oxide, as follows:

$$2Cu + O_2 \rightarrow 2CuO \quad \text{or} \quad 2[Cu^{2+} + O^{2-}]$$

The copper atoms in becoming Cu^{2+} ions, have *lost* their outer electrons on oxidation and, correspondingly, the oxygen atoms have *gained* outer electrons in becoming O^{2-} ions. Thus, a general rule could be written as follows: *if electrons have been lost by an atom the change is one of oxidation, whereas if electrons have been gained the change is one of reduction*. Redox reactions are of great importance in the analysis of minerals by 'wet' chemical techniques, and also in the production of metals from their ores.

1.11 The electrochemical series of metals

Metal elements can be arranged in an order which expresses the ease with which they part with electrons to form compounds. This order is called the **electrochemical series** (or reactivity series), and the order for some of the more common metals is given below:

K Na Ca Mg Al Zn Fe Sn Pb Cu Hg Ag Au

Potassium, the most active metal, begins the series, and gold, the least active, ends it. The complete series shows a graduation in chemical and other properties, from potassium to gold. The recognition of the electro-chemical series ties together many chemical phenomena which are of importance in the occurrence and treatment of minerals. The more active metals are powerful reducing agents, combine readily with oxygen, and corrode and tarnish readily, whereas the non-reactive metals are extremely stable and show the reverse of the properties already outlined.

The hydroxides and carbonates of the most active metals, such as K and Na, are soluble in water, but calcium carbonate, magnesium hydroxide and magnesium carbonate are insoluble. The reaction of metals in acids varies from the most active, which dissolve easily in sulphuric and hydrochloric acids, to the least active, such as gold and silver, which are unaffected by these acids. All the metals in the list above, with the exception of gold and silver, also dissolve in nitric acid.

The readiness with which compounds of metals decompose upon heating increases from potassium to gold, a circumstance affecting the production of metals from their ores. Metals occurring early in the series, from potassium to aluminium, do not occur **native**, or as uncombined or free elements, in nature and these metals are obtained by electrolysis of their molten compounds. Zinc, iron, tin and lead are rarely found native, and are produced by reduction of their ores with carbon. The less active metals, copper and mercury, often occur native, and these metals (including lead), can be readily obtained without a reducing agent, by simply sintering or heating their ores. The most inactive metals, gold and silver, occur mostly native; that is, as free metals.

1.12 Chemical analysis

The determination of the constituent elements of a compound is called **analysis**. Analysis can be either **qualitative**, in which the nature of the constituents contained in a compound is determined, or **quantitative**, in which the proportions of the constituents are determined.

In a **qualitative analysis** the recognition of the constituents hinges upon the fact that certain bases and acids produce well marked phenomena in the presence of known substances or preparations called **reagents**. The characteristic effect produced by a reagent is called a **reaction**. Thus, for example, hydrochloric acid is a reagent which, when added to a clear solution containing salts of lead, silver or mercury, produces a dense, white precipitate consisting of the chlorides of these metals. This reaction, which denotes the presence of one or more of the metals in the original solution, must be supplemented by others to determine precisely which of the metals is present.

A rock or ore can also be ground to a powder and subjected to analysis by **X-ray diffraction**, to reveal which minerals are present; this technique is particularly suitable for non-silicates, especially metallic ores – oxides, sulphides etc. (see also Chs 2 & 7).

The **quantitive analysis** of a mineral (or rock or ore) can be carried out by a chemist, skilled in 'rapid wet methods' of analysis, who reports his results as weight percentages of the elements or, more commonly, of the oxides which constitute the mineral (or rock or ore). Nowadays, most chemical analyses of rocks or minerals are carried out using **X-ray fluorescence** techniques, in which a ground sample of the mineral, previously fused to a bead, is inserted into the path of a beam of 'hard' X-rays. The various elements fluoresce in the path of the X-ray beam, and the X-ray fluorescence **spectrometer** can be used to determine both the element producing a particular fluorescence and also its amount. The results determine the elemental amounts, including oxygen, and these can then be recalculated into the normal oxide percentages which are used for the *major* elements. Minor and trace element amounts in analyses are given in parts per million (ppm) of the element, and never recalculated to oxides. The method of recalculating the formula of a mineral from simple analyses is shown below for two examples.

(1) The composition of the common copper ore chalcopyrite is given in the table below. The atomic weights in the second column (from Table 1.2) are divided into the weight percentages to obtain the atomic proportions in column three. The atomic ratios in column four follow from column three, and are seen to be 1 copper and 1 iron to 2 sulphur, which gives a formula of $CuFeS_2$:

Element	Weight %	Atomic weight	Atomic proportions	Atomic ratio
Cu	34.5	63.55	0.5429	1
Fe	30.5	55.85	0.5461	1
S	35.0	32.06	1.0917	2

(2) An analysis of the common silicate mineral **orthoclase**, or **K-feldspar** (from Deer *et al.* 1966), is given in the table below, in column one. The oxide percentages are divided by the 'atomic' or formula weights of column two to produce the atomic proportions shown in column

three, and the 'atomic' ratios of 1 potash to 1 alumina to six silica follow in column four. It should be noted that the ratios are not exact since the mineral is probably not 'pure':

Oxide	Weight %	Atomic weight	Atomic proportions	Atomic ratio
SiO_2	64.66	60.08	1.0762	6
Al_2O_3	19.72	101.96	0.1934	1
CaO	0.34	56.08	0.0061)	
Na_2O	3.42	61.98	0.0484)	1
K_2O	11.72	94.20	0.1244)	

The formula is therefore $[K_2O.Na_2O.CaO]. Al_2O_3. 6SiO_2$ or, written another way, $[(K,Na)_2Ca)\ Al_2Si_6O_{16}]$, or $[K(Na,Ca)Al\ Si_3O_8]$, which is the usual way the formula for orthoclase or K-feldspar is written.

Apart from the 'dry methods of analysis by X-ray techniques, analysis by the **blowpipe** can sometimes be an invaluable technique for determining the elements present in a mineral or ore. Although this method is rather dated now, there is sufficient interest in it for the method to be given in detail in Appendix A, with the results of such tests accompanying many of the mineral descriptions given in Chapters 8 and 9.

2

Physical properties of minerals

2.1 Introduction

All minerals possess certain physical properties, which are considered in some detail in this chapter in the following order.

(1) Characters depending upon light, such as colour, streak, lustre, transparency, translucency, phosphorescence and fluorescence. Optical properties required in the recognition of minerals in thin section under the microscope are dealt with in a later chapter.
(2) Characters depending upon certain senses, such as those of taste, odour and feel.
(3) Characters depending upon the atomic structure and state of aggregation, such as form, pseudomorphism, hardness, tenacity, fracture, cleavage and surface tension effects.

Crystallography, the study of the regular pattern of faces and interfacial angles of a crystal, and the internal structure of the mineral to which it is related, is considered in the next chapter.

(4) The specific gravity of minerals.
(5) Characters depending upon heat, such as fusibility.
(6) Characters depending upon magnetism, electricity and radioactivity.

2.2 Characters dependent upon light

Colour

The colour shown by a mineral depends upon the absorption of some, and the reflection of others, of the coloured rays or vibrations which constitute ordinary white light. When a mineral absorbs virtually all light vibrations and reflects virtually none, it appears *black*; but when it reflects all the vibrations of the different colours which comprise white light, it appears *white*. If, for example, the mineral reflects the red vibrations of ordinary light and absorbs all the others, it appears *red*.

The colour of a mineral is often its most striking property. However, for purposes of identification the colours exhibited by minerals vary greatly; even minerals belonging to the same mineral group may possess quite different colours. Thus, for example, in the quartz group of minerals, the mineral quartz (SiO_2) is usually colourless or white; but some of its varieties will appear brown (the variety citrine), or pink (rose quartz), or purple (amethyst), due to the amount of certain trace elements included in the mineral's composition. The two precious varieties of the mineral beryl, emerald and aquamarine, also exhibit different colours – green and blue respectively, due also to a slight variation in the trace element chemistry of beryl. A very few minerals will also show colour variation within a single crystal, either arranged in a regular fashion with different colour bands (as in tourmaline), or in patches within the mineral (as in fluorite).

From the above, it is obvious that the true colour of a mineral depends upon the nature and arrangement of its constituent ions. Thus minerals containing Al, Ca, Na, K, Zr, Ba and Sr as their main ions are generally light coloured or colourless, while those with Fe, Ti, Mn, Cr, Co, Ni, V and Cu are usually coloured – often deeply. Different types of bonding of the carbon atoms are responsible for the colour differences between diamond (colourless) and graphite (black). The valency of an ion can affect the colour of a mineral; thus those with Fe^{2+} are commonly green, whereas those with Fe^{3+} are commonly yellow, red or brown. If both Fe^{2+} and Fe^{3+} are present in any appreciable quantity the mineral is usually black because of charge-transfer effects.

Some minerals, when rotated or observed from different directions, display a changing series of prismatic colours similar to those seen in a rainbow or when looking through a glass prism, called a **play of colours**. This is best shown by diamond, quartz and other colourless minerals, and is produced by the splitting up of a ray of white light into its coloured constituents as it enters and leaves the mineral. **Change of colour** is a somewhat similar phenomenon extending over broader surfaces, the succession of colours being produced as the mineral is turned. Certain varieties of the mineral plagioclase feldspar show these changes of colour very well, particularly a feldspar which is a major constituent of the igneous rock *larvikite* from southern Norway. Polished slabs of this rock are widely used as cladding panels by the building trade. The change in colour is caused by the interference of light reflected from thin plates of other minerals enclosed or exsolved in parallel planes within the feldspar crystals. Such a phenomenon is similar to that of **schiller**, an almost metallic lustre exhibited by the orthopyroxene minerals, particularly hypersthene, where reflection of colours takes place on mineral plates exsolved on parallel planes within the crystal.

Some minerals display **iridescence**, a play of colours due to the interference of rays of light either by minute globules of water trapped in the outer layers of the crystal lattice, as in *opal* (also called **opalescence**), or by distortions in the atomic lattice, as in *labradorite*.

Some minerals **tarnish** on the surface when exposed to air, and may exhibit iridescent colours. Tarnishing may result either from oxidation, or from the chemical action of sulphur and other agents present in the atmosphere. Tarnish may be distinguished from the true colour of the mineral by scratching or chipping the mineral, when the superficial nature of the tarnish is revealed. Chalcopyrite ($CuFeS_2$) and bornite (Cu_5FeS_4) often tarnish to an iridescent mixture of colours, the latter mineral sometimes being known as 'peacock ore'.

Some crystals display different colours when rotated in plane-polarized transmitted light under a microscope. This property, known as **pleochroism**, is considered with the special optical properties (see p. 107).

Streak

The streak of a mineral is the colour of its powder, and may be quite different from that of the mineral in mass. Streak is particularly important with metallic ore minerals; for example, *grey* galena (PbS) gives a *black* streak, and brassy yellow pyrite (FeS_2) also give a black streak. Streak is obtained by scratching or rubbing the mineral across a piece of fired but unglazed procelain, called a **streak-plate**.

Lustre

The lustre of minerals differs both in intensity and kind, depending upon the amount and type of reflection of light that takes place at the surfaces of the minerals. The following descriptive terms are used:

Metallic   The ordinary lustre of metals. When feebly displayed this type of lustre is termed **submetallic**, and when not displayed at all is termed **dull**. Gold, iron pyrite and galena have a metallic lustre; chromite and cuprite have a submetallic lustre; and massive magnetite usually has a dull lustre.

Vitreous   The lustre of broken glass. When less well developed it is termed **subvitreous**, and when not developed at all is also called **dull**.

These lustres are particularly well displayed among the silicates, the carbonates, the sulphates and the halides, and other non-metallic minerals. Quartz shows vitreous lustre, whereas the amphiboles (hornblende, etc.) and the pyroxene augite are usually subvitreous or dull.

Resinous   The lustre of resin. This is well displayed by opal, amber and the zinc ore sphalerite, or blende.

Pearly   The lustre of a pearl. It is shown by surfaces parallel to which the mineral is separated into thin plates, similar to the conditions of a pile of thin glass sheets, such as cover glasses on microscope slides. Talc, brucite and selenite show pearly lustre.

Silky   The lustre of silk. This lustre is peculiar to minerals possessing a fibrous structure. The fibrous variety of gypsum known as satin-spar and the fibrous varieties of asbestos called amianthus are good examples of mineral with silky lustres.

Adamantine   The lustre of diamond. Crystals of the tin ore cassiterite exhibit an adamantine lustre.

The lustre of minerals may be of ranging **degrees of intensity**, depending on the amount of light reflected from their surfaces. When the surface of a mineral is so brilliant that it reflects objects distinctly, as a mirror would do, it is said to be **splendent**. When the surface is less brilliant and objects are reflected indistinctly, it is described as being **shining**. When the surface has no lustre at all it is described as **dull** (as above).

The various crystal faces of a single mineral may show different kinds and degrees of lustre.

Transparency and translucency

A mineral is **transparent** when the outline of an object seen through it is sharp and distinct: a clear crystal of quartz is a good example. A mineral is **subtransparent** when an object seen through it is indistinct. A mineral which, although capable of transmitting light, cannot be seen through, is termed **translucent**. In many cases the edges of the mineral transmit more light than the main body and appear much brighter (for example, the common black flint from the Chalk of the South of England). This condition is very common among minerals. If no light is transmitted the mineral is said to be **opaque**. This condition refers, of course, to the mineral in **hand-specimen**, since many opaque minerals, especially among the silicates, become translucent when cut into very thin slices, or even transparent when reduced to thin sections suitable for microscope examination. Some minerals, mostly oxides and sulphides such as magnetite, hematite, ilmenite, pyrite and galena, are always opaque, even in thin section.

Phosphorescence and fluorescence

Phosphorescence is the property by which some substances emit light after having been subjected to certain conditions such as heating, rubbing, or exposure either to radiation or to ultraviolet light. Some varieties of fluorite, when powdered and heated on an iron plate, display a bright phosphorescence. When rubbed together in a dark room, pieces of quartz emit a phosphorescent light. Exposure to sunlight or even ordinary diffused light can produce a phosphorescence from many minerals, seen by transferring them rapidly to a dark room. Diamond, ruby and a few other minerals exhibit a brilliant phosphorescence after exposure to X-rays. Willemite (Zn_2SiO_4) phosphoresces when exposed to X-rays, a property employed in some mines to make sure that this mineral has been completely extracted from its ore.

Some minerals emit light *whilst* exposed to certain electrical radiations. This phenomenon is best displayed by fluorite, and for this reason is called **fluorescence**.

2.3   Taste, odour and feel

Taste

The characters of minerals dependent upon taste are only perceptible when the minerals are soluble in water. The following terms are used: **saline**, the taste of common salt; **alkaline**, that of potash and soda; **cooling**, that of nitre or potassium chlorate; **astringent**, that of green vitriol (hydrated iron sulphate); **sweetish astringent**, that of alum; **bitter**, that of Epsom salts (hydrous magnesium sulphate); and **sour**, that of sulphuric acid.

Odour

Some minerals have characteristic odours when struck, rubbed, breathed upon or heated. The following terms are used:

Alliaceous   That of garlic, given off when arsenic compounds are heated
Horse-radish   That of decaying horse-radish, given off when selenium
    compounds are heated.

Sulphurous That of burning sulphur, given off by iron pyrite when struck, or by many sulphides when heated.

Fetid That of rotten eggs, given off by heating or rubbing certain varieties of quartz or limestone.

Argillaceous or clayey That of clay when breathed upon.

Feel

Smooth, **greasy** or **unctuous**, **harsh**, **meagre** or **rough** are kinds of feel of minerals that may help in their identification. Certain minerals **adhere to the tongue**.

2.4 State of aggregation

Oxygen, nitrogen and carbon dioxide are examples of *natural gases*; and water, mercury and oil are examples of *natural liquids*.

With the exception of mercury and the natural mineral oils, all the minerals we deal with are found naturally in the *solid* state, and the properties dependent on their state of aggregation are now considered.

Form

Under favourable conditions minerals assume a definite **crystal form**. The requirements are that the crystal has been free to grow outwards into the solution or melt from which it formed, neither obstructed by other solid matter nor hindered by a shortage of the constituents needed for growth. In such an environment, the mineral develops as a crystal with a regular pattern of faces and angles between adjoining faces which are characteristic of a particular mineral. The study of this regularity of form, and of the internal structure to which it is related, is called **crystallography**, and is dealt with in the next chapter. The following general descriptive terms are associated with the crystal characters of minerals:

Crystallized A term denoting that the mineral occurs as well developed crystals. Most of the beautiful mineral specimens in museums are of crystallized minerals.

Crystalline A term denoting that no definite crystals are developed, but that a confused aggregate of imperfect crystal grains have formed, interfering with one another during their growth.

Cryptocrystalline   A term denoting that the mineral possesses traces of crystalline structure.

Amorphous   A term used to describe the complete absence of crystalline structure, a condition common in natural rock glasses but rare in minerals.

Habit

The development of an individual crystal, or an aggregate of crystals, to produce a particular external shape, depends upon the conditions during formation. One such environment may give long needle-like crystals, whereas another may produce short platy crystals; and it is quite possible for the same mineral to have several different habits. Descriptive terms for mineral habits are split into (a) those for individual crystals, and (b) those for aggregates of crystals, and these are now described.

Individual crystals

Acicular   Fine needle-like crystals, as in natrolite.

Bladed   Shaped like a knife blade or lath-like; a form commonly displayed by kyanite.

Fibrous   Consisting of fine thread-like strands, as shown by the variety of gypsum called satin-spar, and also by asbestos.

Foliated or **foliaceous**   Consisting of thin and separate lamellae or leaves, as is shown by the mica group minerals and other sheet silicates.

Lamellar   Consisting of separable plates or leaves, as with wollastonite.

Prismatic   Elongation of the crystal in one direction, as in the feldspars, the pyroxenes and the common hornblendes,

Reticulated   Crystals in a cross-mesh pattern, like a net, as in rutile needles found within crystals of quartz.

Scaly   In small plates, as in tridymite.

Tabular   Broad, flat, thin crystals, as in wollastonite and sanidine feldspar.

Crystal aggregates

These may be aggregates of crystals, of which individuals can be seen with the naked eye; or massive aggregates of minerals in which individual crystals are too small to be seen with the naked eye:

Amygdaloidal   Almond-shaped aggregates, common in the zeolites, in which the minerals occupy vesicles or gas holes in lava flows.

Botryoidal Spherical aggregations resembling a bunch of grapes, as in azurite and prehnite.

Columnar Massive aggregates in slender columns, as is seen in stalactites and stalagmites, usually with the mineral calcite.

Concretionary and nodular Spherical, ellipsoidal or irregular masses, as in flint nodules.

Dendritic and arborescent Massive aggregates in tree-like or moss-like shapes, usually with the mineral being deposited in crevasses or narrow planes, as with the dendrites of manganese oxide.

Granular Coarse or fine grains. Evenly sized granular aggregates of minerals, such as olivine in the ultrabasic rock dunite, are often termed **saccharoidal** because of their resemblance to lumps of sugar.

Lenticular Flattened balls or pellets, shown by many concretionary and nodular minerals.

Mammilated Large mutually interfering spheroidal surfaces, as in malachite.

Radiating or **divergent** Fibres arranged around a central point, as in barite and in many concretions.

Reniform Kidney-shaped, the rounded outer surfaces of massive mineral aggregates resembling those of kidneys, and perfectly displayed by the variety of hematite called kidney-iron ore.

Stellate Fibres radiating from a centre to produce star-like shapes, as in wavellite.

Wiry or **filiform** Thin wires, often twisted like the strands of a rope, as in native silver and copper.

Pseudomorphism

Pseudomorphism is the assumption by a mineral of a form belonging to another mineral. **Pseudomorphs** may be formed in several ways:

(1) By **investment** or **incrustation**, produced by depositing a coating of one mineral on to the crystals of another; for example, a coating of quartz on fluorite cyrstals.

(2) By **infiltration**, when the cavity previously occupied by one mineral is refilled by deposition in it of a different mineral by the infiltration of a solution.

(3) By **replacement**, from slow and gradual substitution of particles of new and different mineral matter for the original mineral particules which are removed by solution. This type of pseudomorphism differs

from the preceding one in that substitution takes place before the previous mineral has vacated the space it occupied.

(4) By **alteration**, due to a gradual chemical change which crystals sometimes undergo, their composition becoming so altered that they are no longer the same minerals, although they possess the same forms; for example, the alteration of olivine to serpentine.

Pseudomorphs may often be recognized by a lack of sharpness in the edges of the crystals, while their surfaces usually have a dull and somewhat granular or earthy appearance.

Polymorphism

Two or more minerals may possess quite different physical properties, such as colour, form, hardness, specific gravity, etc., and yet may have identical chemical compositions. Such minerals represent a series of **polymorphs**. In a polymorphous series of minerals the atomic lattices are different so that their physical properties also differ.

Good examples of polymorphs are calcite and aragonite (**dimorphous minerals**); while graphite and diamond, which exhibit different forms of crystals, differ in hardness and specific gravity, and have markedly different optical properties. The three Al_2SiO_5 polymorphs, andalusite, kyanite and sillimanite, also have different physical properties; for example, their crystal forms differ and their specific gravities or densities are 3.13–3.16, 3.58–3.65, and 3.23–3.27 respectively; and their optical properties are also quite different. From experimental phase equilibria investigations, it is known that each of the three polymorphs exists under different conditions of temperature and pressure. High-density polymorphs, such as kyanite and diamond, are invariably favoured by high-pressure conditions, and therefore great depth.

Polytypism

Polytypes are minerals with the same chemical formulae and the same structural sub-units, which are stacked in different ways. Thus there are numerous polytypes of silicon carbide, depending upon the number of close-packed repeat layers. **Aristotypes** are related to polytypes and include the sheet silicates, which show different stacking arrangements of sheets of atoms, with the general formulae $[Si_4O_{10}]$ for the silicate layers, and either $[Al(OH)_3]$ or $[Mg,Fe(OH)_2]$ for the linking sheets. These sheets are stacked along the crystallographic c axis, to produce minerals in which similar atomic lattice arrangements are repeated: (a) at 7 Å for the kandites (including kaolin) and the serpentines; (b) at 10 Å for the mica group, the smectites (including montmorillonite) and the illites; and (c) at 14 Å for the chlorite group.

Cleavage and parting

The tendency of many minerals to split along certain definite planes, the **cleavage planes**, is closely related to crystalline form and the internal atomic structure. In each mineral with cleavages, the directions of the cleavage planes are parallel to either a particular face or to a set of faces (representing a **crystal form**) in which the mineral will crystallize. Cleavages represent planes in the atomic lattice along which preferential splitting will take place. Thus, with the sheet silicates (see polytypism above), the cleavages occur along planes between sheets or multiple sheets (layers) of atoms. Along these planes the adjoining sheets or layers are weakly bonded, usually by monovalent cations (K^+, Na^+, Rb^+, etc.) and easily cleave along these directions. In sheet silicates, such as the micas, the layers of atoms are always arranged parallel to the basal plane of the mineral, and the cleavages are also parallel to these layers; that is, parallel to the basal plane of the mineral. Such a cleavage is called a **basal cleavage**, and *all* sheet silicates possess this basal cleavage.

The pyroxenes and amphiboles are chain silicates in which the Si and O atoms are linked in chains with the compositions $[(SiO_3)_n]$ and $[(Si_4O_{11})_n]$ respectively. These chains are elongated parallel to the c axes (that is, parallel to the prism zones of the minerals) and cleavages in both minerals occur between the stacked chains of Si–O atoms. Both minerals possess two cleavages which are parallel to the prism zones of the minerals, and the minerals are said to possess **prismatic cleavages**.

Certain rocks, such as slate, split readily into thin sheets and are said to be cleaved, but this property of slaty cleavage, as it is called, is the result of recrystallization of the rock under pressure and consequent mineral re-alignment, and has no connection with mineral cleavage.

Minerals may show several cleavages, which are described by stating the crystallographic direction of each cleavage, and also the degree of perfection of each cleavage plane. For example, the mineral zoisite has two cleavages, and these would be described as follows: *perfect prismatic cleavage always present; poor basal cleavage sometimes present.* Cleavage may be described, in order of quality, as **perfect** or **eminent**, **good**, **distinct**, **poor**, **indistinct**, etc. Fluorite, galena, calcite and all the micas have perfect cleavages. Fluorite commonly crystallizes in cubes, and cleaves along planes which will truncate the corners of the cube, these cleavage planes being parallel to the faces of a regular octahedron. Fluorite is said to have a perfect {111} or octahedral cleavage, whereas galena, which has a {100} or cubic cleavage, cleaves parallel to the faces of a cube. Calcite, no matter what its habit, possesses a {10$\bar{1}$0} cleavage, which produces rhombohedral cleavage fragments when the mineral is shattered or crushed.

Glide-planes and **secondary twinning** are related to cleavage, and are produced in a mineral by pressure. For example, during the preparation of a thin section of calcite for examination under the microscope, the pressure of grinding the mineral down to the required thickness may cause it to show an excellent cleavage and also some secondary twinning (twinning is discussed later). The secondary twin planes and the glide planes are often planes along which the mineral separates fairly readily; such planes are called **partings**. A parting is similar to a very poor cleavage, and is described in the same way that we describe cleavage. The mineral diopside (a pyroxene) possesses several partings.

Fracture

The character of the **fracture** displayed on the broken or chipped surfaces of a mineral is an important property. The fracture surface is **not** the smooth surface of a cleavage plane, but is an irregular surface usually totally independent of cleavage. While fracture is an important diagnostic character, and a new one may also reveal the true colour of a mineral, it is unwise to break or chip good crystals, as crystalline form is a far more valuable and constant character by which to recognize a mineral than either its colour or its fracture. The following terms are used to describe fracture:

Conchoidal   The mineral breaks with a curved concave or convex fracture. This often shows concentric and gradually diminishing undulations towards the point of impact, somewhat resembling the growth lines on a shell. Conchoidal fracture is well shown by quartz, flint and natural rock glasses, particularly obsidian.

Even   The fracture surface is flattish, as in chert.

Uneven   The fracture surface is rough by reason of minute elevations and depressions. Most minerals have an uneven fracture.

Hackly   The surface is studded with sharp and jagged elevations, as in cast iron when it is broken.

Earthy   The dull fracture surface of chalk, meerschaum, etc.

Table 2.1 Mohs' scale of hardness.

Hardness	Standard mineral
1	talc
2	gypsum
3	calcite
4	fluorite
5	apatite
6	Feldspar (K-feldspar)
7	quartz
8	topaz
9	corundum
10	diamond

Hardness

Hardness varies very greatly in minerals. Its determination is one of the most important tests used in the identification of minerals, and it may be carried out in several ways.

Hardness may be tested by rubbing the specimen over a fine-cut file and noting the amount of powder, and also the degree of noise, produced in the operation. The less the powder and the greater the noise, the harder is the mineral. A soft mineral yields much powder and little noise. The amount of powder and the noise are compared with those produced by the minerals of the set used as standard samples for hardness tests. The scale in general use, and known by the name of **Mohs' scale of hardness**, is given in Table 2.1. The intervals on this scale are about equal, except for that between corundum (9) and diamond (10), which is estimated to be about ten times as great; this is, if corundum is 9 then diamond is about 100 on the same scale.

Window glass may be used in an emergency as a substitute for apatite, and flint for quartz. A penknife has a hardness of from 5 to 6.5, depending upon the quality of the steel; a copper coin has a hardness of about 3; and a fingernail has a hardness of about 2.5, although they can vary in hardness from one person to another.

The hardness test may also be made by trying to scratch the minerals listed in Mohs' scale with the mineral under examination. If, for example, the mineral scratches K-feldspar but not quartz, it has a hardness of between 6 and 7.

Several precautions should be observed in testing hardness. A definite scratch must be produced in the softer mineral, and this is best seen by blowing away (or licking away, if the observer cares to) the powder produced by the scratching action, and then examining the spot with a hand-lens. A softer mineral drawn across a harder one often produces a whitish stripe which may be mistaken for a scratch in the harder mineral; in the same way, an attempt to scratch harder minerals with a knife blade produces a steel mark on them. Granular specimens may give a kind of scratch by loosening and removing. Finally, it must be remembered that the hardness test has to be performed on a fresh mineral surface, and not one which is coated with decomposition products and the like. During a hardness test, the colour of the mineral powder produced by the scratch should be observed, since this will give the **streak** of the mineral – particularly useful with ore minerals.

Hardness, like many other physical properties, depends upon the atomic structure of the mineral. Amongst other factors it increases with the density of packing in the structure, and because of this, hardness may vary along different directions on the crystal. This difference is usually very small, but in the mineral kyanite, hardness varies between 7 and 5; an old name for kyanite being 'disthene', from the Greek words meaning 'two strengths'.

Tenacity

This is a measure of how a mineral deforms when it is crushed or bent; that is, subjected to some form of deformation. The following terms are used to describe tenacity:

Sectile   The mineral can be cut with a knife, and the resulting slice breaks up under a hammer; examples include graphite, steatite and gypsum.

Malleable   A slice cut from the mineral can be hammered out into thin flat sheets; examples include native gold, silver and copper.

Flexible   The mineral, or thin plates or laminae from it, can be bent but does not return to its original position (that is, it remains bent) when the pressure is removed; examples include talc, chlorite and selenite.

Elastic   The mineral, or thin plates or laminae from it, can be bent and returns to its original position after the pressure is removed; examples include the micas.

Brittle   The mineral crumbles or shatters easily; examples include iron pyrite, apatite and fluorite.

Other descriptive terms for tenancity include **ductile**, used when the mineral can be drawn out into thin wires.

Surface tension effects

The difference in the power of various liquids to adhere to different minerals has formed the basis for numerous processes of ore separation and concentration. The **surface tension** between various metallic sulphides and a selected liquid is greater than that between the gangue minerals quartz, calcite, etc., and the same liquid. In the original *Elmore Process*, a paste of sulphide and gangue was mixed with oil and water and agitated; the oil separated into a layer above the water and carried the sulphides with it. More or less the same principle underlies the method of extracting diamonds from their kimberlite matrix, by causing them to adhere to grease upon a moving belt.

The various **flotation processes** depend upon surface tension. In these, bubbles of air or gas attach themselves to, for example, finely powdered zinc blende or sphalerite, agitated in a liquid containing oil or other suitable organic compounds, and float this mineral to the surface, leaving other sulphides and gangue material at the bottom of the liquid. By varying the conditions of flotation, clean separations of various ore minerals can be produced, and in this way the working of mixed ores has become economically possible.

2.5   Specific gravity

Specific gravity, or density

The **specific gravity** of a body is the ratio of the weight of the body to that of an equal volume of water. The weight of water varies with temperature, and this has to be considered in exact work. However, in the general practice of determinative mineralogy, this correction can be neglected. In selecting material for the determination of specific gravity (called **SG** for short), it is necessary to obtain as pure a sample as possible, and also one free from alteration products, inclusions and the like.

The specific gravity of minerals depends upon the atomic weight of the constituent elements and the way their atoms are packed in the crystal lattice. The first of these controls is illustrated by three mineral sulphates which have the same type of atomic lattice.

Mineral	Formula	Atomic weight of cation		Specific gravity
celestite	$SrSO_4$	Sr	87.63	4.0
barite	$BaSO_4$	Ba	137.36	4.5
anglesite	$PbSO_4$	Pb	207.21	6.3

The influence of the style of packing is well shown by the two carbon minerals, diamond and graphite, with specific gravities of 2.54 and 2.3 respectively; and the three SiO_2 minerals, quartz, tridymite and cristobalite, with specific gravities of 2.65, 2.26 and 2.38 respectively.

Specific gravity is of great importance in the identification of minerals, and the student should become familiar with the relative weights of roughly equal-sized pieces of the common minerals. With the obvious exception of minerals containing appreciable amounts of heavy cations such as Pb, Ba, etc., the non-metallic minerals (silicates, sulphates, carbonates, halides, nitrates, etc.) have specific gravities varying from about 2.5 to over 3.5, whereas the metallic minerals, including ores (sulphides, oxides, etc.) and native elements have specific gravities varying from about 5.0 to over 8.

The main principle employed in most determinations of the SG is that the loss in weight of a body immersed in water is the weight of a volume of water equal to that of the body. If W_a is the weight of the body in air, and W_w its weight in water, then $W_a - W_w$ is the weight of water displaced by the body. The specific gravity is given by the relation $W_a/(W_a - W_w)$.

Methods of determining specific gravity

The following are the main methods of determining specific gravities in mineralogy, the particular method chosen depending usually upon the size and character of the specimen under examination.

(1) The normal **chemical balance**, used for fragments of a solid mineral about the size of a walnut.
(2) **Walkers's steelyard** apparatus, used for large specimens.
(3) **Jolly's spring balance**, used for very small specimens.
(4) By measuring the **displaced water**, for the rapid determination of the approximate SG for a number of specimens of a mineral.
(5) The **pycnometer** or **specific gravity bottle**, used for friable minerals, mineral grains or liquids.
(6) **Heavy liquids**, used mainly for the separation of mineral mixtures into their pure components according to their specific gravities, but also for approximate determinations of specific gravity of mineral grains. For this latter use, the diffusion column and Westphal Balance may be employed.

Determination of SG with the chemical balance
The mineral is weighed on a good, non-digital, chemical balance. It is then suspended by a thread from one arm of the balance and immersed in a

beaker of water standing on a wooden bridge placed over the scale pan of the balance. Bubbles of air sticking to the mineral are removed by a small brush, and the weight of the mineral immersed in water is obtained. The SG of the mineral is given by dividing its weight in air by the difference between its weights in air and water.

Walker's steelyard

This apparatus consists of a long graduated beam which is pivoted near one end and counterbalanced by a heavy weight suspended from the short arm. The specimen is suspended from the graduated beam and moved along it until it exactly counterbalances the heavy weight, the level position of the beam being shown by a mark on a slot in an upright through which the end of the graduated beam passes. The reading, a, is taken. The specimen is then immersed in a beaker of water and the procedure gone through again until a second reading, b, is obtained. The readings a and b are inversely proportional to the weights of the body in air and in water respectively. Then:

$$SG = 1/a \text{ divided by } 1/a - 1/b = b/(b - a)$$

that is, the SG is obtained by dividing the second reading, b, by the difference between the first and second readings, $b - a$.

Jolly's spring balance

This instrument consists of a spring suspended vertically against a graduated scale. To the lower end of the spring are attached two scale pans, one below the other, the lower scale pan always being immersed in water. The reading, a, of the bottom of the spring is obtained with both scale pans empty. A small fragment of the mineral is placed in the upper pan and reading b obtained. The specimen is transferred to the lower pan and a third reading, c is taken. The quantity $b - a$ is proportional to the weight of the mineral in air, and $b - c$ to the loss of weight in water, so that:

$$SG = (b - a) / (b - c)$$

Measurement of the displaced water

The SG of a large number of pieces of the same mineral may be obtained rapidly and reasonably accurately by half filling a graduated cylinder of a suitable size with water, placing the previously weighed specimens into the cylinder, and noting the increase in volume. The weight in grammes of the mineral in air, divided by the increase in volume in millilitres (ml), gives the specific gravity of the mineral.

The pycnometer, or specific gravity bottle

The pycnometer is used to obtain the SG of liquids or of small mineral grains or gemstones, or porous or friable material. It is a small glass bottle fitted with a stopper through which there is a fine opening. When filled up to a certain mark, or to the top of the stopper, the bottle contains a known volume of liquid, so that by weighing the bottle empty and then filled with liquid, the specific gravity of the liquid can be obtained. If the volume of the bottle is not known, the SG of a liquid may be determined by weighing the bottle empty, then filled with water, and finally filled with liquid, so that the SG of the liquid is the weight of liquid divided by that of the water, since their volumes are the same.

When determining the SG of mineral grains, the mineral is first weighed. The bottle is then filled with distilled water. The filled bottle and the mineral are weighed together and their combined weight obtained. The mineral is then placed in the bottle and displaces an equal volume of water, and the bottle is again weighed. The weight of water displaced is given by subtracting the two readings. The specific gravity of the mineral is obtained by dividing the weight of the mineral by the weight of water it displaces.

The use of heavy liquids

If a mixture of two minerals of different specific gravities is placed in a liquid whose SG lies *between* those of the two minerals, the heavier mineral sinks in the liquid and the lighter one floats, thus achieving a more or less complete separation of the two minerals. The SG of a liquid can be adjusted to a value at which a particular mineral will neither float nor sink, that is, the specific gravities of mineral and liquid are the same, and, by knowing the SG of the liquid, the SG of the mineral can be obtained. These two principles are the basis of the use of heavy liquids in mineralogy and petrology.

The liquids most suitable for normal separations are *bromoform* or *tetrabromoethane*, with an SG of 2.89; *methylene iodide*, with an SG of 3.30; and *Clerici's solution*, with an SG of 4.20.

Note that the use of Clerici's solution, which is a saturated solution in water of equal parts of thallium formate and malonate, is not recommended nowadays, since it is quite poisonous and can be absorbed by the body through the skin. Many other heavy liquids are available, but these are frequently poisonous or difficult to use.

Heavy liquids are used in the purification of minerals for analysis, in separating a rock into its constituent minerals, and in separating out the small amount of high-SG minerals (the heavy mineral fraction) in some rocks. In all applications of this technique, the minerals or rocks are reduced in size by crushing, etc., until single grain-sized particles are present. Dust is washed off and the prepared material is placed in the heavy liquid in a **separating funnel**. This usually consists of a small glass tube constricted in the middle (rather like a figure 8), open at the top, and capable of being held in a centrifuge. The centrifuge action causes the heavier minerals to sink into the lower chamber of the tube, with the lighter material on the top. A metal rod with a small, conical-shaped, plastic bung at one end is inserted into the tube, usually rotating the rod while this is done, until the bung closes the central neck of the tube (between the upper and lower chambers). The lighter mineral fraction can then be poured out into a filter funnel for recovery and, later, the heavier fraction can be treated in the same way.

To determine the SG of a mineral using heavy liquids, several methods may be employed. In the first, the heavy liquid is diluted until the mineral is suspended in it (that is, both SGs are the same). The SG of the liquid (and therefore of the mineral) is determined either by a **pycnometer** (if there is a large amount of liquid), or by a **Westphal Balance** (if there is not much liquid). In the latter apparatus a sinker is immersed in the liquid and balanced by riders on a graduate arm, from which the SG can usually be read off directly.

To test the SG of small samples, the **diffusion column** is used. Two perfectly mixable liquids of different specific gravities are placed in a graduated tube without mixing, and allowed to stand for a day or more until regular diffusion of the two liquids has taken place, resulting in a column of liquid in which the SG varies from top to bottom. Small fragments of known SG are placed in the liquid and, coming to rest at particular points of the column, serve as indices. A small quantity of the finely powdered unknown sample is introduced into the column, and its mineral constituents separate into bands with different specific gravities which can be calculated by their position relative to the indices. The **Berman Torsion Balance** can be used to obtain the SG of very small mineral samples.

2.6 Characters dependent upon heat

Fusibility

The relative fusibility of some minerals is a useful character as an aid in their determination by the blowpipe. A scale of six minerals, of which the temperature of fusion was supposed to increase by equal steps, was suggested by Von Kobell. These minerals and their approximate temperature of fusion are *stibnite* (525°C), *natrolite* (965°C), *almandine garnet* (1050°C), *actinolite* (1200°C), *plagioclase* (1300°C), and *olivine* (1400+°C).

2.7 Characters dependent upon magnetism, electricity and radioactivity

Magnetism

Magnetite (Fe_3O_4), and to a lesser extent *pyrrhotite* ($Fe_{1-x}S$), are the only minerals affected by an ordinary bar magnet, but a larger number of minerals are affected by the electromagnet. Minerals containing iron are generally magnetic, but this is not always the case, and the degree of magnetism displayed by a mineral does not, in all cases, depend upon the iron content. Some minerals, such as monazite and other cerium-bearing minerals, which contain no iron, may be sufficiently magnetic to be electromagnetically separated from non-magnetic minerals. Electro-magnetic separation is an important ore-dressing process. By varying the power of the electromagnet, minerals of varying magnetism can be separated from one another (for example, the separations of magnetite from apatite, etc., pyrite from blende, siderite from blende, wolframite from cassiterite, and monazite from magnetite and garnet). Sometimes it is necessary to roast the ore to convert feebly magnetic minerals, such as pyrite and siderite, into strongly magnetic ones. A small electromagnetic separator is used in the laboratory to separate the heavy mineral residues, from heavy liquid separations, into magnetic and non-magnetic fractions. The magnetic properties of some common minerals are as follows:

highly magnetic	magnetite, pyrrhotite
moderately magnetic	siderite, almandine, wolframite, ilmenite, hematite, chromite
weakly magnetic	tourmaline, spinel group, monazit
non-magnetic	quartz, calcite, feldspars, corundum, blende or sphalerite, cassiterite

Bodies of magnetite or pyrrhotite, the latter often accompanied by nickel-bearing minerals, may be located by mapping the variations in the magnetic field of an area, using an **airborne magnetometer**. When certain rocks crystallize, the magnetic particles in them become orientated in the Earth's magnetic field existing at that particular time and place, once the rocks have cooled below the **Curie point**. The directions of these earlier magnetic fields, and the positions occupied by the North pole at these times, can be determined on rock samples (often in the form of rock cores) in the laboratory, and this study, called **palaeomagnetics** or **palaeomagnetism**, is of great importance in studying the movement of the large Earth 'plates' during geological time, an important part of the subject known as **plate tectonics**.

Electricity

Minerals vary in their capacity for conducting electricity. Good conductors are relatively few in number, and include those minerals with a metallic lustre, such as native metals and sulphides, with the notable exception of sphalerite (zinc blende) which has a resinous or oily (non-metallic) lustre. Good conductors can be separated in the laboratory from bad ones by their attraction to a glass rod which possesses an electrostatic charge, previously induced by rubbing the rod with a silk cloth. On a larger scale, variations in conductivites can be applied by dropping a finely crushed and dried ore on to a rotating iron cyclinder which is electrically charged. Good conductors become charged and are repelled from the cylinder, whereas bad conductors are repelled to a much lesser degree. The result is that the shower of ore is separated out into a number of showers, which can each be collected separately; for example, sphalerite (a bad conductor) can be separated from pyrite (a good conductor). Very fine diamonds are also separated from their finely crushed matrix using a similar process.

Certain minerals develop an electrical charge when subjected either to a temperature change (**pyroelectric** minerals), or to a change in stress (**piezoelectric** minerals). Tourmaline is an example of a pyroelectric mineral. When a single-ended crystal of tourmaline is heated it becomes negatively charged at its sharp end (the end with terminal faces), and positively charged at the blunt end. Quartz is a piezoelectric mineral, and suitably orientated thin plates of quartz respond to extremely small variations in directed pressure. Such plates are used for frequency control in communication systems.

Radioactivity

Many minerals contain elements which are subject to radioactive decay throughout geological time. Radioactive minerals are rare, mainly occurring as salts of uranium and thorium, and can be detected in the field by using a **Geiger counter** or a **scintillometer**. However, many minerals contain certain radioactive isotopes which are subject to decay when the **unstable parent isotopes** break down into **stable daughter products**. The more common pairs of elements are given below, with the atomic weight of the element given above its symbol:

Parent isotope		Daughter product		Half-life
thorium,	^{232}Th	lead,	^{208}Pb	4 510 Ma
uranium,	^{235}U	lead,	^{207}Pb	713 Ma
uranium,	^{238}U	lead,	^{206}Pb	4 510 Ma
rubidium,	^{87}Rb	strontium,	^{87}Sr	50 000 Ma
potassium,	^{40}K	argon,	^{40}Ar	11 850 Ma
potassium,	^{40}K	calcium,	^{40}Ca	1 470 Ma
carbon,	^{14}C	nitrogen,	^{14}N	5 730 years

Decay takes place at a determined and constant rate for each unstable isotope. The half-life for each decay is given above; that is, the time at which one half of the original unstable isotope has been transformed into its stable daughter product. The above decay processes are used to determine the age of a mineral (the **radiometric age**) from the equation:

$$t = (1/\lambda) \log_e [(D/P) + 1]$$

where t is the age, λ is the decay constant, D is the concentration of daughter atoms in the mineral or rock due to radioactive decay, and P is the concentration of parent atoms in the mineral or rock (see also Henderson 1982).

A **mass spectrometer** is used to analyse for the stable and unstable isotopes in a mineral. For example, the parent–daughter pair samarium (^{147}Sa) and neodymium (^{143}Nd) have recently been used with great success to determine the age of Precambrian rocks greater than 3000 Ma old.

3
The elements of crystallography

3.1 Introduction

The ancient Greeks noticed that quartz always occurred in forms having a characteristic shape and being bounded by flat faces. From the transparency of quartz, plus the presence of inclusions in it, it was thought that quartz resulted from the freezing water under intense cold, and the name *krustallos* meaning 'clear ice' was given to it. Numerous other minerals were known to exist, each of which also occurred in forms bounded by flat faces, and the name *krustallos* came to mean any mineral showing such forms. These forms are **crystals**, and their study is called **crystallography**.

A crystal can be defined as *a homogeneous solid bounded by naturally formed plane faces which can be related to a regular internal arrangement of atoms.* Crystals may be formed either by solidification from liquid or gaseous states or by precipitation from solutions, all of which processes can be called **crystallization**.

It is the regular internal arrangement of atoms which really defines whether or not a substance is crystalline. For example, in the natural state many diamonds are found which do not possess flat faces. However, their optical and physical properties, and their internal atomic structure are unique to that of the mineral diamond. On the other hand, many jewellers produce 'paste imitations' (glass copies) of famous diamonds which, despite their complete resemblance to the real thing externally, lack the regular internal structure; in other words, these paste imitations are *not* crystalline. In this book, however, we are concerned with the identification of minerals, so that the **external form** or **morphology** of crystals demands most attention. Nevertheless, it is important to consider some of the aspects of the internal structure of crystals at this point, although detailed accounts of the *crystal chemistry* of selected minerals, and of silicate minerals in particular, will be given in later chapters.

3.2   The internal structure of minerals

Diffraction of X-rays by crystals

The French mineralogist Haüy suggested in 1782 that crystals could be constructed from tiny cleavage-fragments of the mineral (his *molécule integrante*) and the stacking arrangements of these fragments within a particular mineral would give rise to the different forms that the mineral might exhibit. In many minerals it was not possible to obtain the necessary cleavage fragment; some, such as quartz and garnet, do not possess any cleaveage, and others possess only one, so that a cleavage fragment is impossible to obtain. However, Haüy's work led to the concept of a mineral possessing a **structural unit** which would contain a complete representation of the substance of the whole crystal. Later, in the 20th century, when the atomic lattice had been proposed and studied, the concept of a tiny **unit cell** containing the essential information on the complete mineral structure was an important part of this concept. The internal structure of a mineral contains atoms, or groups of atoms, in a regular arrangement, and a three-dimensional pattern is repeated many times. The array of points at which the pattern repeats constitutes a **space lattice** (Bravais 1848), and the unit of pattern is called the **unit cell**. The true nature of crystal structure was not revealed until 1912, when Laue and other workers used crystals as three-dimensional diffraction gratings for X-rays. In 1913 the first structural analysis of crystals of sodium chloride were made by Sir W. H. Bragg and Sir L. Bragg, two British physicists.

X-rays were found to have very short wavelengths, which are comparable with the actual distance between atoms in the mineral lattice. When a beam of X-rays falls on to a crystal the crystal 'scatters' the X-rays but, by analogy with optical diffraction, in certain directions the individual scattered wavelets may recombine in phase to produce a strong, reinforced but deviated beam, a **diffracted beam**. As with optical refraction and reflection (sections 4.3 and 4.4), reinforcement occurs when the angle of incidence, the angle of 'reflection' and the X-ray beam all lie in the same plane, and the angles of incidence and reflection are equal. Such a 'reflection' would take place from a *single* row of atoms, but the X-ray beam will actually penetrate the crystal in such a way that successive rows of atoms, parallel to the first, will also reflect and be required to reinforce the reflected beam. In such a diffracted beam reinforcement will take place when the **path difference** for two successive rays is equal to $2d \sin \theta$, where d is the **spacing** between successive planes of atoms, and θ is the complement of the optical angle of incidence and reflection, that is 90° minus the angle of incidence.

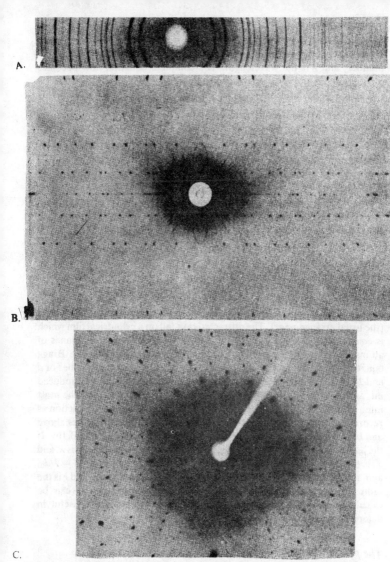

Figure 3.1 Quartz structure revealed by three X-ray techniques. (a) Powder photograph; (b) rotation diagram, crystal rotated about the c axis; (c) Laue diagram, looking down the c axis.

Furthermore, the path difference *must* equal $n\lambda$, where n is a whole number, and λ is the wavelength. The complete equation now becomes $n\lambda = 2d \sin \theta$, and is called the **Bragg equation**.

Various procedures are used for the structural analysis of minerals by X-ray techniques.

The Laue method

In the **Laue** method a beam of 'white' X-rays is used, covering a range of wavelengths. The diffracted X-rays fall on to a photographic plate and a symmetrical pattern of spots results. Laue photographs are difficult to interpret, and a number of other, more practical procedures have since been developed using monochromatic X-rays, but the Laue diagrams (see Fig. 3.1c) are still useful for the study of crystal symmetry.

Rotation photographs

The **rotation** method uses tiny, well formed crystals which can be set up with a zone axis vertical, about which one can then be either rotated or oscillated while in the path of a horizontal beam of monochromatic X-rays (the usual arrangement). The diffracted rays fall on a cylindrical film which is coaxial with the axis of rotation, so that various families of planes of atoms are, in turn, brought into angular positions such that the Bragg equation ($n\lambda = 2d \sin \theta$) is momentarily satisfied by particular values of d and 2θ for the family of planes mentioned. Diffracted beams are produced on, and at each side of, the horizontal plane through the beam. The most intense beam occurs on the horizontal plane, where the diffraction is recorded as a series of dark spots on the **equator line.** Parallel lines above and below the equator line represent successive **layer lines** (Fig. 3.1b). If the angle of elevation of the beam is β, the wavelength of the beam is λ, and d_0 is the length of side of the unit cell in this direction, then: $\sin \beta = \lambda/d_0$, and, if h_1 is the height of the first layer in the actual photograph and r is the radius of the cylindrical camera, then: $\tan \beta = h_1/r$, and d_0 can be calculated from the two equations. The rotation method is useful in determining the size of a **unit cell (UC)**.

The powder photograph and graph

The **powder** method is the most important X-ray technique, since it is commonly used by many geologists as a simple method for identifying minerals, particularly ores and other non-silicates. In this technique a beam of monochromatic X-rays falls either on to a tiny cylinder of finely ground mineral powder (the finer the powder the better), held in an amorphous cement or inside a **Lindemann glass tube**, or on to the powder pressed into an opening in a small metal frame. The cylinder of powder is inserted into a **powder camera**, whereas the pressed powder 'film' is used in a **goniometer** linked to a chart recorder. At the present time, many researchers use a beam of 'white radiation' with an energy dispersive detector in power X-ray work.

The powder camera

When the beam hits the cylinder of powder, diffraction takes place on the planes of atoms in an 'infinite' number of randomly orientated crystal particles. The diffracted X-rays will lie on a series of cones coaxial with the path of the incident beam. Initially, the diffracted rays were received on a flat photographic plate as a series of concentric dark rings but, since this plate could not receive rays deviated through large angles and since only a portion of each (diffraction) ring is needed, modern cameras (e.g. the **Debye–Scherrer** camera) are squat cylinders and the photograph is taken on a narrow strip of film laid around the inside of the camera. When the film has been exposed, it is developed and the powder pattern appears, as shown in Figure 3.1a. The powder lines appear as pairs of curves which are mirror images about the hole in the film through which the beam emerges. The complete distance s between two lines across the hole in the film is measured, and if r is the radius of the camera, then the d-spacing of *any* set of planes hkl in a *cubic* lattice is given by:

$$d_{hkl} = a_0/(h^2 + k^2 + l^2)^{1/2}$$

where a_0 is the length of edge of the cubic unit cell. Inserting this in the Bragg equation, $n\lambda = 2d \sin \theta$, and squaring, we obtain:

$$\sin 2\theta = n^2\lambda^2 \times (h^2 + k^2 + l^2)/4a_0^2$$

Since every mineral has a particular atomic structure which gives a characteristic powder photograph, the powder photograph can be matched with one of a standard set and the mineral identified.

The graph recorder

In this apparatus the powder is pressed into a frame and placed in a goniometer. The sample can rotate in the goniometer through different degrees of 2θ so that different planes of atoms are able to diffract the beam as the angle changes. As different planes 'reflect' the beam, the planes which reflect are translated into **peaks** on continuously moving graph paper. The most strongly reflecting planes produce the tallest peaks, and it is a simple task to select the ten strongest peaks in descending order of intensity. The reflecting or 'glancing' angles of 2θ for each peak can be read directly from the graph, and these can be used in the Bragg equation to obtain the d-spacing of the reflecting plane for a particular angle. A monochromatic X-ray beam (say CuK_α) is used so that the $n\lambda$ part of the equation is a constant. Then since θ is also known the d-spacings of the ten most intense reflecting planes of atoms can be calculated. Graphs for the solution of Bragg's equation exist for different wavelengths of monochromatic X-ray beams and for different values of 2θ. The results are compared with data cards for thousands of minerals prepared by the Joint Committee for Powder Data of the United States of America, and an identification can usually be made. This technique is very good for non-silicates, particularly ores and other minerals which are often quite difficult to identify by normal optical techniques.

The unit cell

Every crystal is an orderly assemblage consisting of atoms or groups of atoms arranged in a three-dimensional pattern which is repeated throughout the crystal. The **unit cell** is the smallest *complete* unit of pattern, and the whole crystal structure can be thought of as unit cells stacked together. The **symmetry** of this unit determines the external symmetry of the crystal. If the atoms or ions are represented by points, their arrangement in the crystal can be shown by the framework of the **space lattice**.

Sodium chloride (NaCl), a **cubic** mineral, is an excellent example of a mineral with a simple space lattice. In the structure, monovalent sodium ions (Na^{1+}) are bounded to chlorine ions (Cl^-) by ionic bonds. The unit cell consists of a cube with Na ions at the corners (**coigns**) of the cube and also at the centres of faces, and Cl ions at the mid-points of the edges of the cube and also in the centre. The result is as shown in Figure 3.2a. The unit cell could also be interpreted as a stack of eight smaller cubes, each having a sodium or chlorine ion at the corners.

If the ions are drawn to their exact size relative to each other, the Cl^- ion is very much larger than the Na^+ ion, at 1.81 Å compared to 0.97 Å; 1 Å (**angstrom**) equals 10^{-10}m, or 10 Å equals 1 nm (nanometre). The unit cell can therefore be considered to resemble more closely Figure 3.2b, where these two different-sized ions have been packed together in the way

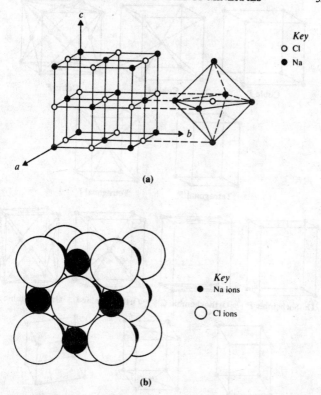

Figure 3.2   The crystal structure of sodium chloride. (a) The unit cell of sodium chloride, showing the octahedral arrangement of six sodium (Na^+) ions around one chloride (Cl^-) ion. (b) The same structure with ions drawn to their exact sizes relative to each other.

depicted in Figure 3.2a. The length of the unit cell edge in sodium chloride has been determined at 5.6402 Å (56.402 nm).

From the sodium chloride unit cell some general points arise. First, the central Cl ion is in six-fold **coordination**. This type of coordination, where one ion (in this case a negative chlorine) is bonded to six others (six positive sodiums), is also called **octahedral coordination** (see also Tables 1.5 and 1.6), since a regular octahedron can be inscribed about the immediate sodium ions surrounding the chlorine ion, as is shown in Figure 3.2a.

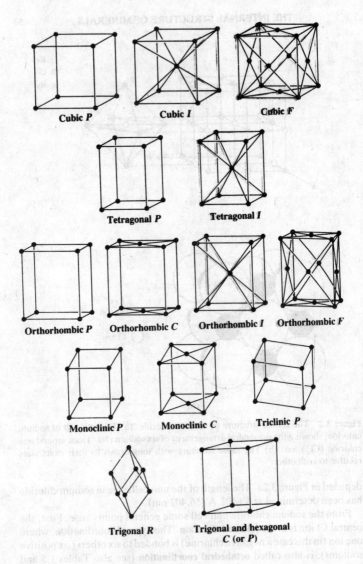

Figure 3.3 The 14 space lattices.

Secondly, the three edges of the unit cell can be used as **axes of reference** to denote the position of any plane in the crystal latttice. These axes, lettered *a*, *b* and *c* in Figure 3.2a (or sometimes known as *x*, *y* and *z*) are of *fundamental* importance in examining the morphology of crystals, and are known as the **crystallographic axes**.

Types of unit cells

There are *seven* major types of unit cell. These are the **cubic**, **tetragonal**, **orthorhombic**, **monoclinic**, **triclinic**, **trigonal** and **hexagonal**. Some of these are further subdivided to give 14 in all. This gives 14 unit cells and, hence, 14 types of space lattice that can be constructed from them, sometimes called the 14 **Bravais lattices** (named after Auguste Bravais, a French physicist in the early 18th century who worked on crystal structures). The main types are given in Figure 3.3, and are as follows:

Cubic Three arrangements occur: (1) *P space lattice – primitive* cube unit, points at the corners of a cube; (2) *I space lattice – body-centred* cube unit, points at the corners of a cube and one in the middle; (3) *F space lattice – face-centred* cube unit, points at the corners and in the centre of each face.

Tetragonal Two arrangements occur: *P* and *I space lattices – primitive* and *body-centred* tetragonal units. A tetragonal unit is a right square prism with the *a* and *b* edge lengths equal and *c* different.

Orthorhombic Four arrangements occur: *P*, *I*, *F* are as above, the *C* type has points centred on a pair of opposite faces (any pair of opposite faces will do). An orthorhombic unit is a rectangular prism with each edge a different length. The convention is that the length along the *a* axis is called *a*, that along the *b* axis is called *b*, and that along the *c* axis is called *c*.

Monoclinic Two arrangements occur: *P* and *C space lattices*. *P* is as above; the *C* type has points on the top and bottom faces, defined by the *a* and *b* axes. The monoclinic unit is a rectangular prism in which the angle between the *a* axis and the *c* axes (called β) is *not* a right angle. *The reader is referred to Figure 3.4 for a more detailed view of the axial arrangements in the seven major types of unit cell or lattice.*

Triclinic Only one arrangement occurs: *P space lattice*, as above. The triclinic unit consists of a rectangular prism with *none* of the angles between the three axes, $a\hat{}c$, $a\hat{}b$ and $b\hat{}c$, equal to a right angle. The convention is that the angles between the axes are denoted as follows: $a\hat{}b = \gamma$, $a\hat{}c = \beta$ and $b\hat{}c = \alpha$.

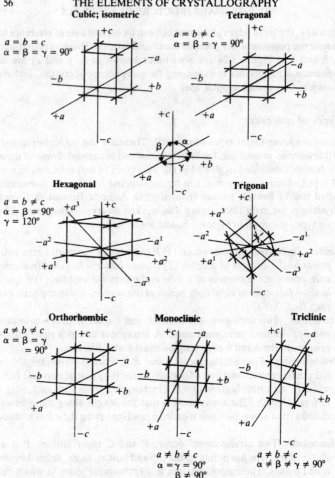

Figure 3.4 The arrangement and nomenclature of the crystallographic axes, related to the 14 unit cells shown in Figure 3.3.

Trigonal and **hexagonal** Two arrangements occur: *trigonal R* and *hexagonal* (or *trigonal*) *P*. In the trigonal system the most appropriate shape of unit cell is a rhombohedron and the only arrangement possible is one where the points occur at its corners – this is the trigonal *R* space lattice.

The hexagonal or trigonal *P* space lattice is based on a *rhombus* with an angle of 60°.

All 14 unit cells described above are illustrated in Figure 3.3.

The seven major types of unit cell or space lattice can be referred to the various sets of axes used to depict each type, and this grouping forms the basis of the classification of crystals into the **seven crystal systems**. There are *six* sets of axes in all, since the hexagonal and trigonal types use exactly the same arrangement of axes, and these are shown in Figure 3.4.

3.3   The nature of the crystalline state

Crystal symmetry

In many crystals the internal atomic symmetry is manifest in the external crystal form, which may consist of a set of naturally formed plane faces. In many crystals a regularity of arrangement of these plane **faces** occurs, and a careful study of such a crystal will reveal **elements of symmetry**:

(1) Many crystal faces mutually intersect in **edges** which are parallel. A **set of faces** which meets in parallel edges is called a **zone**, and the common edge direction is called the **zone axis**.

(2) When similar faces occur in parallel pairs on opposite sides of a crystal, the crystal, is said to possess a **centre of symmetry**. *All* faces of the crystal must occur in parallel pairs for the crystal to possess a centre. Thus a regular cube *does* possess a centre, but a cube with one corner removed does *not*. A regular tetrahedron (a four-faced crystal) does not possess a centre since a face is always opposite a point. Most naturally occurring single-ended prismatic crystals likewise do not possess a centre.

(3) Crystals may be bilaterally symmetrical about a plane (or planes), which are known as **planes of symmetry**. Such planes divide the crystal not just into two equal halves but also into two equal halves which are **mirror images** of each other. Planes of symmetry can also be called **reflection planes**.

(4) An **axis of symmetry** is defined as a line through the crystal such that after rotation through $360°/n$ (where $n \neq 1$) the crystal assumes a congruent position; an identical view being seen again; n represents the *degree* of the axis. There are *four* axes of symmetry:

   (1) If $n = 2$, the same view occurs every 180° on rotation, and the axis is termed a **diad axis**.

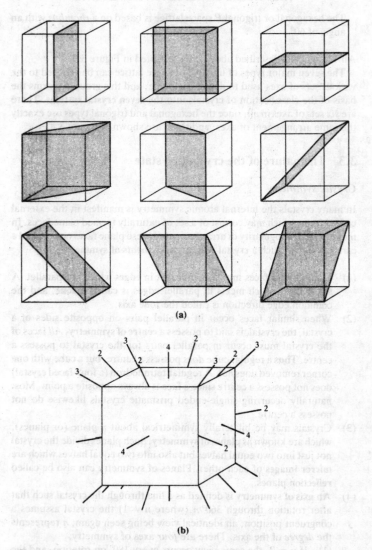

Figure 3.5 The elements of symmetry of a cube. (a) The nine planes of symmetry. (b) The various axes of symmetry: 2 = diad; 3 = triad; 4 = tetrad.

(2) If $n = 3$, the same view occurs every 120° on rotation, and the axis is termed a **triad axis**.

(3) If $n = 4$, the same view occurs every 90° on rotation, and the axis is termed a **tetrad axis**.

(4) If $n = 6$, the same view occurs every 60° on rotation, and the axis is termed a **hexad axis**.

If a cube is examined for elements of symmetry, the presence of a centre of symmetry is easily identified. In addition, a cube possesses nine planes of symmetry, three parallel to the faces, and six diagonal, as can be seen in Figure 3.5a. Finally, a cube possesses 13 axes of symmetry; three tetrads through the centres of opposite faces, four triads between opposite corners, and six diads through centres of opposite face edges (Fig. 3.5b). Thus, in summary, the full crystallographic symmetry of a *cube* is:

1 centre, 9 planes, 3 tetrads, 4 triads and 6 diads

Other crystals may possess different symmetry elements. For example, careful examination of a regular *tetrahedron* shows its symmetry to be:

no centre, 6 planes, 4 triads and 3 diads

and a *perfect* natural crystal of gypsum has a much simpler symmetry of

a centre, 1 plane and 1 diad

On the basis of careful examination of the elements of symmetry of large numbers of natural crystals, it was found that *all* crystals can be grouped into seven major divisions, the seven **crystal systems**. Each of these systems can also be subdivided into a number of **crystal classes**. These will be discussed in detail later (Section 4.6).

Form and Habit

Form is the assemblage of faces of the same type, necessitated by the symmetry, that a crystal must possess. For example, the six faces of a perfect cube are all equal squares, and, if we describe *one* face precisely and also satisfy the elements of symmetry that a cube possesses, that one face is sufficient to describe the **cubic form** that the mineral displays. Similarly, a regular octahedron shows only *one* form, since all the eight

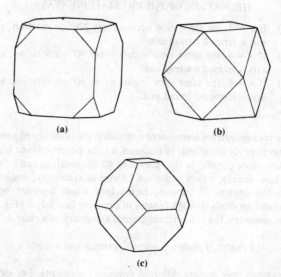

(a)

(b)

(c)

Figure 3.6 Cube modified by octahedron. (a) Cube with coigns modified by octahedral faces. (b) Cube modified by octahedron. (c) Octahedron modified by cube.

triangular faces are identical. Frequently, however, a crystal may exhibit faces of several forms at once. Figure 3.6b shows a regular cube modified by octahedral faces which truncate the corners, or coigns. In this case, although the crystal possesses six cubic faces and eight octahedral ones, in fact only two forms are present, a cube and an octahedron. Thus the word 'form' should be avoided in mineralogical studies if we merely want to convey a sense of shape. Instead, the precise definition of **form** is *the assemblage of faces required by the symmetry when only one face is given.*

So far, we have considered only perfect crystals, with forms developing so that all faces of the same form have the same size and shape. A departure from this can be seen in Figure 3.6, where faces of two forms already described, the cube and the octahedron, are developed to different degrees. In Figure 3.6a, the octahedral faces are small and the crystal resembles a cube with its corners cut off, whereas in Figure 3.6c, the octahedral faces are large and the crystal now resembles an octahedron with its 'apices' cut off. The important point is that all three figures represent the different development of the *same two forms*. This different development is called the **habit**, and can be defined as *the general aspect or shape that a crystal possesses which is produced by the relative development of the different forms*. Terms for the description of habit are given in Chapter 2 (Section 2.2). The same mineral may display several habits which may be caused by differences in temperature or pressure during crystallization, or by additions of certain trace elements, or by other, unknown, factors.

Law of constancy of angle

In 1669, Steno, a Danish scientist, published a dissertation in which measurements on quartz crystals were described and illustrated. Steno examined the **interfacial angles** in a variety of quartz crystals, from perfect examples to some which were highly distorted. Prismatic crystals of quartz were sliced at right angles to their length and the angles between adjoining faces measured, after their shape had been accurately traced on to a sheet of paper. He found that, whatever the shape of the quartz crystal, the angle between adjoining faces was always 120°. A similar result was obtained for other interfacial angles between corresponding faces on quartz crystals with different habits. This study led to the **Law of Constancy of Angle** which states: *In all crystals of the same substance, the angles between corresponding faces have a constant value.*

Goniometry

The accurate measurement of interfacial angles was of obvious importance in crystallographic studies of minerals, and the method adopted by Steno, described above, was not of sufficient accuracy to measure precise interfacial angles on most natural crystals. In 1780, the **contact goniometer** was invented. In this instrument, two flat scale bars are clamped together and the angle between them read off on a graduated scale, like a protractor. This method, although more accurate, suffered in that the crystals measured had to be quite large. However, most natural crystals are very small and would be incapable of being measured in such an instrument. Therefore, another much more accurate instrument was required which was capable of measuring the interfacial angles on extremely small crystals very precisely. Such an instrument, invented in 1809 by Wollaston, an English chemist, was the **reflecting** or **optical goniometer**.

A typical reflecting goniometer (a **horizontal-circle goniometer**) has a horizontal, circular, graduated stage which can be rotated about a vertical axis. The crystal is set up with a zone axis (set of faces with parallel edges)

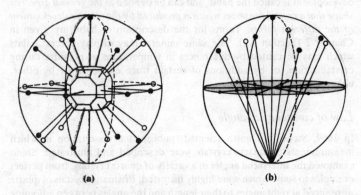

(a) **(b)**

Figure 3.7 (a) The spherical projection of the normals to the faces of a crystal. (b) The stereographic projection of the upper hemisphere of Figure 3.7a. The points made by the faces on the enclosing sphere are projected through the equatorial plane to the point P; the stereographic projection of a face is the point made on the equatorial plane (shaded).

vertical in the centre of the stage. A beam of light meets the crystal from a light source, the reflection from a face is picked up in a telescope, and the angular reading on the graduated stage noted. The crystal is rotated about the zone axis until the next face on the zone is in a position to reflect the beam, and the new reading taken. The difference between the first and second readings is *not* the angle between the two adjoining faces, but is *the angle between normals* (perpendiculars) to the two adjoining faces; that is, the actual angle measured is 180° minus the angle between adjoining faces. With such an instrument, the size and shape of individual crystal faces are incidental and thus, for any crystal, 'interfacial' angular measurements can be obtained for *all* the faces occurring in zones.

3.4 Stereographic projection

Interfacial angles (that is, angles between face normals as measured on an optical goniometer) are used in an accurate method of crystal representation called **stereographic projection**. A set of normals to all faces of a crystal are extended so that they meet the surface of an imaginary sphere enclosing the crystal, and which has its centre at the point of origin of the face normals (Fig. 3.7a). Each point of intersection between a face normal

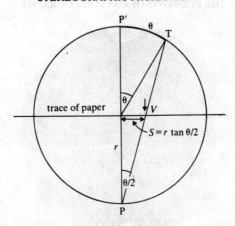

Figure 3.8 The derivation of the value of S in stereographic projection.

and the surface of the sphere is called the **pole** to that particular face. Each face of the crystal possesses such a pole. This projection is still a three-dimensional one, but in 1839, W. H. Miller, a Welsh crystallographer, suggested a means of producing a two-dimensional plot from the projection which would still preserve the angular relationship between faces. The **equatorial plane**, sometimes called the **primitive circle**, of the enclosing sphere, is used as the eventual two-dimensional surface plot. Each **pole** on the sphere's surface is projected on to the equatorial plane by joining it to the 'south pole' (bottom point) of the enclosing sphere, or, if the pole to a face is in the *lower* hemisphere of the sphere, the projection is made by joining it to the 'north pole' (upper point) of the enclosing sphere (Fig. 3.7b). In each case, the intersection of a join with the equatorial plane is the stereographic projection of the pole to a particular crystal face. All the faces of the crystal are treated in the same way, and the stereographic plot of all the faces is called a **stereogram**. Note that the poles to vertical crystal faces already plot on to the circumference of the equatorial plane and these remain unchanged in the stereogram. The distance S from the centre of the stereogram to the projection of a face on the equatorial plane is given by $S = r \tan \theta/2$, where r is the radius of the enclosing sphere, and θ is the interfacial angle between the vertical line joining the north and south 'poles' of the sphere and the pole to a particular face (Fig. 3.8). A plane passing through the centre of the sphere intersects the surface of the enclosing sphere in a *great circle*, which projects on to the stereogram in

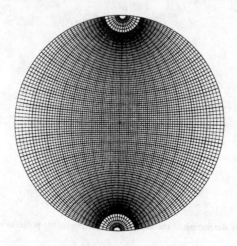

Figure 3.9   The stereographic, or Wulff, net.

different ways depending upon its inclination to the equatorial plane. If the great circle is *vertical* it plots as a straight line; if *horizontal* it plots as the equatorial circumference; and if *inclined* it plots as an arc whose chord is a diameter of the equatorial circle. The most helpful device in stereogram constructions is the **stereographic net**, an example of which is shown in Figure 3.9. On such a net, called a **Wulff net** after its inventor, great circles inclined at 2° intervals and small circles of varying angular radii are drawn, and stereographic plots of crystal faces can be made instantly using the angular distances shown on the net.

Since a stereogram accurately depicts angular relationships, the elements of symmetry for a particular crystal can also be inserted, as well as the stereographic projection of each crystal face. The result is a simple pictorial representation of the crystallographic data for a mineral, and stereograms for every mineral can be prepared. Furthermore, stereograms can be used to accurately depict the differences between the seven crystal systems and their subsidiary classes, and this technique will be used later in this chapter, but detailed descriptions of the use of stereographic projections and stereograms in crystallography are beyond the scope of this book.

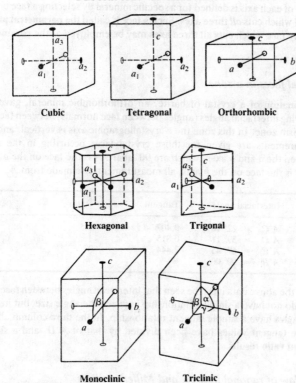

Figure 3.10 Crystallographic axes in the seven crystal systems.

3.5 Description of crystals

Crystallographic axes

Crystallographic axes should be chosen which are related to elements of symmetry. For most crystals three axes are required, the axial lengths and interaxial angles depending upon the crystal system to which the crystal belongs. The three axes intersect at the **origin**. These axes were mentioned earlier this chapter (see p. 55), and are depicted in Figure 3.10. In an orthorhombic crystal, for example, the crystallographic axes are at right angles to each other, and each is of a different length ($a \neq b \neq c$). The length of each axis is defined for a specific mineral by selecting a face on the crystal which cuts *all* three axes. Such a face is called the **parametral plane**, and *any* face which cuts all three axes may be employed as the parametral plane.

Crystal faces in a zone

Examination of a crystal of barite, an orthorhombic mineral, gave the following interfacial angles (angles between face normals) between faces in the prism zone. In this zone the *c* crystallographic axis is vertical, and the measurements are given for those crystal faces occurring in the zone between the *a* and *b* axes, which are 90° apart. *A* is the face on the *a* axis, and *B* is the face on the *b* axis; all measurements are made from *A*:

Interfacial angle		Tangent	Ratio
$A\hat{\ }C$	= 22° 10′	0.407	0.5
$A\hat{\ }D$	= 39° 11′	0.815	1
$A\hat{\ }E$	= 67° 45′	2.444	3
$A\hat{\ }B$	= 90° 00′	∞	∞

From the above data it can be seen that interfacial angles between faces in a zone do not have a simple relationship between the angle size, but instead the angles have a simple tangent relationship. In the third column above, all the tangent values have been divided by that of $A\hat{\ }D$, and a simple **tangent ratio** results.

The law of rational indices, and Miller's indices

In an orthohombic crystal a face is chosen as the **parametral plane** since that face, or that face produced, meets the three crystallographic axes *a*, *b* and *c*. The intercepts of the face on the three axes are *a*, *b* and *c*. The intercept on the *b* axis (*b*) is divided into the intercepts and the result is given as a **ratio** as follows:

$$a/b : b/b : c/b \quad \text{or} \quad a/b : 1 : c/b$$

This is called the **axial ratio** of a crystal and each mineral possesses a *unique* axial ratio.

The **Law of Rational Indices** or, more accurately, the law of rational ratios of intercepts, states that: *The indices of any crystal face are always*

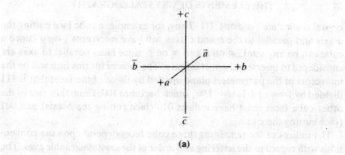

(a)

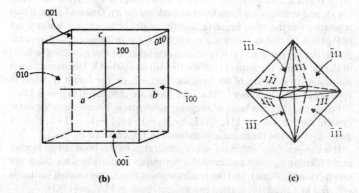

(b)   (c)

Figure 3.11 (a) The lettering of crystallographic axes. (b) The indices of a cube. (c) The indices of an octahedron.

rational numbers, and are determined by dividing the intercepts of any face into the intercepts of the parametral plane, and clearing fractions. Thus any face chosen as the parametral plane, such as the one mentioned above with intercepts *a*, *b* and *c*, will have **indices**:

$$a/a, \quad b/b, \quad c/c, \quad \text{or} \quad 111$$

and such a face is said to be a face of **unit form**. The indices (111) are also known as the **Millerian symbol** of the face.

In a cubic crystal the axes are at right angles and the same lengths ($a = b = c$) so that the parametral plane intersects each axis at the same distance from the origin; if, say, this distance is *a*, then the axial ratio of a cubic crystal is $a/a : a/a : a/a$ or 1:1:1. Thus, for example, a cube face cutting the a axis and parallel to the b and c axes will have intercepts 1 on a (since a cube can be any size), ∞ on b and ∞ on c, since faces parallel to axes are considered to intersect them at infinity. The indices for this face will be the intercepts of the parametral plane divided by those of the face; that is 111 divided by $1\infty\infty$ or $1/1, 1/\infty, 1/\infty$, which becomes 100. From this, two of the other cube faces must have indices 010 (face cutting the b axis) and 001 (face cutting the c axis).

The indices of the remaining three cube faces depend upon the conventions with regard to the **lettering** and **order of the crystallographic axes**. The vertical axis is always called c (or z), the axis running from right to left is b (or y), and that running from front to back is a (or x). One end of each axis is **positive** and the other **negative**, and the angles between axes are as given in Figure 3.10; that is, $a\hat{\ }b = \gamma, a\hat{\ }c = \beta$, and $b\hat{\ }c = \alpha$. These conventions are illustrated in Figure 3.11a. It is now obvious that the three remaining faces of the cube have the indices $(-100), (0-10)$ and $(00-1)$. The minus sign is usually written on top of the symbol; thus -1, or _bar 1_ as it is called in crystallography, becomes $\bar{1}$. The cube faces are indexed in Figure 3.11b.

From this, the eight faces of a regular octahedron, belonging to the cubic system, can be indexed 111, $-111, 1-11, -1-11, 11-1, -11-1, 1-1-1$ and $-1-1-1$ and these are also indexed in Figure 3.11c.

It is not possible, within the space constraints of this book, to go further into Millerian symbols and crystal descriptions (especially since these are seven crystal systems), and the reader wishing to take his reading further is referred to a specialist text on the subject, such as Phillips (1963).

3.6 The crystal systems

Although there are seven crystal systems and 32 crystal classes, many of these classes have no mineral representative, or are represented either by very rare minerals or by chemical compounds. Most common minerals can usually be assigned to one of 15 classes, and it is these that will be examined in detail in this section, although a summary of the others will also be given. In the **classification of crystals** that follows, the crystal systems and their classes are dealt with in order from lowest class (minimum number of symmetry elements) to highest class (always called **holosymmetric**, which has the maximum number of symmetry elements). In Section 3.2, types of unit cells were referred to six sets of axis (see also Fig. 3.4). All crystal forms, of whatever symmetry, that can be referred to the _same set_ of crystallographic axes belong to the same crystal system. The six (or seven) sets of axes are shown in Figure 3.10. The elements of symmetry for each crystal class are given in the descriptions that follow, accompanied by a sketch to illustrate their disposition relative to the crystallographic axes.

Cubic system

Three axes at right angles, and with $a = b = c$:

(1) Tetrahedral pentagonal dodecahedral class: no centre, no planes, 4 triads and 3 diads.

(2) **Didodecahedral class:** a centre, 3 planes, 4 triads and 3 diads – iron pyrite (FeS_2), and many nitrates.

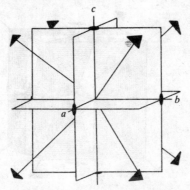

(3) **Hexatetrahedral class:** no centre, 6 planes, 4 triads and 3 diads – sphalerite or blende (ZnS), and some tellurides and phosphates.

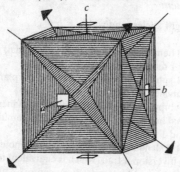

(4) Pentagonal icositetrahedral class: no centre, no planes, 3 tetrads, 4 triads and 6 diads – cuprite was thought to belong to this class, but this is doubtful.

(5) **Cubic holosymmetric** (hexoctahedral) **class:** a centre, 9 planes, 3 tetrads, 4 triads and 6 diads – examples include **free metals** such as gold, silver, copper, lead, platinum and iron, halite (NaCl), galena (PbS), fluorite (CaF_2), spinels (including magnetite Fe_3O_4) and the silicate mineral garnet ($X_3Y_2[SiO_4]_3$).

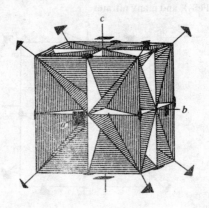

Tetragonal system

Three axes at right angles, and with $a = b \neq c$:

(a) Tetragonal hemimorphic (tetragonal pyramidal) class: no centre, no planes and 1 tetrad.

(2) Tetragonal sphenoidal class: no centre, no planes and 1 inversion tetrad (equivalent to a diad).

(3) Tetragonal bipyramidal class: a centre, no planes and 1 tetrad.

(4) Ditetragonal hemimorphic (ditetragonal pyramidal) class: no centre, 2 planes and 1 tetrad.

(5) **Tetragonal scalenohedral** (tetragonal bisphenoidal) **class:** no centre, 2 planes, 1 inversion tetrad (which is equivalent to a diad) and 2 diads (the symmetry could be written: no centre, 2 planes and 3 diads) – chalcophyrite ($CuFeS_2$), and the melilite group of silicate minerals.

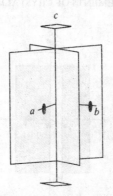

(6) Tetragonal trapezohedral class: no centre, no planes, 1 tetrad and 4 diads.

(7) **Tetragonal holosymmetric** (ditetragonal bipyramidal) **class:** a centre, 5 planes, 1 tetrad and 4 diads – rutile (TiO_2), cassiterite (SnO_2), and the silicate zircon ($ZrSiO_4$).

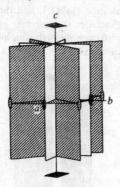

Orthorhombic system

Three axes at right angles, and with $a \neq b \neq c$:

(1) **Orthorhombic hemimorphic** (orthorhombic pyramidal) **class:** no centre, 2 planes and 1 diad – examples include the silicates natrolite ($Na_2Al_2Si_3O_{10}.2H_2O$) and hemimorphite (or smithsonite, $Zn_4(OH)_2$ $Si_2O_7.H_2O$).

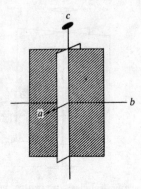

(2) Orthorhombic sphenoidal class: no centre, no planes and 3 diads – some sulphates and chromates belong to this class.

(3) **Orthorhombic holosymmetric** (orthorhombic bipyramidal) **class:** a centre, 3 planes and 3 diads – examples include the sulphates barite ($BaSO_4$) and celestine ($SrSO_4$), and the sulphide stibnite (Sb_2S_3), while silicates include the olivine group ($[Mg,Fe]_2SiO_4$), the orthorhombic pyroxenes, the orthoamphiboles, staurolite, cordierite, andalusite and sillimanite.

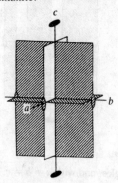

Monoclinic system

Three axes, with angle β between a and c, and with $a \neq b \neq c$:

(1) Monoclinic hemimorphic (monoclinic sphenoidal) class: no centre, no planes and 1 diad.
(2) Monoclinic clinohedral (monoclinic domatic) class: no centre, no planes and 1 inverse diad which is equivalent to 1 plane – rare, but includes the clay mineral kaolin, which belongs to the kandite group.
(3) **Monoclinic holosymmetric** (monoclinic prismatic) **class**: a centre, 1 plane and 1 diad – a *frequently encountered* class with gypsum ($CaSO_4$), and many silicates including the mica group, the clinopyroxenes, the monoclinic amphiboles, K-feldspar, the epidote group and the chlorite group.

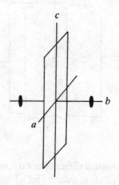

Triclinic system

Three axes, none at right angles, with $a\hat{\ }c = \beta$, $a\hat{\ }b = \gamma$, $b\hat{\ }c = \alpha$. The axes are also unequal with $a \neq b \neq c$:

(1) Asymmetric (triclinic pedial) class: no centre, no planes and no axes.
(2) **Triclinic holosymmetric** (triclinic pinacoidal) **class**: a centre, no planes and no axes – a common class including the plagioclase feldspars, kyanite and perhaps chloritoid.

Trigonal system

Four axes, three horizontal at 120° and one vertical. The three horizontal axes $a_1 = a_2 = a_3$ are different in length from c:

(1) Trigonal hemimorphic (trigonal pyramidal) class: no centre, no planes and 1 triad.

(2) Rhombohedral class: no centre, no planes and 1 inversion triad (equivalent to a centre and 1 triad) – although an uncommon class, the important carbonate dolomite ($[Ca, Mg][CO_3]_2$) belongs to it.

(3) **Ditrigonal hemimorphic** (ditrigonal pyramidal) **class**: no centre, 3 planes and 1 triad – tourmaline ($[Na,Ca][Mg,Fe^{2+},Al,Li] B_3[Al,Fe^{3+}]_6 O_{27}[OH,F]_2$) is the most important member of this class.

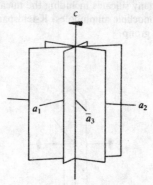

(4) **Trigonal trapezohedral class**: no centre, no planes, 1 triad and 3 diads – the important mineral quartz (SiO_2) and cinnabar (HgS).

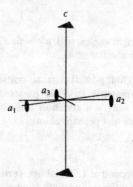

(5) Trigonal holosymmetric (ditrigonal scalenohedral) class: a centre, 3 planes, 1 triad and 3 diads – examples include the carbonates (calcite, siderite, rhodochrosite), hematite (Fe_2O_3) and brucite ($Mg[OH]_2$).

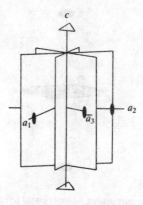

Hexagonal system

Axes identical to those of the trigonal system:

(1) **Hexagonal hemimorphic** (hexagonal pyramidal) **class:** no centre, no planes and 1 hexad – nepheline ($NaAlSiO_4$) is the most important mineral.

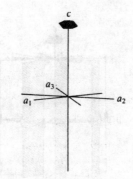

(2) Trigonal bipyramidal class: no centre, no planes and 1 inversion hexad (equivalent to 1 triad axis and 1 plane, normal to the axis).
(3) **Hexagonal bipyramidal class:** a centre, 1 plane and 1 hexad – apatite ($[Ca,F]Ca_4[PO_4]_3$) is the most important mineral.

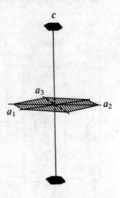

(4) Dihexagonal hemimorphic (dihexagonal pyramidal) class: no centre, 6 planes and 1 hexad.

(5) Ditrigonal bipyramidal class: no centre, 3 planes and 1 inversion hexad (equivalent to a triad axis normal to a plane of symmetry). This gives: a centre, 4 planes, 1 triad and 3 diads.

(6) Hexagonal trapezohedral: no centre, no planes, 1 hexad and 6 diads.

(7) **Hexagonal holosymmetric** (dihexagonal bipyramidal) **class:** a centre, 7 planes, 1 hexad and 6 diads – examples are rare but include beryl ($BeAl_2Si_6O_{18}$).

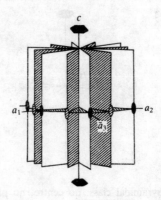

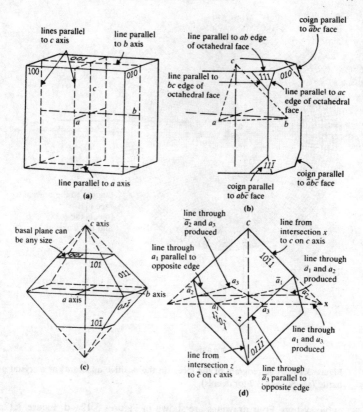

Figure 3.12 (a) A cube. (b) Part of a cube with coigns modified by octahedral faces. (c) A barite crystal. (d) A rhombohedron.

3.7 Crystal Drawings

Drawings of crystals are made on an axial cross in **clinographic projection**. A typical cross is shown in Figure 3.11a, which represents three mutually perpendicular axes. The cross needs to be amended if the angles between the axes are not right angles, as occurs in the *monoclinic* and *triclinic* crystal systems. Furthermore, the cross in Figure 3.11a is of three axes of equal lengths, and needs to be amended for crystals which do not belong to the

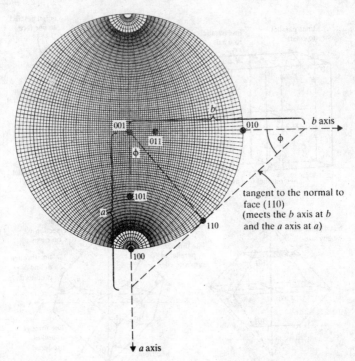

Figure 3.13 A stereogram of the faces (in the positive quadrant) of a crystal of barite (see Section 3.7 for details).

cubic system. Four drawings are shown in Figures 3.12a–d. Figure 3.12a shows a cube with the construction lines left in; the visible faces have been indexed. Figure 3.12b shows the same cube with its corners, or coigns, modified by octahedral faces, and again the construction lines have been left in. Figure 3.12c shows a crystal of barite drawn on axes which have been modified according to the axial ratio of barite, namely 0.8 : 1 : 1.3. In this the *a* axis has been reduced to 0.8 of its length, the *b* axis is of unit length, and the *c* axis has been increased to 1.3 of its length. The forms include {101} {011}, and {001}, and again the construction lines have been left in. In the final diagram, Figure 3.12d, the axial cross has been amended to enable *hexagonal* and *trigonal* crystals to be drawn on them. This necessitates three horizontal axes (all of equal length and 120° apart) being constructed, instead of the usual two, and a rhombohedron of the form {10–11} has been drawn, with the construction lines left in as before.

The methods of crystal drawing outlined above are rarely used or taught today, but they indicate very well the technique of **parallel perspective**. In this, parallel faces remain parallel, and parallel edges (such as in a zone) also remain parallel. More detailed information on crystal drawings, and on **orthographic** and **clinographic** projections can be obtained from a specialist book on crystallography, such as Phillips (1963).

3.8 Simple uses of crystal stereograms

In Figure 3.13, a stereogram of a simple orthorhombic crystal is shown, with various faces (100), (010), (001), (110), (011), (101) and the parametral plane (111) all shown in the positive quadrant. Since face (110) is on the circumference, a line from the origin to (110) is the face normal to that face, and the line at right angles to the normal on the circumference (the tangent to the point (110) when produced) will meet the a and b axes produced, at intersections a on the a axis, and b on the b axis, as shown in Figure 3.13. Then $a/b = \tan\theta = \tan$ [angle between (100) and (110)], an angle which can be directly measured on the circumference of the stereogram. Similarly, using the same methods, $c/b = \tan\pi = \tan$ [angle between (001) and (011)]. Since we are dealing with ratios, b can be $= 1$, and, therefore, $a = \tan(100)\char`^(110)$, and $c = \tan(001)\char`^(011)$. These angles can be read from the stereogram, and from them the **axial ratio** of a mineral can be obtained. If no face (011) exists, or the angle is unknown, the face (101) can be used, since $c/a = \tan\omega = \tan(001)\char`^(101)$. The value of a is already known and is substituted in the equation in order to obtain c. To illustrate this, the following interfacial angles were obtained from measurements on a crystal of barite with a simple optical goniometer:

(100)ˆ(110) = 39°	11′	(100)ˆ(010) = 90°	00′
(001)ˆ(011) = 52°	44′	(001)ˆ(010) = 90°	00′
(001)ˆ(101) = 58°	12′	(100)ˆ(001) = 90°	00′

From the above actual measurements, $a/b = \tan 39°\,11' = 0.815$, and $c/b = \tan 52°\,44' = 1.314$. Therefore the axial ratio for barite is:

$$a : b : c = 0.815 : 1 : 1.314$$

In this example the calculation has been restricted to that for a simple *orthorhombic* crystal with three axes at right angles. In the *monoclinic* and *triclinic* crystal systems such a calculation has to take account of the fact that the angles between the crystallographic axes are not always right angles. However, inter-axial angle data for particular minerals are usually commonly available. The aim here is to illustrate the use that can be made of a simple mineral stereogram, but it is not the intention to go into stereographic calculations in great detail, and the interested reader is, once again, referred to the specialist text for detailed information.

4

The optical properties
of minerals

4.1 Introduction

Certain light-dependent characters of minerals have already been considered. This chapter deals with the optical properties of those minerals which make up the common rocks of the Earth's crust, the so-called **rock-forming minerals**. A study of these rock-forming minerals entails their examination under the microscope, either as tiny grains or as thin (~0.03 mm) slices of rocks or minerals suitable for examination in **transmitted light**. In certain microscope studies, particularly of ores and opaque minerals, the thin slice of rock or ore is polished and examined in **reflected light**. The descriptions of microscopes and thin-section preparation are deferred until after the principles of optics that underlie the study have been considered.

4.2 The nature of light

Light is an electromagnetic vibration, but for the purpose of transmitted- and reflected-light microscopy, it can be considered as being simply the transfer of energy by vibrating 'particles' along a path from the source to the observer. White light consists of many **rays** ranging in wave length, through the visible spectrum, from 380 nm to 770 nm. It is convenient to consider the idealized case of a single ray of monochromatic light (that is, light of a single wavelength). A **wave** is generated by the vibration of particles lying along the path of the ray. If the light is non-polarized, the particles vibrate at random in a plane *normal* to the direction of the ray. The transverse vibrations in a ray can be considered to take place in *all* possible directions perpendicular to the direction of propogation. If, however, the light is linearly or **plane polarized** by means of a polarizing filter, then the particles simply vibrate up and down along the line x–y (see Fig. 4.1). A **wavelength** is the shortest distance between two points in

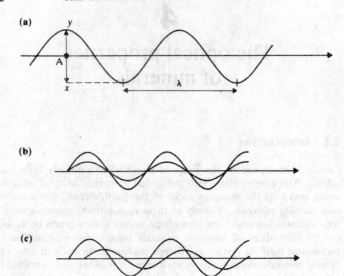

Figure 4.1 Monochromatic light of wavelength λ (a). Two waves of the same wavelength but different intensity, (b) in phase and (c) out of phase.

exactly similar positions on a wave and moving in the same direction. Two waves are said to be **in phase** when they are of equal wavelength and their positions of zero amplitude occur at exactly the same time. Light of the same wavelength and the same or different intensity (amplitude) may either be *in phase* or *out of phase*, as shown in Figure 4.1. If two *coherent* rays (originating at the same moment of time from the same source) are combined in phase, they are added together and their intensity is most enhanced. If, however, the rays are slightly out of phase, the enhancement is reduced; and if the rays have the same amplitude and a phase difference of one half of a wavelength, the vibration will be cancelled out and the amplitude will be zero.

In transmitted-light microscopy, plane-polarized white light travels up the microscope axis, which is at right angles to the plane of the rock thin section lying on the **microscope stage**. On entering an anisotropic crystalline substance rotated from the extinction position, the light can be considered to be separated into two components which travel with different velocities through the crystal. On leaving the crystal the two components may be out of phase and the **path difference** will vary for different wavelengths of light. This complexity in the light leaving the crystal is only apparent when the **analyser** is inserted and **interference colours** are generated.

4.3 Reflection

When a ray of light strikes a surface separating one medium from another, it will be, in part, *reflected* back into the medium through which it originally came *provided that*

(1) the angle of incidence, (i), equals the angle of reflection, r; and
(2) the incident ray, the reflected ray, and the normal to the surface between the two media at the point of reflection must lie in the same plane.

4.4 Refraction

When a ray of light strikes a surface separating two transparent media, it will also in part be *refracted* into the second medium (provided that the two media possess different properties). Two laws govern refraction:

(1) The sines of the angles of the incident ray ($\sin i$), and the refracted ray ($\sin r$), measured from the perpendicular to the surface at the point of contact, always bear a definite ratio to one another; and
(2) as with reflection, the incident ray, the refracted ray, the reflected ray, and the normal to the surface between the two media at the point of contact must lie in the same plane.

Refractive index

The refractive index of a medium is defined as the ratio of the velocity of light in the medium to that *in vacuo*. It varies with wavelength (see below), but the variation is usually small for transparent minerals, so refractive indices for 'white light' are usually quoted.

If V_1 and V_2 are the velocities of light in two different **isotropic** media (see Fig. 4.2), with i the angle of incidence and r the angle of refraction, then:

$$V_1/V_2 = bc/b\cdot c = b\cdot c \sin i/b\cdot c \sin r = \sin i/\sin r = \text{refractive index}$$

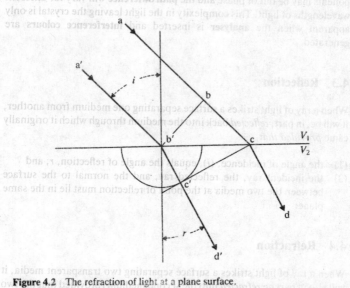

Figure 4.2 The refraction of light at a plane surface.

That is, in isotropic media the ratio between the velocities in the two media is equal to the ratio between the sine of the angle of incidence and the sine of the angle of refraction (*Snell's Law*). The refractive index is a constant and, as shown above, is the refractive index of the medium with light velocity V_2 with respect to the adjacent medium with velocity V_1. The refractive index is usually denoted by the letter **n**, but is frequently abbreviated to **RI**.

For a specific wavelength of light, the refractive index of a medium is inversely proportional to the velocity of light through the medium; that is, *the RI is proportional to 1/V*. The RI also increases as the wavelength of light decreases. Since the wavelength of red light is greater than that of blue, the refractive index of a medium in red light is smaller than the refractive index in blue light. White light entering a medium is split into the colours of the spectrum, with blue nearest the normal, since it has the greatest RI, and red furthest away. This breaking up of white light passing through a medium, due to the component wavelengths having different RIs, is called **dispersion**.

Methods of determining the refractive index of a medium

Refractometers and the critical angle

Typically, when rays of light travel from a source through a medium of *higher* RI into a medium of *lower* RI, they are partly refracted at the interface and partly reflected back into the higher RI medium. There is a particular angle of incidence called the **critical angle**, at which the ray is neither reflected nor refracted but moves along the interface between the two media. In this case the angle of refraction is 90°, but if the ray direction is reversed then the angle of incidence (i) becomes 90°, and RI = sin i/sin r; that is RI = sin 90°/sin r = 1/sin r. Thus the refractive index equals the reciprocal of the sine of critical angle. This technique is used in instruments called **refractometers** to find the RI of certain substances, particularly gemstones. In these instruments, the mineral or substance whose RI is to be determined is placed on a hemisphere of glass with a thin film of liquid, whose RI is intermediate between those of the hemisphere and the mineral, between them. A drop of the liquid is placed on the flat surface of the hemisphere and light from a mirror is refracted at the critical angle (see Fig. 4.3), which is measured by a telescope observing the emergent rays and showing a field of view that is half light and half dark. The RI of the hemisphere is given by 1/sin β, where β is the critical angle for air. If the RI

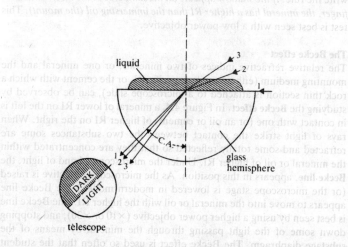

Figure 4.3 The passage of light through a refractometer.

of the hemisphere $= N$, and if n is the RI of the mineral, then $\sin i/\sin r = V$ mineral$/V$ hemisphere $= N/n$; and since $\sin i = \sin 90° = 1$, $n = N \sin r = N \sin \beta$, where β is the critical angle.

Refractometers of this type include the **Abbe refractometer** and, for gems, the **Herbert Smith** and **Tully** refractometers, where the boundary of the light and dark fields, and hence the critical angle, is read directly from a scale.

Inclined illumination

The method of **inclined illumination** is particularly useful for minerals immersed in oils. In this, a finger or card is placed below the microscope stage so that a portion of the light passing through the mineral grain is cut off at one side, and the light rays strike the contact surface between the mineral and the mount (immersing oil) obliquely. If they pass from a mineral of higher RI into a mount of lower RI, they are concentrated by refraction and internal reflection and form a light band along the inner edge of the mineral grain. If they pass from the mount of lower RI into the mineral of higher RI, they are spread out by refraction and reflection and produce a shadow along the inner edge of the mineral grain. In this case, if the finger or card is put in from the right, a shadow appears on the left side of the mineral. However, the microscope objective *reverses* the position of the image, so that we may write the rule: *If the shadow appears on the same side as the inserted card or finger, the mineral has a higher RI than the immersing oil (the mount)*. This test is best seen with a low-power objective.

The Becke effect

The relative refractive indices of two minerals, or one mineral and the mounting medium (either an oil of specific RI, or the cement with which a rock thin section is attached to a microscope slide), can be observed by studying the **Becke effect**. In Figure 4.4, a mineral of lower RI on the left is in contact with one (or an oil or cement) of higher RI on the right. When rays of light strike the contact between the two substances some are refracted and some totally reflected so that they are concentrated within the mineral or oil of higher RI. Under the microscope a band of light, the **Becke line**, appears in this position. As the microscope objective is raised (or the microscope stage is lowered in modern models) the Becke line appears to move into the mineral or oil with the higher RI. The Becke line is best seen by using a higher power objective ($\times 10$ or $\times 30$), and stopping down some of the light passing through the mineral by means of the substage diaphragm. The Becke effect is used so often that the student should observe the following rule: *As the objective lens is raised (or the*

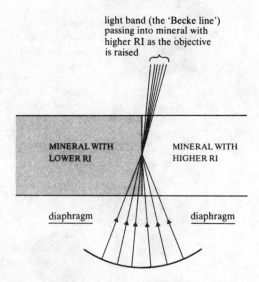

light band (the 'Becke line') passing into mineral with higher RI as the objective is raised

MINERAL WITH LOWER RI

MINERAL WITH HIGHER RI

diaphragm diaphragm

Figure 4.4 The Becke effect.

stage is lowered), the Becke line moves into the substance of higher refractive index.

The immersion method

For an accurate determination of the RI of mineral grains the **immersion method** is used. The grain is placed in an oil of known RI and the relative refractive indices noted (that is, whether higher or lower) using the Becke line method. If, for example, the mineral has a higher RI than the oil used, another oil of slightly higher RI is used, and the technique is repeated until, finally, an oil is used which is itself just higher than the mineral under examination. The RI of the mineral then must lie between the last oil used and the second last. For greater precision these last two oils can be mixed together in varying proportions until the refractive indices of the mineral and oil are identical (the Becke line appears to split into reddish and bluish lines which move in the opposite directions). The RI of the oil can then be determined precisely using a refractometer, particularly the **Leitz jelly refractometer**, which is simple and fast to use.

Suitable immersion oils include kerosene (1.448), clove oil (1.530), α-monobromnapthalene (1.658), methylene iodide (1.740), and

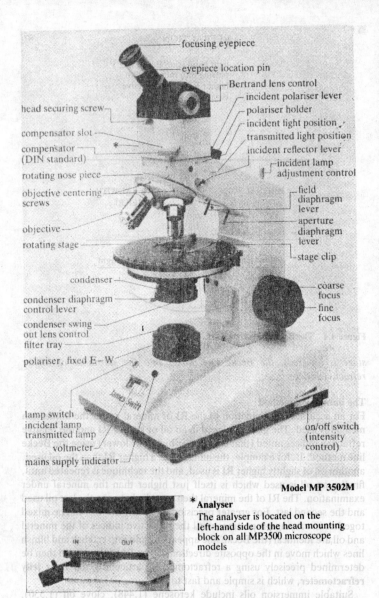

- focusing eyepiece
- eyepiece location pin
- Bertrand lens control
- incident polariser lever
- polariser holder
- incident light position
- transmitted light position
- incident reflector lever
- incident lamp adjustment control
- field diaphragm lever
- aperture diaphragm lever
- stage clip

- head securing screw
- compensator slot
- compensator (DIN standard)
- rotating nose piece
- objective centering screws
- objective
- rotating stage
- condenser
- condenser diaphragm control lever
- condenser swing out lens control
- filter tray
- polariser, fixed E – W

- coarse focus
- fine focus

- lamp switch
- incident lamp
- transmitted lamp
- voltmeter
- mains supply indicator

- on/off switch (intensity control)

Model MP 3502M

* Analyser
The analyser is located on the left-hand side of the head mounting block on all MP3500 microscope models

Figure 4.5 The Swift Student polarizing microscope (photograph courtesy of Swift Ltd).

methylene iodide saturated with sulphur (1.778). Sets of immersion oils, *many of which are toxic and must be handled with care*, whose stated refractive indices increase by regular steps, can be purchased.

The liquids used in this technique have a greater dispersion than have the minerals; that is, they have greater differences in refractive indices for different wavelengths of light. Thus a stage can be reached when the mineral and oil have the same RI for yellow light but not for blue light ($RI_{min} < RI_{oil}$) or red light ($RI_{min} > RI_{oil}$). At this point **colour fringes** appear at the edge when the **shadow** method (inclined illumination) is used. For accurate work, however, **monochromatic sodium light** (yellow light) is best, and the Becke line method is used.

4.5 The petrological microscope

A diagram of a typical transmitted-light microscope, used in the study of **thin sections** of rocks and minerals, is shown in Figure 4.5. The essential parts of the instrument are indicated in this diagram. The microscope consists simply of a barrel with an **eyepiece** or **ocular** at the top containing the microscope **crosswires**, and with an **objective lens** at the bottom. Both objectives and eyepieces can be changed for others with different properties. Just above the objective lens there is usually a slot in the microscope barrel, into which **accessory plates** can be inserted. The microscope barrel sits above the **microscope stage**, which is capable of rotation and of being locked at any point. A **vernier** attached to the stage is employed for precise angular measurements during a rotation. Below the stage is a lens system (see below for details), and at the foot a **light source**. In its simplest form this is a mirror capable of directing light from a lamp up through the microscope, but most modern microscopes possess a built-in light source. This is usually a bulb with a tungsten filament giving a yellowish tint to the light, which can be rectified by inserting a blue filter to produce a 'daylight' colour. The microscope is **focused** either by moving the microscope stage up or down (newer models), or by moving the upper barrel of the microscope up or down (older models). In either case both coarse and fine adjusting knobs are present. Other important microscope fittings are described below.

Polarizer and analyser (plane-polarized light and crossed polars)

Two polarizing films are fitted, a lower one or **polarizer** is held within a lens system located below the microscope stage, and an upper one or **analyser** is located above the microscope stage. On passing through the polarizer, light is polarized so as to vibrate in a single plane (usually east–west). This is called **plane-polarized light**, or abbreviated to **PPL**. The analyser is similar to the polarizer but oriented at right angles to the polarizer. When the analyser is inserted with the polarizer in position, and with no intervening mineral specimen, the analyser receives light vibrating in an east–west direction from the polarizer but cannot transmit it; thus the light is absorbed and the field of view is dark. The above arrangement of analyser and polarizer is referred to as **crossed polars**, or abbreviated to **XP**.

Substage diaphragms

One or two diaphragms may be located below the stage. A **field diaphragm** is used to reduce the area of light entering the thin section (but this is rarely present), and an **aperture diaphragm** can be closed to increase resolution. This diaphragm is particularly useful when examining Becke lines.

Condensing lens

This is also called the **condenser** or **convergent lens**. It is a small hemispherical lens attached to a swivel bar so that it can be inserted into the train of light. It is used to produce a cone of light and to give optimum resolution for the high-powered objectives used (×40, etc.). This lens is always used when the Bertrand lens is inserted. The entire lens system below the microscope stage, namely the polarizer, diaphragm and condensing lens, can often be racked upwards or downwards. Some microscopes which do not possess a condensing lens need, instead, to have the substage lens system racked upwards until it is just below the surface of the stage.

The Bertrand lens

This lens is used when examining **interference figures**. When it is inserted into the upper microscope tube an interference figure is produced which fills the field of view, provided that the condensing lens is also in position. *If the microscope does not possess a Bertrand lens the entire eyepiece should be removed in order to examine interference figures.*

4.6 Isotropic and anisotropic substances

Isotropic substances transmit light with equal velocity in all directions. A **ray velocity surface** represents the surface composed of all points reached by light travelling along all possible rays from a point source within a crystal in a given time. In isotropic crystals, the ray velocity surface is a *sphere*. Isotropic substances include glass, almost all fluids, and all minerals crystallizing in the *cubic* system. Another representation of the RI of a transparent medium is called an **indicatrix**. The isotropic indicatrix is also a sphere with a radius equal to r, (proportional to $1/V$, where V is the velocity of light in any direction in the substance). The ray velocity surface and the indicatrix are different but complementary representations of the RI variations for the substance.

Anisotropic crystals transmit light with different velocities in different directions, and the ray velocity surface and indicatrix of an anisotropic crystal are *ellipsoids*, which may be of two principal geometric types, **biaxial** and **uniaxial**. Anisotropic substances include all minerals crystallizing in the *orthorhombic, monoclinic* and *triclinic* systems (biaxial minerals), and also the *tetragonal, trigonal* and *hexagonal* systems (uniaxial minerals).

4.7 Isotropic minerals

As described above, the isotropic indicatrix is a sphere in which all radii equal n (the RI for a specific wavelength of light), and the wave corresponding to a particular radius of the sphere travels at right angles to the radius, with a velocity proportional to $1/n$.

4.8 Uniaxial minerals

In uniaxial minerals light is polarized so as to vibrate in two mutually perpendicular planes. Light transmitted through the mineral has a velocity which depends upon its direction of propagation. There is one direction along which all monochromatic light moves with the same velocity. This direction is parallel to the c crystallographic axis and is called the **optic axis**. Since there is only *one* optic axis, crystals with these optical properties are described as **uniaxial**.

For each wavelength of light, uniaxial crystals have *two* **principal** refractive indices, from which it follows that light travelling in any direction, except along the optic axis, consists of two mutually perpendicular, plane-polarized components with different velocities. The **uniaxial indicatrix** (Fig. 4.6) is an imaginary surface showing the variation in the refractive indices of light waves in their directions of vibration; each radius vector

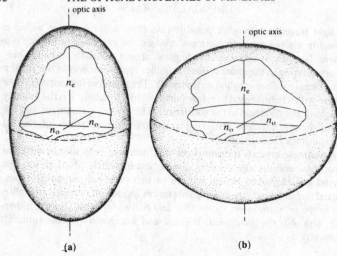

Figure 4.6 The positive (a) and negative (b) uniaxial indicatrix.

represents a vibration direction and the radius length is the RI of a wave vibrating parallel to it. Waves vibrating parallel to the equatorial radii, all of which are designated n_0, are called **ordinary** or **O waves**; and rays transmitting light parallel to the equatorial radii are called **ordinary** or **O rays**. Waves vibrating in a plane containing the optic axis (**a principal section**) have refractive indices and velocities depending upon their direction of propagation, and are called **extraordinary** or **E waves**. The RI of a wave vibrating parallel to the optic axis, designated n_c, is at a maximum or minimum depending upon whether the crystal is positive or negative. A wave vibrating in a principle section and travelling in a random direction has an RI between n_0 and n_e. Rays transmitting light vibrating in a principal section are called **extraordinary** or **E rays**:

In uniaxial minerals, light is polarized so as to vibrate in two mutually perpendicular planes, and along the principal axes this gives:

(1)   For light travelling parallel to the optic axis (*c* crystallographic axis), all light has a constant RI and a velocity proportional to $1/n_0$

(2)   For light travelling along an equatorial radius, there are two mutually perpendicular components; the component vibrating in the principal section has an index of n_e and a velocity proportional to $1/n_e$; whereas the other component vibrates perpendicular to the principal plane

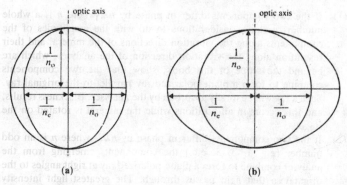

Figure 4.7 Principal sections in positive (a) and negative (b) uniaxial ray velocity surfaces.

(that is, in the equatorial plane), with an RI of n_0 and a velocity proportional to $1/n_0$.

From the above, ray velocity surfaces can be constructed for **positive** uniaxial crystals, in which $n_e > n_0$, and for **negative** uniaxial crystals, in which $n_e < n_0$, and these are shown in Figure 4.7. In the figure the wavefront of the ordinary ray is a sphere, whereas the wavefront of the extraordinary ray, the velocity of which varies with its direction, is an ellipsoid.

Negative uniaxial minerals include calcite, nepheline, beryl, tourmaline and vesuvianite. Positive uniaxial minerals include quartz, zircon and cassiterite.

Uniaxial crystals under crossed polars

Light entering a uniaxial mineral in any direction, *except parallel to the optic axis*, is split into two mutually perpendicular components, each travelling with a different velocity; that is, one component will be relatively **fast** and the other **slow**; these two components will differ *in phase* on leaving the mineral. Consider a beam of monochromatic light entering the mineral thin section from the substage **polarizer**, and vibrating in an east–west direction. The beam emerges from the mineral resolved into two mutually perpendicular components which differ in phase, and this phase difference is preserved when the components emerge from the **analyser**, situated above the microscope stage, and which resolves light into components vibrating in a north–south direction. The two components **interfere** on leaving the analyser.

(1) If the two components differ in phase by $n\lambda$, where n is a whole number, certain considerations to do with the amplitudes of the components along the vibration directions of the mineral and their resolution along the vibration direction of the analyser, which are beyond the scope of this book, show that the two components combine to form a plane-polarized ray parallel to the original ray. Such a ray will be totally absorbed by the analyser. Blackness results, and this occurs in all positions while the mineral is rotated on the stage.

(2) If the two components differ in phase by $n\lambda/2$, where n is an odd number (e.g. 1, 3, 5, etc.) the components emerging from the analyser combine to form a plane-polarized ray at right angles to the original so that light passes through. The greatest light intensity occurs in the 45° position; that is, when the two polarized light components produced by the mineral are at 45° to the position of either the analyser (north–south) or the polarizer (east–west). However, this condition of maximum illumination does not hold throughout a complete rotation of the microscope stage. When the two components of light produced by the mineral are parallel to the polarizer and analyser, the resultant waves will have a zero amplitude and blackness will result. Thus a mineral will extinguish *four times* during a complete rotation of the microscope stage (called the **extinction positions**). Any phase difference of components from the analyser results in a certain amount of light getting through, but the intensity of the light decreases as the phase difference approaches zero.

(3) If the phase difference is any other fraction of λ, the two rays will not combine to form a plane-polarized ray and the light of this wavelength will neither be entirely transmitted nor entirely absorbed by the analyser.

Birefringence and Newton's scale of interference colours

If two ways, with velocities V_1 and V_2 and refractive indices n_1 and n_2, traverse a mineral plate of thickness M, in times of t_1 and t_2 respectively, then $t_1 = M/V_1 = Mn_1$ (since velocity is inversely proportional to RI), and, in the same way, $t_2 = Mn_2$, so that $t_2 - t_1 = M(n_2 - n_1)$. That is, the **relative retardation** of the two rays is equal to the *thickness* multiplied by the *difference* in refractive index. This latter quantity $(n_2 - n_1)$ is called the **birefringence** of the mineral, and is represented in texts by the symbol δ.

If, under crossed polars, white light is passed through a crystal fragment of constant thickness, a phase difference of $(n/2)\lambda$ for some wavelength results, and the colour for that wavelength is seen. One wavelength, that of the complementary colour, has a phase difference of $n\lambda$ (where n is a whole number) and is removed from the white light by the analyser; that is, it is extinguished. Other wavelengths are partly transmitted. The colour produced is called the **interference colour** of the mineral. Such an interference colour will not change during rotation of the microscope stage but will only vary in intensity. The crystallographic orientation of the mineral fragment with respect to the microscope axis has an important bearing on the interference colour obtained. If the mineral has its optic axis parallel to the microscope axis, *no* interference colour is produced because all light travelling through the crystal has the same velocity and no path or phase difference results. A *maximum interference colour* is produced by a mineral fragment orientated so that its optic axis is at right angles to the microscope axis, because, in this orientation, light passing through the mineral is resolved into components with a maximum path difference. A fragment with intermediate orientation shows an interference colour somewhere between the maximum and minimum colours.

If λ is the wavelength of the monochromatic light used, and P is the phase difference, then:

$$P = \text{retardation}/\lambda = M(n_2 - n_1)/\lambda$$

From the above expression, in a wedge of a mineral, where there is a constant difference between the refractive indices of the two components (rays) traversing the wedge and where the thickness of the wedge varies rom zero to some finite value, the phase difference must vary. When such a mineral wedge (say, of quartz) is examined under crossed polars in monochromatic light, it shows alternating dark and light bands corresponding to phase differences of 0 λ (dark), $1/2\lambda$ (light), 1 λ (dark), $3/2$ λ (light), and so on.

If, for example, monochromatic sodium light is used, light bands are seen where the path difference is $(n/2)\lambda$, where $n = 1,3,5$, etc., and dark bands occur where the path difference is $n\lambda$, where $n = 0,1,2,3$, etc. The wavelength of sodium light is 580 nm (1nm = 10^{-9}m); thus yellow bands occur at 580/2, $3 \times 580/2$ $5 \times 500/2$, etc., and dark bands at 0, 1×580, 2×580, 3×580, etc.

White light consists of light waves with wavelengths ranging from about 390 nm (blues) to 770 nm (reds). As the quartz wedge is inserted thin end first into the microscope, the colours change from black to grey, white, yellow and then a characteristic red, which marks the highest colour of the **first-order spectrum**. The **second-order spectrum** is more sharply separated into its component colours, with violet followed by indigo, then blue, green yellow, orange, again ending at another red. Interference effects are much more pronounced in the **third-order spectrum**, and some colours are removed, the colours of this order being indigo, green-blue, yellow, red and violet. The fourth, fifth and higher orders show pale green and pinks and, at a higher birefringence, a peculiar white colour resulting from a mixing of the component colours of white light; the interference colour seen is called **white of a higher order**. It is important that the student should familiarize himself with the actual colours seen in the different orders by inserting a quartz wedge into the microscope under crossed polars, with nothing on the microscope stage, and noting the colours seen in the different spectrums. The complete sequence of colours is called Newton's scale of interference colours, but this is usually abbreviated to **Newton's scale**

Anomalous interference colours

Several minerals, including chlorite and zoisite, exhibit interference colours which are *not* present on Newton's scale, such colours being called **anomalous**. The most common anomalous colours are a dark blue, or **Berlin blue**, and a **buff-coloured Berlin brown**, which are seen under crossed polars. This phenomenon depends upon the dispersion of light by the minerals in question, and will not be considered further in this book (but see the discussion of refractive index on p. 84).

Compensation and the determination of interference colour

Light passing through a uniaxial mineral plate is resolved into two mutually perpendicular components, one *fast* and the other *slow*. Under crossed polars the mineral can be rotated into any of the four extinction positions, where the E and O rays are parallel to the polarizer and analyser respectively, that is north–south and east–west. Which ray is fast depends upon whether the mineral is positive or negative, since the components represent the extraordinary and the ordinary rays. The mineral plate is rotated through 45° and the quartz wedge inserted into the appropriate slot on the microscope (usually just above the objective lens). The **quartz wedge** used in microscope studies *is always length slow*; that is, the wedge is cut parallel to the prism zone of a quartz crystal. If the wedge is inserted along the slow component of the mineral it is clear that the effect is one of thickening the plate or of increasing retardation, so that the interference colours increase as the wedge is pushed in. This effect is called **addition**. The mineral is then rotated backwards through the extinction position and for a further 45°, so that the other component is in a position along which the quartz wedge can be inserted. This time the effect is one of thinning or decreasing retardation, so that the interference colours decrease as the wedge is pushed in, until a point is reached when the mineral interference colour is exactly 'neutralized' by the wedge. At this point the phase difference is zero and darkness is produced on the mineral plate. This effect is called **compensation**. When darkness is obtained, the wedge is withdrawn and the number of **orders** of interference colours seen are counted until the wedge is completely removed. This gives the interference colour for a particular mineral plate. It is usual to investigate several plates of the same mineral and to take the *highest interferene colour* obtained as representing the interference colour of a mineral. This technique is particularly important in minerals with high orders of interference colours, such as the olivine group and the epidote group.

Microscope accessory plates

Apart from the quartz wedge, the most important accessory plate is the **first order red** or **sensitive tint plate**, which is cut so that the interference colour displayed by it under crossed polars is red of the first order spectrum (first order red). Thus a length slow, first order red plate, inserted along the slow component of a mineral plate showing first order white interference colour, will have the interference colour of the mineral *added* to that of the plate, and a *blue* colour of the second order results; whereas if it is inserted along the fast component the interference colour of the mineral is *subtracted* from that of the plate and a *yellow* colour of the first order results. Always ascertain whether the plate is length slow or length fast, as this is extremely important in certain microscope techniques discussed later.

Interference figures in uniaxial minerals and their sign

The microscope is set up with crossed polars, the substage condensing lens in position, and the Bertrand lens inserted. If the microscope does not have a Bertrand lens, the entire eyepiece should be removed. A high power objective lens, ×40 or ×45, is also required.

Isotropic minerals (cubic minerals such as garnet) show no interference figures, whereas anisotropic minerals display interference figures consisting of isogyres and isochromatic curves. **Isogyres** are black or grey areas which may or may not change position as the microscope stage is rotated. **Isochromatic curves** are colour bands or areas which are systematically distributed with respect to the isogyres and are seen in highly birefringent minerals.

In uniaxial crystals, to obtain an interference figure, an **isotropic section** is required; that is, a mineral section cut at right angles to the *c* crystallographic axis (cut at right angles to the optic axis) so that the mineral grain appears black under crossed polars throughout a complete 360° rotation of the stage.

The **condensing lens** is a hemispherical lens which, when inserted into the light train of the microscope, changes the inclination of the incident light to the mineral section on the stage. Thus, at the centre, light passes straight through and the crystal still behaves as an isotropic section. However, as the incident light gets further away from the centre its inclination increases until, at the edge of the lens, light is passing almost horizontally along the mineral section. Path differences occur among the components of light emerging from the crystal plate, and the *loci* of similar path differences; for example, 1λ, 2λ and 3λ appear as concentric rings of darkness, and $(1/2)\lambda$, $(3/2)\lambda$ and $(5/2)\lambda$ appear as rings of bright light for monochromatic 'ight. This is because the substage condensing lens is hemispherical and produces 'cones' of light of equal inclination. The central point, where light passes through as before, is dark with a path difference of 0λ. When all other wavelengths are considerd, the resulting **interference figure** consists of alternate concentric circles of colour corresponding to Newton's chart. These are called **isochromatic curves**, and upon these circles a black cross is superimposed, the arms of which are parallel to the planes of polarization of the polarizer and analyser. The arms of this black cross are called **isogyres** (see Fig. 4.8).

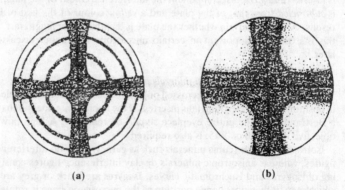

(a)   **(b)**

Figure 4.8 Uniaxial interference figures, (a) with and (b) without isochromatic curves.

Light passing straight through the centre of the condensing lens passes through the mineral plate, travelling parallel to the optic axis, and behaves as if the crystal was isotropic – remember that we are using an isotropic section of a uniaxial mineral in this study. However, the cone of light from the outer edge of the condensing lens passes into the mineral plate almost horizontally, and is resolved into two mutually perpendicular components. Since light is travelling virtually at right angles to a **basal section** (the basal section of a uniaxial mineral is an isotropic section), it is passing into the crystal along the direction of an equatorial radius. One component has an RI of n_e and a velocity proportional to $1/n_e$, whereas the other has an RI of n_0 and a velocity proportional to $1/n_0$. The extraordinary component points towards the c axis (that is, it points radially towards the centre of the black cross), and the ordinary component is at right angles to this (that is, tangential to, and concentric with, the isochromatic curves).

Determination of mineral sign

After an isotropic section of the mineral being studied has been found, the microscope is set up in the conoscopic mode; that is, with crossed polars, high-power objective and Bertrand lens inserted (or, if not present, the entire eyepiece or ocular removed) and with the substage condensing lens in position. An interference figure is then obtained and the black cross, if not centred, placed in the bottom left-hand corner of the field of view by rotating the stage. A length slow first order red plate is inserted (towards the centre of the cross) so that it is superimposed parallel to the component of light emerging radially from the mineral plate which has a velocity proportional to $1/n_e$. If $1/n_e$ is slow (compared to $1/n_0$), that is in the uniaxial indicatrix $n_e > n_0$, then $1/n_e$ is added to the slow direction of the first order red plate, and a second-order *blue* colour is observed near the centre of the isogyre in the top right or north-east quadrant. The mineral is seen to be **positive**. If $1/n_e$ was fast (in the indicatrix $n_e < n_0$), then $1/n_e$ is subtracted from the interference colour of the first order red plate and a *yellow* colour results. In this case, the mineral would be **negative**.

4.9 Biaxial minerals

Minerals in the orthorhombic, monoclinic and triclinic crystal systems are **biaxial** and are characterized by possessing three principal refractive indices at right angles to each other, and two directions normal to which light vibrates with the same velocity in all directions; that is, there are *two* optic axes (hence *bi*axial, compared with *uni*axial crystals which have only

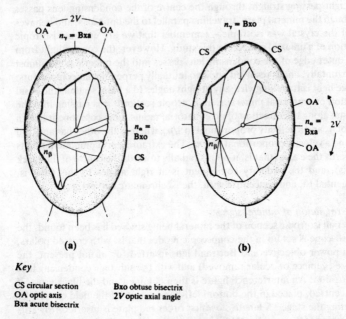

(a)　　　　　　　　　**(b)**

Key

CS circular section　　　　　　Bxo obtuse bisectrix
OA optic axis　　　　　　　　　2V optic axial angle
Bxa acute bisectrix

Figure 4.9 The positive (a) and negative (b) biaxial indicatrix.

one). The **biaxial indicatrix** is an ellipsoid with the three major semi-axes denoted n_α, n_β and n_γ. The three **principal sections** of the biaxial indicatrix are, therefore, ellipses. In the indicatrix $n_\gamma > n_\beta > n_\alpha$. From Figure 4.9a it can be seen that n_β is between n_α and n_γ in length. In the principal section containing n_α and n_γ, two radial lines occur which are also *exactly* n_β in length. Thus two **circular sections** can be constructed, each containing one of these two radii, and the n_β semi-axis in the horizontal principal section in the diagram. The perpendiculars to *both* these circular sections also lie in the principal section containing n_α and n_γ. Light travelling along these perpendiculars (that is, at right angles to the circular sections) behaves as if the crystal were isotropic, and these directions represent the **two optic axes** of a biaxial mineral. Since these two optic axes lie on the principal section containing n_α and n_γ, this section is called the **optic axial plane**, or **OAP** for short. The angle between the two optic axes is called the **optic axial angle**, or **2V**. The optical axial angle is *bisected* by a principal semi-axis, either n_α or n_γ, and this semi-axis is called the **acute bisectrix**, or **Bxa**; the other

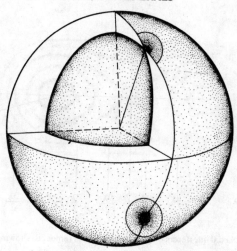

Figure 4.10 Ray velocity surfaces in three dimensions.

semi-axis is the **obtuse bisectrix**, or **Bxo**. By definition, in a positive biaxial mineral (Fig. 49a), *Bxa is always* n_γ, whereas in a negative biaxial crystal (Fig. 49b), *Bxa is always* n_α. Note that the third semi-axis, namely n_β, is always perpendicular to the OAP and is called the **optic normal**.

Plane-polarized light travelling through a biaxial crystal is resolved into two mutually perpendicular components, as happens in *all* anisotropic minerals. Light moving parallel to the vertical semi-axis n_γ is resolved into the two components with velocities proportional to $(1/n_\beta)$ and $(1/n_\alpha)$. Similarly, along the two horizontal semi-axes, the velocity components are proportional to $(1/n_\beta)$ and $(1/n_\gamma)$ along n_α, and $(1/n_\alpha)$ and $(1/n_\gamma)$ along n_β. Thus ray velocity surfaces can be constructed (as we did for the uniaxial indicatrix) and Figure 4.10 shows these for a positive biaxial crystal. In the ray velocity surface four 'dimples' appear, representing the four intersection points that exist on the surface (Fig. 4.10). In all biaxial crystals, $n_\gamma >$ $n_\beta > n_\alpha$, and therefore $(1/n_\gamma) < (1/n_\beta) < (1/n_\alpha)$. The difference between positive and negative biaxial crystals is a measure of the size of n_β, and whether it is nearer n_α or n_γ in size. This fact dictates the position of the circular sections in the indicatrix, and whether n_α or n_γ is the acute bisectrix (Bxa).

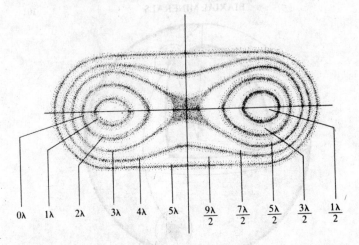

Figure 4.11 A biaxial interference figure, with isogyres removed, showing isochromatic curves.

Interference figures in biaxial minerals

The microscope is set up in the usual mode for interference figures; that is, with crossed polars, Bertrand lens in position (or, if not present, the eyepiece removed), with a high-power objective ($\times 45$ or more) in place and with the substage condensing lens in position. Two specifically orientated sections of minerals are suitable for this study. These are given below.

Mineral section perpendicular to Bxa

The interference figure may show isochromatic curves and black brushes or isogyres, the appearance and behaviour of these on rotation depending mainly upon the $2V$. Once again, it is beyond the scope of this book to discuss the detailed evolution of the isochromatic curves, and we merely state that, in monochromatic light, these curves are again dark and light, the dark curves representing the loci of points of emergence of all components of light with phase or path differences of $n\lambda$, where n is a whole number, and the light curves representing the points of emergence of components which have phase or path differences of $(n/2)\lambda$ where n is an odd number. The dark curves are sometimes known as **Cassinian curves** (see Fig. 4.11). In white light the isochromatic curves represent the same

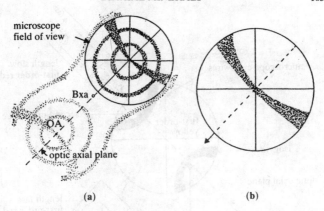

Figure 4.12 A biaxial interference figure, (a) with and (b) without isochromatic curves. In each case, $2V$ is large ($\approx 60°+$).

Newton's scale of colours, increasing outwards from two points in the field of view. The two points mark the emergence of the two optic axes, occurring as two dark points where the path differences are zero. Remember that the field of view of a microscope in the 'conoscopic mode' is such that both optic axes will be in view *only* if the optic axial angle ($2V$) of the mineral under examination is *less than* 45°, although this value depends upon the properties of the objective lens. Isogyres again consist of curves determined by loci of points of emergence of light whose planes of vibration are parallel to, or nearly parallel to, the planes of polarization of the polarizer and the analyser. Isogyres appear as crosses or hyperbolae as the microscope stage is rotated. The isogyres pass through the points of emergence of the two optic axes, and an example of a biaxial interference figure is shown in Figure 4.12. The curvature of the isogyres changes depending upon the value of $2V$; if $2V$ is large, say approaching 90°, the isogyres are virtually straight, whereas if $2V$ is small ($<30°$), they are highly curved. At one extreme, when $2V$ is 0°, as in some biotites, the two isogyres touch and the interference figure appears uniaxial.

Mineral section perpendicular to a single optic axis
This section is important in minerals with large optic axial angles, such as amphiboles, olivines and feldspars. It resembles more or less half of the interference figure described in the preceding paragraph; that is, a *single* isogyre is present, sometimes with the accompanying isochromatic curves.

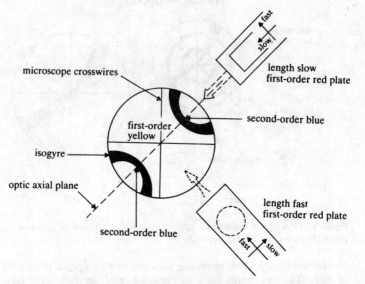

Figure 4.13 The determination of sign in an interference figure from a positive biaxial mineral, using either length slow or length fast first-order red plates.

In this type of figure the curvature of the isogyre is displayed very well, and it is possible to get a good estimate of $2V$ with a little practice.

Determination of sign in biaxial minerals

An interference figure is obtained using one of the techniques described above. Whichever indicatrix section is employed, whether normal to Bxa (*two* optic axes) or normal to an optic axis (*one* optic axis), the microscope stage is rotated so that the OAP (which is vertical in *both* sections) is at an angle of 45° to the microscope crosswires, and an accessory plate is inserted across it, as shown in Figure 4.13. Once again, the full optical explanation is outwith the scope of this book and, although other methods involving differently orientated mineral sections can be employed, the two mineral sections already described above are the most direct and the only ones dealt with here.

Bxa figure

In the 45° position the first-order red plate is inserted so that the slow direction is along the OAP. On the *concave* side of the isogyre the slow direction of the plate is parallel to a component of light whose velocity is proportional to $1/n_{Bxa}$. If the mineral is *positive* the colour *blue* will appear near the centre on the concave side of the isogyre. In all biaxial minerals, $n_\gamma > n_\beta > n_\alpha$ and, conversely, $(1/n_\gamma) < (1/n_\beta) < (1/n_\alpha)$. A positive mineral has n_γ as n_{Bxa}, and therefore $1/n_\gamma$ is slow. Since $1/n_\gamma$ is slow, this component ($1/n_{Bxa}$) is added to the red colour of the first-order plate and a blue colour of the first-order spectrum of colours is obtained. A negative mineral has n_α as n_{Bxa}, and since $1/n_\alpha$ is fast, this component is subtracted from the red colour of the first-order plate and a *yellow* colour indicating a *negative* mineral appears, again on the concave side of the isogyre.

Optic axis figure

The optic axis figure is treated in exactly the same way. The isogyre is rotated until it is pointing to the south-west, i.e. the lower left-hand corner of the field of view, and in this position the OAP is at 45° to the crosswires (running through the isogyre from south-west to north-east). The sign is obtained as described in the preceding paragraph. *When obtaining the optic sign of a biaxial mineral the investigator should always note the curvature of the isogyre, and make an estimate of the size of the 2V of the mineral at the same time.*

In some minerals which have 2V angles approaching 90°, it is extremely difficult to determine the curvature of the isogyre, since a single optic axis shows the isogyre as a straight black 'brush' at right angles to the OAP. In such a case, as would happen, for example, with most Mg-rich olivines from basic igneous intrusions, the information obtained can merely be written as $2V = 90° \pm$ from a simple microscope study. More detailed optical information requires the use of an instrument called a **universal stage** with which the size of the 2V can be measured directly, usually to an accuracy of ±2°.

Determination of extinction angle in biaxial minerals

Extinction angles in minerals are almost always measured to a prismatic cleavage or to a prism face edge. In most cases, a **prismatic section** (that is parallel to *c*) of a mineral is employed in this type of study, and such a section usually shows both the trace of any prismatic cleavage(s) that the mineral possesses (appearing as frequent thin black lines parallel to the *c* axis of the mineral), and also good anisotropic properties under crossed polars.

The microscope is set up in the usual mode for 'normal' thin-section examination, that is with a low-power objective in position and with crossed polars. A suitable prismatic section of a mineral is obtained in the thin section under examination and the mineral is rotated into extinction. At this point, as has been mentioned before, the two mutually perpendicular vibration components of light resolved by the mineral plate are parallel to the polarizer and analyser of the microscope; that is, the two light vibration components, in the extinction position, are orientated north–south and east–west – parallel to the crosswires of the microscope field of view. From previous discussion (in this chapter), one of the components is *slow* and the other is *fast*. From the extinction position, the mineral slice is rotated through 45° and the length slow first order red plate inserted along one of the two components. It is then determined whether addition (the interference colour is increased) or subtraction (the interference colour is decreased takes place). If, say, addition takes place, the light component parallel to the length of the plate will be slow, and the *extinction angle equals the angle between that component and the cleavage*. Such information is usually written:

$$\text{extinction angle} = x° \ (slow \ \hat{} \ cleavage)$$

The actual angle is measured by rotating from the extinction position to the position where the cleavage trace is parallel to the crosswire, and measuring the angle on the microscope stage with the vernier attached. The angle which is less than 45° is usually given. For example, employing the above technique, the extinction angle is found to be slow ^ cleavage =51°. This implies that the *other* component (the *fast* one) has an extinction angle of $90° - 51° = 39°$ with the cleavage. In this case the extinction angle would be written fast ^ cleavage = 39°. The extinction described above is termed **oblique extinction**. However, some minerals (for example, the mica group minerals) go into extinction when the two components resolved by the mineral plate are parallel and normal to the cleavage traces of the mineral. In this case there is *a zero extinction angle*, and the mineral is said to possess **straight extinction**.

(1) Straight extinction is shown by prismatic sections of minerals crystallizing in the tetragonal, trigonal, hexagonal and orthorhombic crystal systems.

(2) Oblique extinction is shown by most prismatic sections of minerals crystallizing in the monoclinic and triclinic crystal systems.

Note that the micas, which are monoclinic minerals, show virtually straight extinction in almost every prismatic section, but it should be remembered that the micas are 'pseudo-hexagonal' minerals.

4.10 Pleochroism

Pleochroism is *exhibited only by coloured minerals*, and a mineral is said to be **pleochroic** if it shows a change in hue or intensity of colour during rotation in plane-polarized light (that is, with the microscope substage polarizer in place). Pleochroism is caused by the different degrees of absorption of light by the mineral in different orientations. For example, in a longitudinal section of biotite, when plane-polarized light from the polarizer enters the mineral and vibrates parallel to the cleavage trace, considerable absorption of light occurs and the biotite appears dark brown in colour. If the biotite section is then rotated through 90° so that light from the polarizer enters the mineral and vibrates normal to the cleavage trace, much less absorption of light occurs and the biotite appears pale yellow.

Isotropic minerals possess the same absorption in all directions, so that all sections of an isotropic (cubic) mineral exhibit the same colour and are non-pleochroic.

Uniaxial minerals possess the same absorption normal to the optic axis, so that sections cut at right angles to the optic axis (basal sections) are non-pleochroic. A prismatic section is the best section with which to examine pleochroism in a coloured uniaxial mineral. In an elongate section of tourmaline, for example, the section is rotated so that its length is parallel to the north–south crosswire. If the polarizer is producing plane-polarized light vibrating in an east–west plane (as is customary in many microscopes), then the light emerging from the mineral section has the colour (wavelength) appropriate to the ordinary ray (see the section on the uniaxial indicatrix for details; p. 91). The mineral is then rotated through 90° until it is lying east–west, and the colour appropriate to the extra-ordinary ray (parallel to the c crystallographic axis) is seen. This gives the complete scheme for a uniaxial mineral such as tourmaline (see also Ch. 9 for specific details).

Biaxial minerals have different light-absorbing properties in all directions, except on sections cut normal to the two optic axes. Light entering these behaves as if the mineral was isotropic, but *only* with respect to the velocity of light and *not* to the selective absorption of wavelengths. Because a biaxial indicatrix has three principal refractive indices, and consequently three principal vibration directions, a **pleochroic scheme** for a biaxial mineral must identify the three main colours (wavelengths) appropriate to the three principal vibration directions. In order to identify each colour precisely *two* orientated sections are needed, and these are:

(1) *A section showing a centred interference figure.* The figure is obtained in the normal way, and set up so that the sign of the mineral can be obtained. In this, the OAP bisects the crosswires of the microscope, crossing the field of view from top right to bottom left. The vibration direction parallel to n_β is at right angles to the OAP (since n_β is the optic normal; see p. 101). The mode of the microscope is changed to normal; that is, the condensing lens and Bertrand lens are removed, and the high-power objective is replaced by a low-power objective, with uncrossed polars. The stage should be *locked* while this is going on! Then the stage is rotated through 45° (anticlockwise) so as to bring the vibration direction, n_β, into alignment with the vibration direction of the substage polarizer. The colour for this is noted (usually called β in mineral descriptions).

(2) *A section showing maximum interference colour.* Under normal microscope mode and with crossed polars, a thin section of the mineral being investigated is found showing maximum interference colour. In this section the *birefringence* will be at maximum. Now the maximum birefringence equals the difference between the maximum and minimum refractive indices of a mineral (see p. 94), and in a biaxial mineral this equals $n_\gamma - n_\alpha$. In such a section the OAP is lying in the plane of the section, and the optic normal is parallel to the microscope axis.

Having found the appropriate section it is rotated into an extinction position, where the two components (equal to the vibration directions parallel to n_γ and n_α) will be parallel to the directions of vibration of the polarizer and analyser. The signs of the two components are then determined (that is, whether fast or slow) using a first-order red plate. A quartz wedge may be needed if the interference colours are high (see p. 96). Then the fast component is identified as the vibration direction parallel to the axis n_α, and is termed α. The slow component is the vibration direction parallel to n_γ, and is called γ. The polars are uncrossed and each component is rotated in turn into the same position (say, east–west) as the vibration direction of the light from the substage polarizer, and the colour appropriate to each noted. This gives the complete pleochroic scheme for a coloured biaxial mineral.

Some minerals, such as cordierite and biotite, contain minute inclusions which may have a surrounding area which is more pleochroic than the rest of the mineral. These areas are called **pleochroic halos**, and are due to alteration of the host mineral by radioactive emanations from the inclusions. Minerals which commonly occur as inclusions showing these effects are zircon, monazite and xenotime.

Pleochroism is of special value in gemstone determination, and for this purpose an instrument called a **dichroiscope** is used. This consists of a cleavage rhomb of calcite (variety Iceland spar), contained in a tube provided with a square aperture at one end and a lens at the other. On looking through the lens at a transparent crystal placed over the aperture at the other end, two images of the aperture can be seen side by side. The images are formed by the two mutually perpendicular components into which light has been resolved after passing through the calcite crystal, namely the ordinary ray and the extraordinary ray. Since these are at right angles to each other, different colours may be seen in the case of a pleochroic mineral, depending upon the orientation of the specimen at the aperture of the instrument.

4.11 Thin sections of rocks and minerals

For the examination of rocks or minerals under the microscope, **thin sections** or slices are required. The modern method is to cut off a thin slice from a specimen using a diamond saw, polish one side to a perfectly smooth, flat surface and attach it to a thin glass slide using some type of cement (always with an RI = 1.540). The slide is then mounted in a machine where the specimen is further reduced in thickness either by milling action or by rotary grinding with different grades of carborundum powder. When the specimen is very thin – almost transparent – it is removed and finished to the correct thickness by hand grinding on a glass plate, until its thickness is approximately 0.03 mm. It is washed and dried, the specimen surface is covered with a thin glass cover slip attached with the same cement as before, and the microscope thin section is now ready for use.

It is possible for a person to prepare a thin section, however, without the modern machinery now in use. This technique involves taking a chip from the specimen under examination, and grinding one side perfectly flat with carborundum powder on a glass plate. Different grades of carborundum powder are used, starting with coarse and continuing with finer powders, remembering to wash the specimen between each application to remove coarser powder. A glass slide (75 mm × 25 mm) is taken and heated while a spot of cement (RI = 1.540, such as Lakeside, Canada balsam, etc.) is placed on it. The correct moment to stop heating balsam is judged by taking up a small quantity on forceps and, by opening them, causing a bridge of balsam to form. If the balsam has been heated sufficiently this bridge will be hard and brittle when cool. The flat side of the rock chip is pressed against the cement on the glass slide, to remove any air bubbles that form. When cool, the chip will be firmly attached to the glass slide. The next part of the operation is the same as the first part; that is, further grinding down of the rock chip, again using carborundum powders, beginning with coarse and ending with the finest powder. Great care is needed during the final stages or the rock chip can be completely rubbed away. The correct thickness is judged by examining the section under a microscope with crossed polars, to see if the polarization colours attributable to some known mineral, such as quartz (grey or white of the first order) are the usual ones. If the thickness *is* correct the thin section is thoroughly washed, and all remaining cement scraped away from around the chip. The section is then covered with a cover slip, which is attached to the section by the same cement as was used at the beginning, again pressing down to remove any air bubbles. Excess cement is removed using methylated spirits. The result is a rock or mineral thin section.

The above process requires much patience and skill for good results to be obtained, but in the field some information can be obtained by first rubbing down the rock chip on a grindstone and then using a whetstone or carborundum file for the final grinding. By these methods, and others which will suggest themselves to the practical person, rock slices can be obtained which, although thick and uneven, can yield information even by means of a simple hand-lens.

4.12 Systematic description of minerals under the petrological microscope

The optical properties of minerals include those determined in plane-polarized light (PPL) and others determined under crossed polars (XP). Examination of minerals in ordinary light (that is, with both polarizer and analyser out of the light path) is not very common nowadays but it can be useful to, say, obtain the true mineral colour, since coloured minerals are not pleochroic in ordinary (unpolarized) light.

Properties in plane-polarized light

Colour Minerals show a wide range of colour, from colourless minerals such as quartz and the feldspars, to coloured minerals such as brown biotite and green hornblende. Many coloured minerals are also pleochroic, and differently orientated sections of these may change colour during a complete rotation of the microscope stage.

Pleochroism Some minerals change colour between two extremes when the stage is rotated. The two extremes are seen *twice* during a complete rotation. Such a mineral is said to be pleochroic, and minerals such as the amphiboles, biotite, staurolite and tourmaline possess this property. Pleochroism is caused by the different degrees of absorption of light by differently orientated sections of the same mineral (see p. 107). Pleochroic halos surrounding certain inclusions are characteristic of certain minerals such as cordierite and biotite (see also p. 109).

Form and **habit** The crystalline *form* of a mineral may be deduced by studying a large number of sections of the mineral, but an excellent crystal outline with well developed faces is decidedly helpful. The shape of a mineral is referred to as its *habit*. Furthermore, a crystal in a rock may appear well shaped (**euhedral**), or with no crystal faces present at all (**anhedral**). For other terms descriptive of habit, the reader is referred to Chapter 2. Some crystals frequently contain inclusions, and in some minerals these are arranged in a characteristic pattern. Leucite contains inclusions in concentric or radial patterns, and andalusite (var. chiastolite) has inclusions in the shape of a cross.

Cleavage Cleavage appears in thin sections as one or more sets of parallel cracks, which tend to be thin, straight and evenly spaced. The number of cleavages seen depends upon the orientation of the mineral section; thus a basal section of a pyroxene will show two cleavage traces, whereas a prismatic section of the same mineral will show only one trace because both cleavages are parallel to prism faces and therefore have parallel traces in a prismatic section. When a cleavage is poorly developed it is called a **parting**. Partings are usually straight and parallel but *not* evenly spaced.

Relief The relief of a mineral is directly related to the refractive index (RI). The relief of a mineral depends upon the difference between the RI of the mineral and the RI of the enclosing cement of the thin section, which is always 1.540. The greater the difference, the rougher or more prominent the appearance of the mineral's surface. When the mineral and cement have about the same RI, the mineral's surface appears smooth. The relationship of the RI of a mineral to that of the enclosing cement, that is whether higher or lower, can be determined using the Becke line method (see p. 86). An arbitrary scheme used in thin section descriptions is as follows:

RI	Description of relief
1.40–1.50	moderate
1.50–1.58	low (similar to cement at 1.540)
1.58–1.67	moderate
1.67–1.76	high
>1.76	very high

Note from the table above that two groups of minerals show moderate relief; one group with RIs less than the cement (negative RIs) and the other with RIs more than the cement (positive RIs).

Alteration   Alteration of minerals in thin section usually appears as turbid or cloudy areas within the mineral, and the alteration products may develop along cracks or cleavages. Under crossed polars an altered mineral frequently shows **aggregate polarization** because the original homogeneous crystal has been altered to a multitude of randomly arranged crystals of the alteration product.

Properties under crossed polars

Isotropism   Minerals belonging to the *cubic* system are isotropic and remain dark under crossed polars whatever the orientation of the section. All other minerals are anisotropic and go into extinction four times during a complete rotation of a mineral section. All anisotropic minerals possess at least one orientation which acts as an isotropic section, that is remains dark during a complete rotation. Such a section is normal to an optic axis.

Birefringence and **interference colour**   The interference colour of aniso-tropic minerals under crossed polars varies depending upon the orienta-tion of the mineral section. Thus, at standard thickness, quartz may vary from grey to white, and olivine may show a wide range of colours. These are colours from Newton's scale of interference colours (see p. 95), which is divided into several orders, as follows:

Order	Colours
first	grey, white, yellow and red
second	violet, blue, green, yellow, orange and red
third	indigo, green, blue, yellow, red and violet
fourth and above	pale pinks and pale greens

The colours of this scale depend upon the thickness of the thin section (which is a constant at 0.03 mm) and the birefringence, which is a measure of the difference between the maximum and minimum refractive indices possessed by an isotropic mineral. This is usually denoted by δ. The maximum interference colour of a mineral will be shown by a mineral section exhibiting a maximum birefringence; that is, with the maximum and minimum refractive indices in the plane of the section. Such a section will not show any of the interference figures described above. Descriptive terms for interference colours are given below:

Maximum birefringence (δ)	Interference colour range	Description
0.000–0.018	first order	low
0.018–0.036	second order	moderate
0.036–0.055	third order	high
>0.055	fourth order and above	very high

If the mineral shows amomalous interference colours (see p. 96), the descriptive term used can be either 'very low' or 'anomalous colours shown'. Insertion of a first-order red plate may help to locate the position of an anomalous colour on Newton's scale if the colour is only slightly anomalous.

Interference figures All minerals except cubic minerals possess interference figures, of which there are two types, namely *uniaxial* and *biaxial*. Uniaxial figures are produced from basal sections of tetragonal, trigonal and hexagonal minerals (see p. 97)). In studying these the microscope is set up with crossed polars, a high-power objective ($\times 40$ or $\times 45$) in position, the substage condensing lens in and either the Bertrand lens in position, or, if the microscope does not have one, the complete eyepiece removed. A black cross (the isogyre) is seen and rotated, if off-centre, into the lower left-hand (south-west) corner of the field of view. The first-order red plate is inserted so that its slow direction is towards the centre of the black cross, and the colour of the north-east quadrant of the cross is noted:

blue indicates that the mineral is *positive*   (denoted +, or +ve)
yellow indicates that the mineral is *negative*   (denoted −, or −ve)

Biaxial figures are produced by suitable sections of orthorhombic, monoclinic and triclinic minerals, especially sections normal to optic axes or, if the 2*V* is small (<40°), a section cut normal to Bxa (see p. 102). The microscope is set up in the mode described in the last paragraph, and the interference figure consists of one or two arcuate black lines (the isogyres) crossing the field of view. The stage is rotated so that the isogyre is in the 45° position (relative to the microscope crosswires), and concave towards the north-east segment of the field of view. In this position the isogyre curvature can indicate the size of the optic axial angle (or 2*V*). A straight isogyre gives a 2*V* of 90°, whereas when the 2*V* is very small both isogyres are seen in the field of view and the interference figure resembles a uniaxial cross (which actually does occur when 2*V* = 0°). The length slow first-order red plate is inserted and the colour noted on the *concave* side of the isogyre:

blue means that the mineral is *positive*   (denoted +, or +ve)
yellow means that the mineral is *negative*   (denoted −, or −ve)

Note that the accessory plate mentioned here has been *length slow*. If, however, the plate is *length fast*, the colours mentioned in this section for sign determination (blue for positive and yellow for negative) will be *reversed*.

Extinction angle   Anisotropic minerals go into extinction *four times* during a complete rotation of the stage. Once in extinction the analyser can be removed and some physical property of the mineral, such as a cleavage or a prism face, can be related to the microscope crosswires. All uniaxial minerals possess straight extinction or parallel extinction, since a prismatic cleavage or prism face is parallel to a crosswire in the extinction position, when examining a prismatic section. Biaxial minerals possess either straight extinction or oblique extinction. Orthorhombic minerals show straight extinction against a prism face or prismatic cleavage when examining a prismatic section. All other biaxial minerals possess oblique extinction where, in the extinction position, the cleavage and crosswire are *not parallel*. The mineral is rotated into extinction and the angle through which the stage must be rotated to bring the cleavage and crosswire parallel is measured. Several grains are examined and the *maximum* extinction angle is usually taken; on no account should the readings be averaged out. When the mineral is in extinction, the fast or slow nature of the two light components should be determined so that the extinction angle can be given as *fast* or *slow to cleavage* (see p. 105).

Twinning This property takes the appearance of two grains of the same mineral when areas with differing extinction orientations and different interference colours have planar contacts. Often, only a single twin plane is seen, but in some minerals (particularly plagioclase feldspars and cordierite) multiple or lamellar twinning occurs with parallel twin planes.

Zoning Concentric compositional variation (**zoning**) within a single mineral may be shown by either changes in mineral colour in PPL or by changes in birefringence, or by changes in extinction orientation. These changes may be abrupt or gradational and commonly occur as a sequence from the core of a mineral (the early-formed part) to its edge (the last-formed part). Plagioclase feldspars, particularly those which occur as phenocrysts in basic extrusive igneous rocks, often show well developed zoning from a Ca-rich core to an Na-rich edge.

4.13   The microscopic investigation of ore minerals

The investigation of ore minerals and other opaque minerals by **reflected-light techniques** is an important branch of mineralogy. The study of deposition and replacement of ore minerals, the order of crystallization of ores in a mineral vein and other allied phenomena are important topics in economic mineralogy, but these are beyond the scope of this book, and only a few remarks are given here to indicate the methods of study employed.

The reflected-light microscope

Opaque minerals are examined with a microscope fitted with a **light source** placed at right angles to the microscope barrel. In front of the tungsten light, various filters for 'daylight' type light or for particular monochromatic wavelengths can be inserted. An **incident illuminator**, comprising either a glass disc, a glass disc and mirror, or a glass prism, is used to reflect light down through the objective lens on to the polished specimen. The reflected light travels back to the eyepiece, passing through the objective and the incident illuminator. Low-power objectives (up to ×10) or special high-power objectives are used. A **polarizer** can be inserted between the light source and the incident illuminator. An **analyser** is located in the microscope barrel above the objective lens and the incident illuminator, and a **Bertrand lens** may also be present. Various **diaphragms** are present near the polarizer to control the amount of light and also the resolution, focus and brightness of the specimen.

Polished section preparation

The specimen under examination may be prepared for the microscope in one of three ways: (1) as a polished specimen in a block of resin; (2) as a polished thin section; and (3) as a polished wafer or a section polished on *both* sides. Preparation involves five stages, namely *cutting* the sample with a saw to obtain a slice; *mounting* the sample on glass or resin; *grinding* the surface or surfaces flat using carborundum powder (various grades) and water; *polishing* the surface using diamond grit and an oily lubricant; and finally *buffing* the surface using gamma-alumina powder and water.

Examination of the polished section

Properties such as **colour, pleochroism, reflectance** (used in comparing minerals, from dark grey, low-reflectance minerals, to bright white, high-reflectance ones), and **bireflectance**, which is a measure of the difference between maximum and minimum reflectance values for minerals, are examined with the polarizer in position. Under crossed polars, properties such as **anisotropy, internal reflections** and **twinning** are important. Finally, properties of **hardness** are extremely useful. One of these is the **Vickers Hardness Number** (or **VHN**), which has a reasonable relationship with Moh's scale of hardness (see Ch. 2), with values ranging from 10 (talc) through 500 (apatite) to 1300 (quartz) and ~2400 (corundum). VHN values are tested using a **Vickers microindentation hardness tester**. Another hardness property is that of **polishing hardness**, causing differences in relief between minerals which can be identified using a technique called the **Kalbe light line** which is similar to the Becke line.

4.14 Microchemical tests

A minute amount of a mineral examined in a polished section can be scraped off, placed upon another glass slide, and then subjected to a series of chemical tests. The reactions to various reagents are studied under the microscope and, from the crystalline form, colour and general character of the crystallization or precipitates produced during the reactions, the nature of the ore minerals present can be determined.

4.15 X-ray diffraction studies of minerals

The identification of many minerals, particularly ore minerals, is best carried out using an **X-ray diffractometer**. In this technique the mineral is ground to less than 100 mesh and placed in a goniometer, in which it is in the path of a beam of X-rays, such as Cu K_α radiation. In the **Bragg equation**:

$$n\lambda = 2d \sin \theta$$

everything is known except d, which is the interatomic distance of a set of planes of atoms in the mineral lattice which is causing the incident beam to reflect at a given angle (θ). Graphs are available for the solution of the equation for various types of radiation used. In any mineral, several planes will produce reflections of the beam, each reflection being at a different angle 2θ and having a different intensity. The ten most intensely reflecting planes are selected for a particular mineral and the d **spacing** for each plane calculated. Such a set of data is peculiar to a particular mineral and the mineral can easily be identified. This technique is particularly good for ore minerals and most non-silicates, but is not always suitable for silicates, because of the amount of elemental substitutions which can occur in the silicate lattice, resulting in changes in lattice dimensions and, therefore, d spacings which are not always diagnostic of a particular silicate mineral.

5

The occurrence of minerals

5.1 Introduction

This chapter gives an outline of the various modes of occurrence of minerals in the Earth's crust. Certain terms to be used in the detailed descriptions of minerals will be explained here.

Rocks are composed of assemblages of **minerals**, and we consider here the character, classification and occurrence of rocks. A brief account of the common rock types and their relationships is given, and mention is made of features of importance in economic mineralogy: this is followed by a summary of the main characters of mineral deposits.

5.2 Classification of rocks

According to their manner of formation, rocks are divided into *three* great classes:

(1) **Igneous rocks** are formed from molten rock, or **magma**, which has originated well below the Earth's surface, has ascended towards the surface, and has consolidated or *crystallized* as solid rock, either on the surface as **lava**, or deep within the Earth's crust, as its temperature fell.

(2) **Sedimentary rocks** are formed from the accumulation and compaction of either: (a) fragments from pre-existing rocks which have been distintegrated by erosion; (b) organic debris such as shell fragments or dead plants; or (c) material dissolved in surface waters (rivers, etc.) or ground water, which is precipitated in conditions of oversaturation; or some combination of these.

(3) **Metamorphic rocks** are formed from pre-existing rocks of *any* type, which have been subjected to increases of temperature or pressure, or both, such that the rocks are changed from the original parental material in appearance, texture and mineralogy.

Table 5.1 Major oxides in igneous rocks and their ranges of composition.

Constituent (oxide)	Range of concentration (wt.%)
SiO_2	30–78
Al_2O_3	3–34
Fe_2O_3	0– 5
FeO	0–15
MgO	0–40
CaO	0–20
Na_2O	0–10
K_2O	0–15

5.3 Igneous rocks

Chemical composition

In spite of there being almost 100 elements, only eight of these are abundant in igneous rocks on the Earth's surface. These, in decreasing order of abundance, are oxygen (O), silicon (Si), aluminium (Al), iron (Fe), calcium (Ca), sodium (Na), potassium (K) and magnesium (Mg). From thousands of chemical analyses carried out on igneous rocks, the range of concentration of the main elements (expressed as oxides) is as shown in Table 5.1.

The major constituents in Table 5.1 generally comprise more than 1% by weight of an igneous or **magmatic** rock. Minor constitutents, which comprise by weight generally between 0.1% and 1%, include H_2O^+ (which represents **combined water**, such as (OH) groups and complete water molecules), H_2O^- (which includes water absorbed on to mineral surfaces), TiO_2, P_2O_5, MnO and the volatile substances CO_2, Cl, F and S. Finally, there are **trace elements** which range in amount from several thousand ppm (parts per million; 10 000 ppm = 1%) to less than 10 ppm. The most important trace elements include nickel (Ni), chromium (Cr), barium (Ba), strontium (Sr) and rubidium (Rb).

All silicate minerals possess the silicate oxyanion $[SiO_4]^{4-}$, which resembles a tetrahedron in outline, with a silicon atom at the centre and four oxygen atoms at the apices. Modern classification of silicate minerals is based upon the **degree of polymerization** of the $[SiO_4]^{4-}$ tetrahedra.

Crystallization of magma

Silicate minerals form when igneous magma crystallizes. The $[SiO_4]^{4-}$ groups combine in different ways to produce different types of minerals, as was mentioned in the previous section. The system of classification of the major silicate minerals depends upon how many oxygens in each tetrahedron are shared with other similar tetrahedra. There are *six* major groups of silicates, as outlined below:

(1) **Nesosilicates** or **orthosilicates** or **island silicates**. These minerals contain independent tetrahedra, which are joined together by other cations (Mg^{2+}, etc.). The minerals of this group include the olivine group, the garnet group, sphene, zircon and topaz.

(2) **Cyclosilicates** or **ring silicates**. These result from tetrahedra sharing two oxygens linked together to form a ring. The minerals of this group include tourmaline and *perhaps* beryl.

(3) **Sorosilicates**. In these groups two tetrahedra share a common oxygen. Minerals in this group include the epidote group and the melilite group.

(4) **Inosilicates** or **chain silicates**. Two major groups of minerals belong to this group. The first is termed **single-chain silicates**, since the tetrahedra join to form single chains of composition $(SiO_3)_n$, and the most important minerals are the pyroxenes (the orthopyroxenes and the clinopyroxenes, including augite). The second is termed **double-chain silicates**, since the tetrahedra join to form double chains of composition $(Si_4O_{11})_n$, and the most important minerals are the amphiboles (including hornblende).

(5) **Phyllosilicates** or **sheet silicates**. In this group three oxygens are shared between tetrahedra, the structure taking the form of a sheet of composition $(Si_4O_{10})_n$. This group includes the mica group (particularly biotite and muscovite), chlorite and serpentine.

(6) **Tektosilicates** or **framework silicates**. Here, all four oxygens in a tetrahedron are shared by other tetrahedra, and a 'framework' structure results. This is a most important group and includes the feldspars (plagioclase and K-feldspar), quartz, the feldspathoids and the zeolite group.

The type of mineral crystallizing from a magma depends both on the composition and on the temperature of the magma. If the magma has high amounts of iron and magnesium (and calcium) present, the **ferromagnesian minerals** (olivine, the pyroxenes and the amphiboles) may be expected to crystallize out, together with calcium-rich plagioclase feldspar, whereas if the magma is high in silica, alumina and akalis, the K and Na **aluminosilicates** (the alkali feldspars and quartz) will crystallize (see Fig. 5.1). Other minerals appear depending upon certain compositional characteristics of

Table 5.2 Order of crystallization of minerals from a magma.

Mineral	Early High temperature X	Late Low temperature Y
olivine		
pyroxene		
amphibole		
mica (biotite)		
plagioclase-feldspar	Ca-rich_____	_____Na-rich
K-feldspar		
quartz		
mica (muscovite)		
	basic rock types	acidic rock types

the magma, for example, if the magma is low in silica but high in alkalis the feldspathoids will appear, and if the magma is high in alumina, alkalis and combined water, low in silica and variable in iron and magnesium, the micas will form.

In a crystallizing magma the **order of crystallization** of the main silicate minerals is important, and this is given in Table 5.2. In Table 5.2, minerals which crystallize early (at high temperatures) from a magma can be ascertained by inserting a vertical line at some point in the table between 'early' and 'late', and seeing which crystallizing minerals it passes through. Thus a line down from point X shows that the minerals olivine, pyroxene and Ca-rich (plagioclase) feldspar will all be crystallizing from a magma at that temperature, whereas at point Y, at a much lower temperature, the magma is crystallizing Na-rich (plagioclase) feldspar, K-feldspar, mica (biotite), quartz and perhaps amphibole.

Magma which is crystallizing minerals at point X is termed a *basic magma*, since it is high in iron, magnesium and calcium, and low in silicon and aluminium, whereas the magma crystallizing at point Y is called an **acidic magma**, since it is high in silicon, aluminium and alkalis, and low in iron, magnesium and calcium. Magma crystallizing between point X and Y is **intermediate** in composition; and magma of a composition to the left of point X (which would be crystallizing olivine, Ca-rich plagioclase feldspar and perhaps pyroxene) would be **ultrabasic** in composition.

Mineralogy of igneous rocks

The four main categories of igneous rocks based on magma composition can be recognized on their silica content as well as their mineralogy, and

Table 5.3 Composition of igneous rocks.

Rock composition	SiO₂ (wt.%)	Minerals present
ultrabasic	<45	Ca-rich plagioclase feldspar, olivine, ±pyroxene
basic	45–55	Ca-rich plagioclase feldspar, pyroxene, ±olivine, ±amphibole
intermediate	55–65	Plagioclase feldspar (Na > Ca), amphibole, mica (biotite), ±quartz, ±pyroxene
acidic	>65	Na-rich plagioclase feldspar, K-feldspar, quartz, mica (biotite and muscovite), ±amphibole

Table 5.3 gives the percentage of silica (by weight), as well as the mineral constitutents.

Textures of igneous rocks

A variety of textures may occur in igneous rocks, each reflecting the physical conditions under which the rock crystallized. With the exception of the pyroclastic rocks and some volcanic and subvolcanic rocks, igneous rocks are composed of interlocking crystals, and are said to possess a **crystalline texture**. The size of the individual crystals is important since this is used as a criterion, together with mineralogy and chemical composition, in the simplest classification of igneous rocks. Crystal size is generally related to the length of time taken by the rock to solidify, and thus crystal size provides a simple and reasonably accurate method of identifying the three main types of igneous rocks:

(1) **Fine-grained rocks (<1 mm)** are rocks which have cooled quickly, and represent the **extrusive igneous rocks**, which include rocks crystallizing on the surface of the Earth, such as **lava flows**.
(2) **Medium-grained rocks (1–5 mm)**; in the opinion of some authors 5 mm is too large a grain size here, and the value of 3 mm has been proposed as an alternative – thus 1–3 mm can be used. These are rocks which have cooled more slowly than the extrusive rocks since they cool below the Earth's surface in minor **intrusions** such as sills, sheets and dykes, and represent the **hypabyssal igneous rocks**.

(3) **Coarse-grained rocks** (**>5 mm**); in the opinion of some authors the crystal size boundary should be 3 mm – thus the size suggested is >3 mm. These are rocks which have cooled deep within the crust of the Earth from large volumes of magma contained within very large intrusions, and erosion throughout geological time has eventually revealed them on the surface. This category includes the **plutonic igneous rocks**.

Other textures which occur in igneous rocks include the following:

Glassy   Found in some extrusive rocks; where a magmatic liquid has been quickly chilled, leaving no time for crystals to form.

Porphyritic   Found in *all* categories, but particularly in extrusive and hypabyssal rocks; in this texture large crystals, usually well formed, occur in a *finer-grained* matrix or groundmass.

Vesicular and **amygdaloidal**   Commonly found in extrusive rocks, but also observed in some hypabyssal rocks; this texture is formed when **volatiles**, present within the magma at depth, change to gases when the pressure is released as the magma reaches the surface, and the rock consolidates with gas bubbles trapped within the rock. This is called a *vesicular* texture, but this can change to an **amygdaloidal** texture if the gas-holes are filled with later minerals deposited from solutions circulating in the rock just before, or at some time after, consolidation.

Ophitic   Almost exclusively occurs in *basic* hypabyssal rocks; this is an interlocking texture of two or three minerals which crystallize more or less simultaneously in a basic hypabyssal rock such as dolerite.

Pegmatitic   Found in late-stage veins from plutonic intrusions; this texture is produced where the concentration of water and other fluxes in the late-stage residue of a plutonic magma lowers the temperature of crystallization of the minerals forming in it (quartz, K-feldspars, etc.), allowing an individual crystal to achieve a size well beyond what is meant by 'coarse-grained'. This texture is found in late-stage **veins** (called pegmatite veins) associated with large plutonic intrusions.

Textures may vary within a single igneous rock unit. Thus a lava flow or a small hypabyssal intrusion may have **chilled margins**, where the liquid rock magma was in contact with colder country rocks, but be much coarser in texture towards the centre of the flow or intrusion, where the crystallization has been slower.

Table 5.4 Classification of igneous rocks.

Composition (SiO_2 wt.%)	Rock categories		
	Extrusive <1 mm	Hypabyssal 1–5 (or 3) mm	Plutonic >5 (or 3) mm
acid (>65)	obsidian☆ rhyolite☆ dacite	quartz- and orthoclase-porphyries	various granites granodiorite
intermediate (65–55)	andesite	plagioclase-porphyries pitchstones☆	diorite tonalite
basic (55–45)	various basalts	various dolerites†	gabbros and norites
ultrabasic (<45)	komatiites and other ultrabasic lava nows	peridotites etc. in minor intrusions	peridotites‡ picrites serpentinites dunites‡

☆ Glassy and partly glassy extrusive rocks.
† Dolerite and diabase are equivalent terms.
‡ These rocks may be partly or wholly serpentinized.

Classification scheme for igneous rocks

The principal rock types for the three main categories of igneous rocks are given in Table 5.4, with a division into the four categories depending upon their SiO_2 content. In addition to these principal rock types, late-stage coarse- and fine-grained rocks associated with plutonic intrusions are pegmatite and aplite respectively, usually found in veins. Table 5.4 gives a classification of the so-called **calc-alkaline igneous rocks** (that is, rocks crystallized from magmas containing appreciable amounts of both calcium and alkalis).

Other magma types include **alkaline magmas**, rich in alkalis and with (K + Na) > Ca, which crystallize to give alkali feldspar-rich rocks such as alkali basalts, trachyte (extrusive) and its equivalents, felsite (hypabyssal) and syenite (plutonic); and **alkaline silica-undersaturated magmas**, rich in alkalis and low in silica, which crystallize to give rock types containing feldspathoids, such as nepheline- and leucite-basalts (extrusive), and their equivalents, phonolites (hypabyssal) and nepheline-syenites (plutonic).

Other important rock types, which are difficult to place in the categories already discussed, include sub-aqueous lava flows of any liquid composition, submarine lava flows, and abyssal tholeiites found in oceanic areas as basalts on the sea floor (see also Section 5.5); **pillow lavas** or **spilites**, which are hydrothermally altered submarine theoliitic basalts occurring in pillowed flows; and alkali basalts which occur, not on the sea floor, but on oceanic islands, guyots, etc.

Pyroclastic rocks

Ash and dust may be ejected from volcanoes during eruptions, and this material may fall on the land or in water to form **tuffs** or deposits of **volcanic ash**. General terms for the ejected material are **tephra** or **ejecta**, leading to deposits of **tephrite**. Occasionally, the tuffs are welded together by residual heat and compaction and these are called **welded tuffs** and, if the tephra from the eruption flows as a turbulent density current down the volcano slope, the result is **ash-flow deposits** or **ignimbrites**, which may or may not be welded. Coarse ejected material may fall back into, or near, the volcanic vent producing, after consolidation, a rock known as an **agglomerate**. The collective name for all these rocks is pyroclastic deposits, which most commonly tend to originate from calc-alkaline or alkaline magmas.

Igneous rock forms

The extrusive rocks form **lava flows** and **pyroclastic deposits** associated with **volcanoes**. The hypabyssal rocks are formed from thin (usually less than 50 m thick) intrusive sheets of magma, which are called **dykes** if they cut the original strata at high angles, and **sills** if they are intruded more or less parallel to the existing strata which is usually subhorizontal. Dykes tend to occur in **swarms**, which occupy cracks resulting from tensions acting tangentially to the surface of the Earth. The plutonic rocks may occur in several types of forms: **laccoliths** represent small 'mushroom-shaped' intrusive bodies, often of intermediate composition; **batholiths** represent huge intrusions of acidic magma, usually granitic or granodioritic in composition, and with smaller intrusive features such as **stocks** and **bosses** associated with them; and finally **lopoliths** (nowadays often called **large intrusive sheets**), which are very thick intrusions of **basic magma**, often crystallizing to give a **layered intrusion**, in which there is a variation in composition from bottom to top, due to the early-formed, high-temperature minerals settling out by various mechanisms at the base of the intrusion, and the later, lower-temperature minerals, crystallizing near the top when the magma finally solidifies. This type of layered intrusion is called a **differentiated** or **fractionated** intrusion.

Igneous rock associations

The outermost shell of the Earth, which is about 100 km thick, is called the **lithosphere** and includes the **crust** and uppermost **mantle**. Parts of the rigid lithosphere (called **plates**) can move on top of the layer beneath called the **asthenosphere**. The study of the movement of these plates is called **plate tectonics**. Plates are bounded by **active ridges**, **trenches**, **passive margins** and **strike-slip faults**. At **mid-ocean ridges**, where two plates spread apart, new **oceanic basalt** is extruded and added to each plate (**constructive plate margin**). Trenches occur where plates converge, and one oceanic plate is thrust under another and the underthrust plate is 'consumed' in a **subduction zone**. Where two plates collide along such a **destructive plate margin**, either **island arcs** may form on the surface or **mountain ranges** may form in the upper crust.

The **continental crust**, as opposed to the oceanic crust, consists of rocks formed at various times throughout the entire history of the Earth. It consists of very old **shield areas**, containing igneous and metamorphic rocks; younger **platform areas**, containing sedimentary rocks and metamorphosed sedimentary rocks overlying the shield areas; and younger fold mountain belts called **mobile belts** of variable age.

The main associations of igneous rocks are now briefly mentioned:

(1) **Calc-alkaline volcanic rocks** include arc volcanoes, with extrusive rocks such as basalt, andesite, dacite and rhyolite occurring , and also pyroclastic deposits and pyroclastic flow deposits, including ignimbrites.

(2) **Calc-alkaline plutonic rocks** are mainly found in continental settings with granitic and granodioritic batholiths, and minor diorites and gabbros. The most common occurrence is along continental margins overriding subducting oceanic lithosphere.

(3) **Continental tholeiitic basaltic rocks** and **ultramafic rocks**. In continental regions **plateau** or **flood basalts** (basalt and rhyolite) occur, with gabbros and ultabasic (ultramafic) rocks (anorthosites, dunites, etc.) being produced in much the same tectonic settings. The layered, differentiated intrusions of different sizes, from huge sheets to small sills, are included in this group.

(4) **Oceanic tholeiitic basaltic** to **ultramafic rocks**. Tholeiitic basaltic magmas occur both in oceanic *and* continental settings as island arc basalts, abyssal tholeiites and flood basalts. There are two types of

Table 5.5 Idealized section through an ophiolite

Ophiolite unit	Thickness (km)
sediment	variable, thin
pillow lavas	0.3–5.0
sheeted dolerite dykes	0.1–4.0
gabbroic rocks	0.5–2.3
ultramafic rocks	0.5–2.0
metamorphosed ultramafic rocks	of variable thickness

associations connected with this group: (a) intraplate volcanoes that are distant from plate boundaries and may appear as islands; and (b) volcanic rocks at mid-ocean ridges, and in marginal or **back-arc basins** behind island arcs. Rocks which occur include sub-alkaline tholeiites (just-saturated, basic rocks) and tholeiitic basalts. Oceanic lithosphere assemblages, called **ophiolites**, are found in an oceanic setting – either a mid-ocean rift or a dilating back-arc basin. An idealized section through a typical ophiolite sequence is shown in Table 5.5.

(5) **Oceanic alkaline rocks** are associated with intraplate magmatic activity, well away from plate boundaries. Rocks exposed on volcanic islands may range from tholeitic to highly alkaline, and from ultrabasic to felsic. Typical rocks include alkali olivine basalts, feldspathoid-bearing basalts and subordinate trachyte and phonolites.

(6) **Continental alkaline rocks** are found associated with areas of continental **rifting**. As well as alkaline lava flows, the magmatism occasionally produces rocks such as carbonatite (the only magmatic rock composed largely of carbonates of Ca, Mg and Na with subordinate feldspar, pyroxene, olivine, etc.) with associated fenite (a **metasomatic rock** produced by sodium metasomatism of the contact rocks into which the carbonatite was intruded), and nephelinite (a soda-rich, undersaturated basic rock found, for example, in the volcanic cone into which the carbonatite was intruded). **Highly potassic rocks**, such as kimberlites (potassium-rich, basic to ultrabasic diatremes, small (<1 km across) carrot-shaped intrusions, containing diamonds), are found emplaced into a stable old continental crust and are derived from well below the Earth's crust in the upper mantle at depth.

5.4 Sedimentary rocks

Sedimentary rocks are formed either from the solid debris and dissolved mineral matter produced by the mechanical and chemical breakdown of pre-existing rocks or, in some cases, from the remains of dead plants and animals. A sedimentary rock may be described according to the type of environment in which it accumulated. A **continental deposit** is laid down on land (including lakes, rivers, etc.) by rivers, ice or wind. An **intermediate deposit** is laid down in an estuary or delta, and a **marine deposit** represents material accumulating on the seashore, in the shallow waters of continental shelves, or in the deep oceans.

Mineralogy of sedimentary rocks

The composition of a sedimentary rock will depend upon: (a) the original source rocks from which the waste material was derived; (b) the chemical and mechanical resistance of each mineral component during transport; and (c) the distance travelled. Resistant minerals such as quartz are common constituents of sedimentary rocks, and some rarer minerals (garnet, rutile and zircon) have similar properties. Feldspar is less resistant, but is so common that it is a major constituent of many sedimentary rocks. Rock fragments may appear as constituents of some sedimentary rocks but these usually imply short distances of transportation. Precipitated minerals forming in the area of accumulation include the carbonates (calcite, dolomite, etc.), the sulphates (gypsum and anhydrite), chlorides, and chalcedonic silica (chert, flint, etc.). Processes which affect a sedimentary rock during and after deposition but before the onset of metamorphism and lithification are called **diagenesis**. Sedimentary rocks may be bonded together by different types of cements such as silica, carbonate or iron oxide. These affect the colour of the rock as well as its weathering properties.

Sedimentary rock textures

The most important textural feature used in the classification of the **terrigenous sedimentary rocks** is **grain size**. Other important textural features indicating mode of transport include the **degree of roundness** of grains, which is dependent upon (a) the distance travelled by grains (river sand increases in roundness down current); and (b) the length of time that the grain is in motion (beach sands are highly rounded, whereas greywackes are angular after rapid transportation by a turbidity current).

Table 5.6 The three divisions of terrigenous sedimentary rocks.

Class	Grain size (mm)	Rock types
rudaceous	>2*	conglomerate, breccia
arenaceous	0.06 –2	various arenites (sandstones), arkoses, and various wackes (including greywacke)
	0.002–0.06	siltstones
argillaceous	<0.002	shales (clay-rocks), mudstones

* More than 30% of the grains *must* be greater than 2 mm in size.

The **degree of sorting** expresses the relative homogeneity of a rock in the range of grain sizes present and indicates the nature of the transporting medium (wind-blown sands are well sorted, whereas greywackes are poorly sorted.

The **sphericity** of individual grains is a property dependent upon the original shape of the grain; garnet has a high sphericity whereas pieces of micas (slate, etc.) have a low sphericity.

Sedimentary rock structures

Most sedimentary rocks display bedding or layering, in which individual layers may be less than 10 mm thick. In **graded bedding** there is normally an upward transition from coarse to fine occurring in the particles making up the rock, related to a waning in current activity. **Cross-bedding** is the internal layering found in current-generated ripples or dunes.

Classification of sedimentary rocks

From the criteria outlined above, it is possible to formulate the following simple classification of the sedimentary rocks:

(1) **Terrigenous (clastic or detrital) sedimentary rocks**. Rocks formed from minerals or rock fragments derived from the breakdown of pre-existing rocks. There are *three* categories, given in Table 5.6. Arenites usually represent quartz-rich, well sorted arenaceous rocks; wackes represent poorly sorted arenaceous rocks, often with clays and volcanic material present; and arkoses represent fairly well sorted, feldspathic arenaceous rocks.

(2) **Chemical sedimentary rocks**. These are formed from the precipitation of salts dissolved in water, and include some of the chalcedonic silicas, and a group of rocks called **evaporites**, which includes the sulphates gypsum and anhydrite, and many halides including halite.

(3) **Organic sedimentary rocks**. These are formed from the skeletal remains of plants and animals, and include peat, coal and oil.

(4) **Limestones and dolomites**. These include rocks consisting of more than 50% carbonate, and include reef limestones, which are framework limestones composed of organic remains such as corals and algae. Rocks associated with this group also include bioclastic limestones (such as shelly limestones and chalk), oolitic limestones (which are rocks composed of spherules of calcite often formed in lagoonal conditions), marls, which are clay–limestone mixtures, travertine and tufa, which are formed by the evaporation of spring waters, and calcilutites or calcitic mudstones, which are fine-grained rocks consisting of carbonates (>50%) and muds.

5.5 Metamorphic rocks

This third major group of rocks consists of rocks in which the original textures, mineralogy and sometimes chemistry have been altered by changes in temperature and/or pressure subsequent to their original formation. The constituent minerals of metamorphic rocks are more or less in equilibrium at the new conditions of temperature and pressure. Typical metamorphic minerals include chlorite, epidote, staurolite, cordierite, garnet, andalusite, kyanite and sillimanite. Under certain conditions these are called **index minerals**.

Metamorphic rocks are affected by increases in pressure and temperature which occur at considerable depths in the Earth's crust. The important factors controlling metamorphism are (a) **confining pressure**, consisting of the pressure due to *depth* in the crust and *fluid* pressure, which is the pressure exerted by fluids in pore spaces; (b) **strain**, which is the shape and volume changes due to *stress* during deformation; and (c) **temperature**, the rate at which temperature changes with depth being known as the **geothermal gradient**.

Metamorphic rocks possess *directional fabrics* in which minerals are arranged in parallel alignment due to conditions of high stress. Platy minerals (micas, etc.) may define a *planar fabric* or •**foliation**, whereas prismatic or elongated minerals may define a *linear fabric* or **lineation**. Commonly used fabric terms are **(slaty) cleavage** in fine-grained rocks

Table 5.7 A textural classification of regional metamorphic rocks.

Metamorphic grade	Rock type	Texture present
low	slate	slaty cleavage
	phyllite	slaty cleavage
medium	schist	schistosity
high	gneiss	banding

schistosity in medium-grained rocks, and **banding** in coarse-grained rocks.

Other types of metamorphism include **thermal metamorphism**, in which rocks have been subjected to increases in temperature only (such as would occur close to an igneous intrusion); this type of metamorphism is also called **contact metamorphism**. Far away from the intrusion the rocks are hardly affected, although in some cases they may show hardening, but close to the intrusion the rocks have undergone recrystallization, and a new, very tough rock is formed in which the minerals are more or less randomly orientated, such a rock being called a **hornfels**. Extensive **metasomatic zones** occur around some intrusions due to movement of material in hydrothermal solutions circulated through the country rock. In limestones such zones are called (metasomatic) **skarns**.

Dynamic metamorphism occurs where stress is the dominant agent acting on the rocks (such as would occur under conditions of strong shearing stress, along fracture belts and low-angle faults or thrusts in the Earth's crust). The most changed rock in such a regime is called a **mylonite**, formed by pervasive ductile flow.

In all metamorphic conditions, progressive increases in temperature and/or pressure will change the **metamorphic grade** of the new rocks formed, and also increase the grain size, which becomes more coarse as the grade increases. Table 5.7 gives a simple classification for rocks produced by **regional metamorphism**.

At the highest grades melting is possible and **migmatites** may form. Low-temperature eutectic melts (granitic) may form in metamorphic rocks at appropriate temperatures, and appear as veins in a migmatite, which is a mixed rock consisting of these veins intermixed with, or invading, the original metamorphic rock material. Metamorphic rocks which have been formed under very high temperatures and pressures (such as occur at the base of the crust or in the upper mantle) are called **granulites**, and are composed almost entirely of anhydrous minerals such as pyroxenes and feldspar, in a granular texture.

Metamorphosed sandstones (quartz arenites) and limestones give rise to **quartzites** and **marbles** respectively.

Metamorphic regimes and rock associations

(1) *Regional metamorphism in orogenic belts.* Metamorphic rocks evolve within orogenic belts during the convergence of lithospheric plates, and are characterized by the presence of **greenschists** (low-grade rocks containing chlorite, epidote group minerals, albitic feldspar, etc.), medium-grade rocks, **schists**, **amphibolites** (containing amphibole, more Ca-rich feldspar, garnet, etc.), **gneisses**, and high-grade anhydrous quartzo feldspathic rocks with pyroxene ± garnet (felsic or mafic **granulites**). Regionally extensive **migmatite complexes** may also occur within some orogenic belts.

(2) *Sub-seafloor metamorphism.* Mineral assemblages in the altered oceanic crust are dominantly of a very low-pressure type, including **zeolite-bearing** rocks (lowest grade), and **greenschists** (slightly higher grade). In this environment, **ophiolite complexes** represent ancient, metamorphosed oceanic crust, the study of which has suggested that seawater circulation in the upper oceanic crust is important. The chemical reactions involved in this process, besides producing metal oxides, sulphides and carbonates, also release silica to add to deep marine cherts. The basic oceanic crust is leached of elements such as Mn, Fe, Co, Cu, Zn, Ag and Au, which are removed as chloride complexes; spilites, occurring as pillow lavas, for example, are depleted in these elements. The eventual deposition of these leached elements in various forms creates ore deposits which are potentially exploitable.

(3) *High-pressure blueschist metamorphic belts.* Submarine pillow lavas and related ophiolite suites are intimately associated with **arc–trench greywackes**, deep marine sediments, and metamorphic belts (blueschist belts) containing high-pressure mineral assemblages. A blueschist belt forms in a subduction zone – ocean trench setting, and derives its name from a blue variety of amphibole called glaucophane. Other typical minerals include jadeite, pumpellyite, garnet, etc. These blueschists have formed deep within the sediments of the trench, the '**accretionary prism**' shown in Figure 5.1. Although formed at deep levels, blueschists are frequently found as **clasts** (fragments) within a sedimentary heterogeneous rock called a **mélange**, which might form by faulting, shearing or some other mechanism. When ancient ophiolites have been investigated on land, they are intimately associated with areas of mélange.

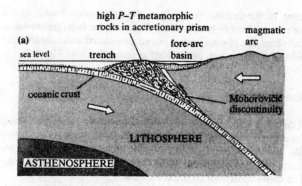

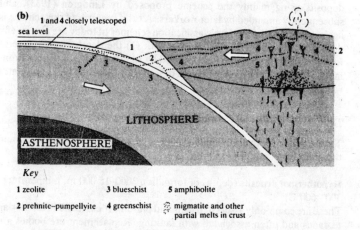

Key

1 zeolite	3 blueschist	5 amphibolite
2 prehnite–pumpellyite	4 greenschist	:::: migmatite and other partial melts in crust

Figure 5.1 Cross sections through a convergent oceanic–continental plate boundary. (a) Arc-trench sediments in an accretionary prism; a mélange may form by faulting, shearing etc. – no mixing process is excluded in its formation. (b) Metamorphic zones in a subduction zone, oceanic trench setting.

(4) *Paired metamorphic belts.* In the Mesozoic–Cenozoic terranes of the circum-Pacific region, **paired** metamorphic belts have been recognized. These involve a high-pressure (high-temperature), blueschist belt associated with ophiolites and trench sediments lying

adjacent to the oceanic trench (i.e. the subduction zone), and a low-pressure (variable-temperature) metamorphic belt (greenschists, amphibolites, etc.) found further inland, associated with calc-alkaline intrusive igneous bodies, particularly granites (see also Fig. 5.2).

5.6 Mineral deposits

A **mineral deposit** is a rock or mineral that is of economic value, and which can be worked profitably. Deposits containing valuable elements are called **ore deposits**. An **ore** is a mixture of the valuable mineral(s), termed the **ore minerals**, and the accompanying, unwanted minerals are termed the **gangue**. The metal content of an ore is its **tenor**.

Previous editions of this book have discussed the classification of ore deposits using mainly the scheme proposed by Lindgren (1933), and subsequently amended by later workers. A brief synopsis of this scheme is again given, but the modern classification schemes of today are also briefly discussed. The following will be outlined: (1) the depth–temperature of formation scheme, which was proposed by Lindgren; and (2) the environmental–rock association scheme, as described by Stanton (1972) and later workers.

Depth–temperature of formation scheme

This is divided into three major groups, and two more minor ones, and these are now dealt with in turn.

Hypothermal deposits (depths of formation 3000–15 000 m, temperature 300°–600°C)

These are commonly associated with acid plutonic rocks in old Precambrian terranes and often associated with faulting. Replacement ore bodies are most common, and often broadly tabular in shape, but veins also occur. The elements found are Au, Sn, Mo, W, Cu, Pb, Zn and As; and the important ore minerals are magnetite, pyrrhotite, cassiterite, arsenopyrite, wolframite and scheelite (with specularite, molybdenite, bornite chalcopyrite, pyrite, galena and gold also found). Common gangue minerals include garnet, biotite, muscovite, topaz, tourmaline, (high-temperature) quartz and Fe-chlorite (but plagioclase, epidote and carbonates also occur). The **wall-rocks** may suffer albitization, **tourmalinization**, perhaps sericitization in feldspars if present, and chloritization, and with development of rutile. Changes with depth are gradual, and the ore bodies are usually coarse-grained.

Mesothermal deposits (depth of formation 1200–4500 m, temperature 200–300°C)
These are associated with intrusive igneous rocks, regional fractures, and faulting. Ore bodies vary from massive to disseminated, and include **stockworks, pipes** and tabular bodies. The elements occurring include Au, Ag, Cu, As, Pb, Zn, Ni, Co, W, Mo, U and others; the important ore minerals being gold, chalcopyrite, bornite, sphalerite, galena, enargite, chalcocite, bournonite, pitchblende, niccolite, cobaltite and tetrahedrite (and with pyrite, argentite and sulphosalts also found). The main gangue minerals include quartz, sericite, chlorite, carbonates and siderite (the high-temperature minerals found in the hypothermal deposits – garnet, tourmaline, etc. – are *not* present), and with albite, epidote and smectite clays also present. The wall-rocks suffer **chloritization, sericitization** or **carbonitization**. Minerals change gradually with depth, and the ore bodies tend to be variable in grain size.

Epithermal deposits (depth of formation near surface to 1500 m, temperature 50–200°C)
These are found in Palaeozoic or younger sedimentary or igneous rocks, associated with high-level intrusive or extrusive igneous rocks and faulting. The orebodies occur as **veins**, with ore chambers developed, and also in pipes and stockworks. Elements present include Pb, Zn, Au, Ag, Hg, Sb, Cu, Se, Bi and U; the important ore minerals being Ag-rich gold, copper, marcasite, sphalerite, galena, cinnabar, stibnite, realgar, orpiment, ruby silvers, argentite and selenides (and with silver, pyrite, chalcopyrite and the tellurides also found). The main gangue minerals include chalcedonic silica, dickite clays and zeolites (and with quartz, Fe-poor chlorite, epidote group minerals, carbonates, fluorite and barite also occurring). Wall-rock alteration is rare, mineralization varies quickly with depth, and the grain size is very variable, with vugs and vein brecciation present.

The two minor groups of deposits include **leptothermal**, to cover those gradational between mesothermal and epithermal in type, and **telethermal**, of which details are given below.

Telethermal deposits (depth of formation near surface, temperature ±100°C)
These occur in sedimentary rocks or lava flows without (apparent) igneous rock present, appearing in discontinuities in the rocks. Elements present are Pb, Zn, Cd and Ge; the important ore minerals being galena, sphalerite and marcasite, etc. (similar to epithermal minerals). Gangue minerals

include calcite and dolomite, etc., and **dolomitization** and **chertification** may affect the wall-rocks. All other features are similar to those of the epithermal deposits.

Environmental – rock association scheme

This is divided into many different types of deposit depending upon their rock associations, and the more important categories are briefly discussed below.

Magmatic deposits of Cr, Fe, Ti and Pt, associated with basic and ultrabasic igneous rocks

Chromite occurs in ultrabasic rocks and anorthosites of either **stratiform (Bushveld)** type, or **podiform (Alpine)** type. In the stratiform types (for example, the Bushveld intrusion of South Africa), the chromite occurs as thin bands of chromitite, towards the base of the huge differentiated basic intrusion. These occur in association with pyroxenites, norites and gabbros in a zone of the Bushveld called the **critical zone**, which also contains the Pt-bearing Merensky Reef. Podiform chromite deposits occur in peridotite masses, which occur in orogenic mobile belts or island arc settings, appearing as lenticular bodies associated with peridotitic and gabbroic rocks.

Platinum group elements occur as primary layer-type deposits in basic plutonic intrusions such as the Bushveld intrusion (**Merensky Reef**) and the Sudbury intrusion, where they occur in layers in the intrusion associated with arsenides, sulphides and antimonides. Platinum may also occur along with chromite in basic masses.

Titanium is mainly obtained from **placer deposits** of rutile. However, primary ilmenite is found associated with anothosite–gabbro igneous complexes and may be a magmatic seggregation deposit.

Iron, in the form of magnetite (often vanadium-rich) layers, can occur in some basic intrusions (e.g. Bushveld).

Magmatic deposits of Cu, Ni, Fe (and Pt), associated with basic and ultrabasic igneous rocks

These are mainly nickel sulphide deposits which, in addition, contain iron, copper and platinum group elements. These occur in a variety of crustal settings, which are outlined below.

Orogenic areas

(a) Bodies coaeval with eugeosynclinal volcanism, including early Precambrian greenstone belts, which are subdivided into a tholeiite suite associated with picrites (with Ni) and anorthosites (without Ni), and a komatiite suite, involving extrusive ultrabasic lava flows as well as intrusive sills, which contains important Ni mineralization.

(b) Syntectonic concentric layered bodies, which are 'ring-type' complexes found particularly in Alaska, do not contain economic nickel deposits.

Non-orogenic areas (cratonic settings)

(a) Large stratiform differentiated plutonic intrusions. These (Bushveld, Sudbury, Stillwater, etc.) are important metal producers and contain great concentrations of nickel ores.

(b) Large sills (intrusive equivalents to extrusive flood basalts).

(c) Medium- and small-sized intrusions, including Skaergaard, alkaline ring complexes and kimberlite pipes.

Carbonatites

Carbonatites are associated with alkaline igneous rocks, and usually occur in stable cratonic regions which have been affected by **rift faulting**, such as the East African Rift Valley. Elements associated with carbonatites include niobium (in pyrochlore), **rare earth elements** (usually written as **REE**) associated with monazite, bastnäsite, etc., Cu, Zr, U and Th. Economic minerals include magnetite, fluorite, barite and strontianite.

Pyrometasomatic deposits (depth of formation a few km, temperature 350–800°C)

These are near or within deep-seated igneous intrusions, which have been emplaced in carbonate rocks, or sometimes schists or gneisses. Ore bodies are irregular, occurring along planar structures, and distribution within the thermal aureole is also irregular. Elements present include Fe, Cu, W, C, Zn, Pb, Mo, Sn, etc., and the main ore minerals found include magnetite, graphite, pyrrhotite, scheelite and wolframite (with specularite, gold chalcopyrite, galena, sphalerite, pyrite, molybdenite and cassiterite also present). Gangue minerals include high-temperature **skarn minerals** such as grossularite, idocrase, epidote, Mg-olivine, diopside–hedenbergite, Ca-rich plagioclase, actinolite and high-temperature quartz. Carbonates may be present. Wall-rocks are changed to skarn, marble and tactite. Zonal arrangements occur with silicates, often barren of ore minerals, at the highest temperatures next to the intrusion, followed by the sequence scheelite – magnetite – cassiterite – base metal sulphides as the temperature decreases. Copper occurs close to the contact, whereas Pb and Zn occur throughout the aureole. Rocks of pyrometasomatic deposits are usually coarse-grained.

Stockwork and disseminated deposits of Cu, Mo and Sn, associated with igneous plutonic intrusions

Deposits of this type are intimately associated with intermediate to acid plutonic intrusives and, in all deposits, the host rocks show extensive hydothermal alteration. The deposits occur either disseminated through the host rock or in a branching network of quartz veinlets which are called a **stockwork**. Copper deposits are usually associated with intrusions of porphyry, and these came to be called **porphyry copper deposits**, even though similar tin and molybdenum deposits are usually known as stockworks. The main types of deposit are now described.

Porphyry copper deposits

These are low-grade disseminated or stockwork deposits of copper containing molybdenum and gold in addition. The typical deposit is a cylindrical stock-like mass about 1.5×2.0 km in areal extent. The central part of the intrusion is porphyritic, often with an outer shell of non-porphyritic medium-grained rock. The most common hosts are acid plutonic rocks (adamellite to tonalite in composition).

Hydrothermal alteration produces a pattern of coaxial zones which are a useful guide in porphyry copper exploration. The four zones of hydrothermal wall-rock alteration, centred on the porphyry stock, are as follows:

(1) The **potassic zone** (the innermost zone), characterized by secondary orthoclase, biotite and chlorite replacing primary K-feldspar, plagioclase and mafic minerals, and with anhydrite also present.
(2) The **phyllic zone**, characterized by quartz, sericite and pyrite, and with minor chlorite, illite and rutile. This zone is itself zoned with sericite in the inner parts and clays further out.
(3) The **argillic zone**, characterized by clay minerals (kaolin in the inner parts nearer to the ore body, and montmorillonite further away), and with pyrite common.
(4) The **propylitic zone**; this, the outermost zone, is characterized by chlorite, pyrite, calcite and epidote.

The boundaries between zones are gradational from tens to hundreds of metres.

Hypogene mineralization takes the form of zones which overlap, but *may not* be coincident with the hydrothermal zones. The main zones are now outlined:

(1) The **innermost alteration zone**, which is coincident with the potassic alteration zone. It is several hundreds of metres across and the main minerals are chalcophrite, pyrite and molybdenite.

(2) The **inner alteration zone**, which closely corresponds with the phyllic alteration zone, and forms the main ore shell, with the minerals present being pyrite, chalcophyrite, molybdenite, bornite (and also chalcocite, sphalerite, magnetite, and enargite).

(3) The **intermediate alteration zone**, which corresponds roughly with the argillic zone, and contains the main ore minerals pyrite, chalocopyrite, bornite (and with traces of other minerals also found in the second zone).

(4) The **outer alteration zone**, which corresponds roughly with the propylitic zone and contains the main mineral pyrite.

In zone (1) the ore minerals are disseminated throughout the zone, but in all other zones the mineralization is commonly in the form of veinlets. Some deposits are brecciated. The majority of porphyry copper deposits are associated with Mesozoic and Cenozoic mountain belts and island arcs.

Porphyry molybdenum deposits

These are very similar to porphyry copper deposits with the host rocks being similar (diorites to granites), and the ore-bodies are associated with simple, multiple or composite intrusions or with dykes or breccia pipes. Hydrothermal alteration similar to that described in the previous section occurs, and the molybdenite ore occurs in quartz veinlets, fissure veins and breccias, and is occasionally disseminated.

Porphry tin deposits

These are mostly uneconomic, and are associated with acid plutonic host rocks. Many features are similar to those of the porphyry copper deposits, with alteration zones, etc.

Stratiform sulphide and oxide deposits, found in sedimentary and volcanic environments

Sedimentary stratiform sulphide deposits

These deposits may occur throughout geological time, from the Precambrian to the Tertiary, and occur in **non-marine** or **deltaic** environments. These stratiform (flat-lensoid) ore deposits are very fine-grained, and organic remains may also be present. Sometimes, a zonation occurs of Cu and Ag (shore) to Pb and finally Zn (in the basin). Examples include: the

European *kupferschiefer*, which is regarded as a syngenetic base metal ore deposit contained in black, organic-rich shales of Permian age, with copper (in the form of the sulphides bornite, chalcocite and chalcopyrite) disseminated through the rocks, and overlain by lead and zinc (as galena and sphalerite); and the Zambian copper belt, where copper occurs in Katangan sediments, mainly shales or dolomite shales, but occasionally arkoses and arenites, of Proterozoic age. Copper is accompanied by iron and cobalt, and mineralization shows a zonal pattern from chalcocite (near the shore) to bornite, then chalcopyrite and finally pyrite (deep water).

Volcanic massive sulphide deposits

There are three classes of deposit, namely: copper, zinc–copper and zinc–lead–copper, all of which contain iron in addition. The most important host rock is rhyolite, and lead deposits are always associated with this rock type. Copper ores are usually associated with mafic volcanics. These deposits have been linked to crustal evolution; thus (iron)–copper–zinc mineralization (with the minerals pyrite, chalcopyrite and sphalerite) may be associated with Archaean greenstone belts, forming from early sulphur-rich mantle material. The other deposits have been formed either by **volcanic exhalitive** or by **sedimentary exhalitive** processes, by the action of **hydrothermal solutions** passing upwards, and through, a stockwork. These solutions may be magmatic in origin *or* may represent circulating sea water. There are three types of deposit:

(1) The **Cyprus type**; cupriferous pyrite bodies associated with basic volcanics, usually part of an ophiolite sequence, and formed at oceanic or back-arc spreading ridges.

(2) The **Besshi type**; copper–zinc deposits associated with mafic volcanics and thick greywacke sequences, and formed at an early stage (calc-alkaline) of island-arc formation.

(3) The **Kuroko type**; copper–lead–zinc deposits associated with felsic volcanics and developed at a later stage in island-arc evolution.

Veins

These include **lodes, pipes, mantas** (flat, blanket or tabular bodies), **fissure veins**, etc., and represent material precipitated from aqueous solutions, which are most probably hydrothermal in origin. Gangue minerals include quartz and calcite, and the main ore minerals include sulphides, native metals such as gold and silver, and oxides in the case of uranium and tin ores.

Strata bound deposits

Base metal deposits in carbonates

Mainly lead–zinc deposits, and with fluorite and barite, these are widely distributed in Palaeozoic and Mesozoic rocks, and associated with carbonate reef rocks and carbonate mudstones deposited at the margins of marine basins in stable cratonic areas. The sulphur for the ore deposits may have been contained in the original sediments as sulphates (laid down in contemporaneous **sabkhas**) and released by later, uprising, mineralizing solutions which deposited the ores in various types of bodies, from stockworks to stratiform deposits.

Base metal deposits (including U and V deposits) in sandstones

These include deposits of Cu, Ag, Pb–Zn, and U ± V, found in terrestrial sediments, often fluviatile and red in colour, laid down in arid conditions. In these desert conditions the uranium-bearing ore bodies occur in conglomerates and sandstones, and contain V, Cu, Ag, Se (selenium) and Mo, in addition to U. The sedimentary hosts, which may have been derived from the weathering of granitic-type rocks, are enclosed by impermeable shales, and the mineralization was probably **epigenetic**, that is, formed after the sedimentary host rocks were deposited.

Sedimentary deposits

These can be divided into two categories. In the first, the sediments have been transported to their area of deposition (**allochthonous** sediments); and in the second, the sediments have formed in their area of deposition (**autochthonous** sediments).

Allochtonous sediments

The most important deposits in this category are **placer deposits** containing the minerals gold, cassiterite, chromite, columbite, copper, garnet, ilmenite, magnetite, monazite and rutile. The more important of these include:

(1) **Residual placers**, which accumulate above a deposit by its weathering *in situ*.
(2) **Alluvial placers**, which accumulate in streams due to the deposition of heavy minerals, at points in the course where the velocity drops sufficiently.
(3) **Beach placers**, which include minerals such as diamond, in addition to those listed above. Placers of this type accumulate on the tidal zone of an unsheltered beach where heavy minerals can separate out.

(4) **Fossil placer** deposits are occasionally recognized, such as the Witwatersrand goldfield of South Africa, where gold is found disseminated throughout the matrices of Precambrian conglomerates.

Autochthonous sediments

The most important deposits in this category are iron and manganese deposits.

Precambrian banded ironstone formations (BIF) are huge ore bodies, thousands of kilometres across and hundreds of metres thick, with the individual layers about 0.5 m thick, and still-finer layering commonly occurring. The banding consists of layers of silica (chert, etc.) alternating with layers of iron minerals such as hematite. There are four main subdivisions (called facies):

(a) **Oxide facies**, consisting of hematite and magnetite with some carbonates and chert, perhaps representing a shallow-water origin.

(2) **Carbonate facies**, consisting of interbanded chert and siderite, probably formed in a shallow-water mud; this facies is gradational into (1) and (4).

(3) **Silicate facies**, consisting of iron silicates (chamosite, glaucophane and greenalite) associated with magnetite, siderite and chert, probably also formed in a shallow-water environment.

(4) **Sulphide facies**, consisting of pyrite argillites containing carbonaceous material, and which formed under reducing conditions.

BIF is usually of Precambrian age, and may form either in a geosynclinal environment or be exhalative in origin, but is characteristic of Archaean greenstone belts. However, the BIF deposits of the Lake Superior region in the United States, which were formed in the early Proterozoic, may have formed by the accumulation of iron-rich precipitates in either a shallow basin or a continental shelf.

Phanerozoic ironstones may be of two types:

(1) The **Clinton type**, consisting of massive red beds of oolitic ironstones (hematite–chamosite–siderite), associated with shales and limestones, and formed in shallow water on continental edges of continental shelves, or in shallow miogeosynclines; usually of Palaeozoic age.

(2) The **Minette type**, consisting of oolitic chamosite and siderite along with limonite, associated with fine-grained clastic sediments containing organic material, and formed in shallow basins; usually of Mesozoic age.

Manganese deposits are divided into two groups:

(1) **Non-volcanogenic**: this group includes (a) manganese ore (oxide or carbonate) associated with sands (containing glauconite), silts and clays in a shelf environment; and (b) manganiferous carbonate, associated with limestones and dolomites and formed either in a geosynclinal environment or on rigid cratonic blocks.

(2) **Volcanogenic–sedimentary**: this group comprises (a) manganese ore associated with greenstone belts and spilite–keratophyre volcanism in a geosynclinal environment; and (b) manganese ores associated with porphyry and trachyte–rhyolite volcanism.

Residual deposits and supergene enrichment

Residual deposits of aluminium

The most important aluminium ore is bauxite, which is formed from the weathering of aluminous rocks with a low iron content. Most of these deposits are post-Tertiary and first form bauxite which is mainly composed of gibbsite. This later changes to boehmite and diaspore.

Iron-rich laterites are uneconomic weathered deposits consisting of hematite and goethite, which may contain appreciable amounts of alumina.

Residual deposits of nickel

These are formed from severe tropical weathering of ultrabasic and ultramafic rocks, such as peridotites and serpentinites. The nickel may form on iron minerals present or appear as garnierite and other nickel minerals in the weathered rock below the lateritic cover. Cobalt may accompany the nickel and is usually contained in the mineral wad or asbolite.

Supergene enrichment

Sulphide deposits frequently show supergene enrichment, and the process has been used to include similar enrichments of oxide and carbonate ores. A typical supergene enrichment of a sulphide vein is shown in the generalized section in Figure 5.2. The weathered upper part of the deposit is called a **gossan**. Below this, the ore body may be **oxidized** and **leached** of its valuable elements down as far as the **water table**. In the **zone of oxidation** the percolating waters may still precipitate secondary oxides, carbonates and hydrated minerals such as cuprite, malachite and azurite. Below the water table and overlying the primary ore occurs the **zone of (supergene)**

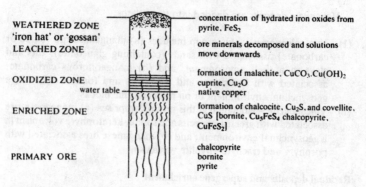

WEATHERED ZONE
'iron hat' or 'gossan' — concentration of hydrated iron oxides from pyrite, FeS_2

LEACHED ZONE — ore minerals decomposed and solutions move downwards

OXIDIZED ZONE — formation of malachite, $CuCO_3.Cu(OH)_2$ cuprite, Cu_2O native copper

water table

ENRICHED ZONE — formation of chalcocite, Cu_2S, and covellite, CuS [bornite, Cu_5FeS_4 chalcopyrite, $CuFeS_2$]

PRIMARY ORE — chalcopyrite bornite pyrite

Figure 5.2 Supergene enrichment in a copper lode

enrichment, in which very high grades of economic ores are deposited. Supergene enrichment is most important in the development of copper ores, but it has been recognized as an important factor in some iron and manganese deposits.

It is obvious from the above brief account of mineral deposits that the modern view requires ore deposits to be linked to their crustal setting and, especially, to their setting in a plate tectonic framework. In the brief descriptions of each ore deposit and their rock associations given here, the crustal environment of their formation has been mentioned as a necessary part of the description, but a detailed discussion of mineralization and ore deposits in space and time is not possible, given the constraints of this book, and the reader is referred to some of the specialist texts on this subject such as Evans (1980), Stanton (1972) and Dixon (1979).

5.7 Earth history

The sedimentary rocks of the crust that contain fossils have been arranged in the order of their deposition, the complete succession being shown in the **stratigraphical column** of Table 5.8, in which the oldest rocks are at the bottom and the youngest at the top. In this column, each named portion indicates a geological **system** of rocks deposited in a **period** of geological time. The oldest system that uses fossils as a dating technique is the **Cambrian**; and rocks older than this are **Precambrian** in age. The Precambrian is divided into three great **aeons** of time: the **Priscoan**, from the origin

Table 5.8 The stratigraphic column and the geological timescale.

Era	Period (system)	Epoch (series)	Age to beginning of period (Ma)	Major orogenic events (Europe)
Cenozoic	Quaternary	Holocene Pleistocene	0.01 1.5–2.0	} Alpine orogeny
	Tertiary	Pliocene Miocene	25	
		Oligocene Eocene Palaeocene	65	
Mesozoic	Cretaceous		144	
	Jurassic		213	
	Triassic		248	
Palaeozoic	Permian		286	} Variscan or Hercynian orogeny
	Carboniferous	Pennsylvanian[a] Mississippian[a]	320 360	
	Devonian		408	} Caledonian orogeny
	Silurian		438	
	Ordovician		505	
	Cambrian		590	
Precambrian				

[a] Pennsylvanian and Mississippian are names of subperiods of the Carboniferous in North America

of the Earth (*c.* 4600 Ma; 1 Ma = 10^6 years) to 4000 Ma; the **Archaean**, from 4000 Ma to 2500 Ma; and the **Proterozoic**, from 2500 Ma to 590 Ma. The entire period after 590 Ma constitutes *one* aeon, the **Phanerozoic**, which itself is divided into three great **eras** of time, each era comprising several systems; these eras are the **Palaeozoic**, the **Mesozoic** and the **Cenozoic**.

Igneous and metamorphic rocks obviously cannot be dated by fossils, although some lava flows can be dated by knowledge of the age of sediments immediately above and below them. However, most igneous and metamorphic rocks are dated by obtaining a **radiometric age**, which uses the process of radioactive decay of certain elements in the rock (see Section 2.6 for a fuller explanation).

In Table 5.8 the lengths of some major orogenic episodes occurring throughout geological time are also given.

6

The classification of minerals

The purpose of this short chapter is to introduce the arrangement of minerals adopted for their description in Chapters 7–9. Minerals may be classified in several ways, nearly all dependent either on their composition or their crystal chemistry.

The chemical classifications usually used begin with the elements and then follow with subdivisions based on the anion groups present. In the scheme employed by the American mineralogist, J. D. Dana, the following subdivisions were made:

   I   **Native elements**
  II **Sulphides** – selenides, tellurides, arsenides, antimonides
 III **Sulpho-salts** – sulpharsenites, sulphantimonites, sulphobismuthites
 IV **Haloids** – chlorides, bromides, iodides, fluorides
  V **Oxides**
 VI **Oxygen salts** – carbonates; silicates, titanates; niobates, tantalates; phosphates, arsenates, vanadates; antimonates; nitrates; borates, uranates; sulphates, chromates, tellurates; tungstates, molybdates
 VII **Salts of organic acids** – oxalates, mellates, etc.
VIII **Hydrocarbon compounds**

Modern variants of this arrangement follow much the same lines. All chemical classifications have some definite relation to crystal structure, and more direct applications of the latter have been made either in terms of crystal symmetry or of structure.

In this book it is considered more advantageous for our purpose to employ both the Dana type of classification (Ch. 8) and a classification based on the useful element or group of elements contained in the mineral. In Chapter 7, therefore, the minerals are assembled simply into economic groups according to elements. For this purpose, we use the Periodic classification of elements which has already been described earlier in this text. The somewhat simplified Periodic Table used here brings together only these elements whose minerals are described in this book. Other elements represented in nature by rare or unimportant minerals are not considered. This method of grouping is helpful both to the mineralogist determining minerals by some analytical technique, and also to the economic geologist, who may wish to know which various ores are associated with a particular element.

The groups and subgroups obtained from Table 1.8 are dealt with in the order shown in that table, and are set out below:

Group	Ia	lithium, sodium, potassium
	Ib	copper, silver, gold
Group	IIa	calcium, strontium, barium (radium)
	IIb	beryllium, magnesium, zinc, cadmium, mercury
Group	IIIb	boron, aluminium
Group	IVa	titanium, zirconium, [cerium], thorium
	IVb	carbon, silicon, tin, lead
Group	Va	vanadium, niobium, tantalum
	Vb	nitrogen, phosphorus, arsenic, antimony, bismuth
Group	VIa	chromium, molybdenum, tungsten, uranium
	VIb	sulphur, selenium, tellurium
Group	VIIa	manganese
	VIIb	fluorine, chlorine, bromine, iodine
Group	VIIIa	iron, cobalt, nickel
	VIIIb	ruthenium, rhodium, palladium, osmium, iridium, platinum

Formal descriptions of all the minerals are given either in Chapter 8, for the non-silicate minerals, using the Dana subdivisions given earlier, or in Chapter 9, for the silicate minerals, using crystal chemical subdivisions.

As in the previous editions of this book, information on elements and industrial minerals is obtained from the *Mining Annual Review* issued by the *Mining Journal*, and it is worth summarizing the arrangement of industrial minerals used in the *Review*, from which the importance, complexity and variety of mineral products in industry is immediately apparent. The arrangement is as follows:

Precious metals: gold, silver, platinum metals.
Older major metals: copper, tin, lead, zinc.

Light metals: aluminium, magnesium, titanium.

Steel industry metals; iron ore, steel, nickel, manganese, chromite, cobalt, molybdenum, tungsten, vanadium, niobium, tantalum.

Fuel minerals: coal, oil, natural gas.

Nuclear metals: uranium, caesium and rubidium, beryllium, rare earths, zirconium and hafnium.

Electronics metals and minerals: mercury, mica, rhenium, indium, cadmium, selenium, tellurium.

Chemical metals and minerals: phosphate rock, potash, salt, sulphur, soda ash, lithium, antimony, boron, kaolin, bismuth, gypsum, fluorspar (fluorite), barytes (barite).

7

Economic grouping of minerals according to elements

Group Ia   Lithium, sodium and potassium

Lithium minerals

Lithium (Li) does not occur in a free state in nature, nor are its compounds very abundant as minerals. The characteristic mode of occurrence of lithium minerals is in pegmatites, where there has been a concentration of this somewhat rare element during the latter stages of consolidation of granite magma. The most common lithium minerals are:

silicates	lithium-bearing micas (lepidolite, zinnwaldite)
	petalite, $LiAlsi_4O_{10}$
	spodumene, $LiAlsi_2O_6$
phosphate	amblygonite, Li (F, OH) $AlPO_4$

Spodumene, petalite, amblygonite and lepidolite are sources of lithium salts. Production figures are not always available, but in 1985 Zimbabwe produced 30 000 t (of Li_2CO_3 equivalents), with the USA (15 000 t) and China also being important producers. In 1990, Chile will start production with a capacity of 30 000 t per annum, and initial production estimated at 10 000 t per annum. The annual world consumption of lithium metal and equivalents in 1985 was estimated at 59 700 million lb (5091 t pure Li metal), with USSR production unknown but estimated at around 4000 t per annum. Lithium carbonate is produced in increasing amounts from brines in the western USA. Lithium salts are used in glass, batteries, television tubes, ceramics, molten carbonate fuel cells and in aluminium lithium alloys for the aerospace industry.

Sodium minerals

Sodium (Na) does not occur native. Metallic sodium is produced by the decomposition of a melt of sodium chloride by an electric current. Sodium is a soft silver white metal, tarnishing easily in air. It decomposes in water, forming sodium hydroxide and hydrogen. It is used in heat-exchange systems.

Sodium forms 2.8% of the Earth's crust. It is a constituent of many very important rock-forming silicates, notably of albite, $NaSlsi_3O_8$, the sodic end-member of the plagioclase feldspars. A less common feldspar, anorthoclase, $(Na,K)AlSi_3O_8$, crystallizes from sodium-rich magmas, as do sodium-pyroxenes, sodium-amphiboles and the feldspathoids, nepheline, sodalite and hauyne. Scapolite, and paragonite the sodium-mica, are other sodium-bearing silicates. Amongst the important group of silicates known as the zeolites, sodium-bearing types are represented by heulandite, analcime, natrolite, stilbite etc. All these and other sodium-bearing members are described on the appropriate pages in the account of the silicates (Ch. 9).

The non-silicate sodium minerals mostly occur as saline residues deposited by the evaporation of enclosed bodies of salt water. Sodium chloride, NaCl, is the compound of the most frequent occurrence in nature, and is procured for industrial purposes either as rock salt or by the evaporation of sea water. NaCl is used mainly for the production of chlorine (see p. 203 for details). Sodium production is given in terms of the amount of soda-ash (hydrous sodium carbonate) produced. In 1985 the total amount of soda-ash produced was 29.3 Mt, of which 9.4 Mt was produced by the USSR and the Eastern bloc, 8.16 Mt by the USA, and 6.25 Mt by Western Europe. Compared with the vast supplies of salt, other sodium compounds are relatively insignificant, though there are important deposits of the carbonate, sulphate and nitrate.

Tests. Sodium salts colour the blowpipe flame intense yellow. After heating before the blowpipe, sodium compounds give an alkaline reaction with litmus; this reaction is, however, also given by salts of the other alkalis and of the alkaline earths.

The sodium minerals, other than silicates, dealt with are:

chloride	rock-salt, halite, NaCl
nitrate	soda-nitre, $NaNO_3$
sulphates	thenardite, Na_2SO_4
	mirabilite, $Na_2SO_4.10H_2O$
	glauberite, $Na_2Ca(SO_4)_2$

carbonates
thermonatrite, $Na_2CO_3.H_2O$
natron, $Na_2CO_3.10H_2O$
trona, $Na_3H(CO_3)_2.H_2O$
gaylussite, $Na_2Ca(CO_3)_2.5H_2O$

To these may be added:

borates
borax, $Na_2B_4O_7.10H_2O$
ulexite, $NaCaB_5O_9.8H_2O$

fluoride
cryolite, Na_3AlF_6

Borax and ulexite are worked for their boron content; similarly, cryolite is an ore of aluminium.

Potassium minerals

Potassium (K) does not occur native. The metal is prepared in a similar fashion to sodium, and its chemical and physical properties are very like those of that metal. Formerly, potassium salts were procured for the most part from vegetable matter, this being burnt and the soluble portion of the ashes dissolved in water. However, the plants have, in the first instance, procured their potassium from soils which have resulted more or less from the decomposition of igneous rocks containing orthoclase feldspar, $KAlSi_3O_8$. Examples of other potash-bearing silicates are leucite, $KAlSi_2O_6$, and muscovite (potash-mica), $KAl_2[Si_3AlO_{10}](OH)_2$ and apophyllite. Potash-bearing zeolites are represented by harmotome. The potash-bearing silicates are described with the other rock-forming silicates.

The extraction of potash from such silicates is a complex and costly process, and more readily accessible supplies of potassium compounds are available in the saline residues (see p. 225). The most interesting deposit of this kind is at Stassfurt, Germany (p. 225). In this, the most important potassium compound is carnallite, $KMgCl_3.6H_2O$, but other potassium minerals occurring there include kainite, $KCl.MgSO_4.3H_2O$, polyhalite, $K_2Ca_2Mg(SO_4)_4.2H_2O$, and sylvine, KCl. A third natural source of potash is found in the mineral alunite, $KAl_3(SO_4)_2(OH)_6$, and in this case also the extraction of potash is not a complicated process. Some deposits of nitre, KNO_3, are of organic origin; nitre occurs also in small amounts in the sodium nitrate deposits of Chile. Sea water contains about 0.04% of potassium salts, and the recovery of such salts supplies a small proportion of the potash production; similarly, the waters of salt lakes, such as the Dead Sea, contain potash salts, and are exploited.

New extraction plants on the Dead Sea have increased the amount of potash recovered to over 1.5 million tonnes (Mt) per annum.

The following gives a summary of the potash production in normal years, considered in terms of the minerals exploited. Italy produces a small quantity of leucite. The overwhelming proportion of the production comes from the saline residues, with an annual output of over 29 Mt in 1985. The major producers are the USSR (10.5 Mt), Canada (6.6 Mt), East Germany (3.5 Mt), West Germany (2.6 Mt), France (1.7 Mt) and the USA (1.2 Mt), with Israel, Spain, Jordan, the UK and Italy all producing between 0.1 and 1.0 Mt per annum. Alunite is produced by Korea, Italy, the USSR, Japan, Australia and Spain. Nitre production is largely from India, where it is of organic origin, and Chile, where it is of inorganic origin in the sodium nitrate deposits.

The most important use of potash salts is as fertilizers, taking 95% of total K_2O production. Other uses are in the manufacture of special glasses, soaps and detergents.

Tests. Potassium compounds give a lilac flame coloration, which is, however, masked by sodium and other elements; the flame should be viewed through blue glass or an indigo prism, whereby elements other than potassium are eliminated. Fused potassium compounds give an alkaline reaction with litmus. For the detection of small quantities of potassium compounds in solution, a few drops of platinic chloride produce in such a solution, after prolonged stirring, a precipitate of minute yellow crystals of potassium platinochloride, K_2PtCl_6.

The potassium minerals, other than silicates, dealt with are:

chlorides	sylvine, KCl
	carnallite, $KMgCl_3.6H_2O$
	kainite, $KCl.MgSO_4.3H_2O$
sulphates	polyhalite, $K_2Ca_2Mg(SO_4)_4.2H_2O$
	alunite, $KAl_3(SO_4)_2(OH)_6$
nitrate	nitre, KNO_3

Group Ib   Copper, silver and gold

Copper minerals

Copper (Cu) is a widely distributed and abundant element in combination, and is also found in the native state. The metal copper has a specific gravity of about 8.9, and melts at about 1100°C. It is a comparatively soft but extremely tough metal, very ductile and malleable when pure and, next to silver, the best conductor of electricity. These properites are a consequence of its cubic close-packed or face-centred cubic structure, as described and illustrated in Section 3.2.

Copper is obtained from its ores (usually sulphide) by an elaborate series of metallurgical operations, commonly consisting of roasting, to expel part of the combined sulphur, fusion in blast or reverberatory furnaces for the production of a concentrated double sulphide of copper and iron called matte, and the conversion of the matte to crude metallic copper in a reverberatory or Bessemer converter. Blast furnaces are sometimes used for ores that are rich in sulphur for the production of copper matte, or for the further concentration of matte from the first reverberatory furnace fusion; the former operation, the smelting of sulphur-rich ores, or pyritic smelting, utilizes the heat produced by the oxidation of the sulphides, with the aid of little or no fuel. Oxidized ores may be reduced in a blast furnace with coal or coke, but are best smelted in admixture with sulphide ores. In the case of native copper, the ore is crushed and the metal, separated from its gangue by dressing, melted in some form of reverberatory furnace. Copper of the necessary purity for use as conducting wire is obtained from the crude metal by electrolysis. Copper is obtained from poor ores and from residues from the pyrites used in the manufacture of sulphuric acid by roasting with common salt, leaching out the soluble copper chloride with water, and deposition of the metal on scrap iron or by electrolysis.

The most important use of copper is in the electrical industry, both as a conductor and for electrical machinery. It is also of great importance in the construction of machinery generally, in the motor-car industry, and in chemical engineering. Copper is extensively used in the manufacture of alloys, such as bronze, gun-metal and bell-metal (copper and tin), brass (copper, zinc and sometimes tin), nickel silver (copper, zinc and nickel), and some others of great technical importance, such as phosphor and manganese bronzes (copper and tin, with small percentages of phosphorus or ferromanganese respectively), silicon bronze, monel metal, etc. Copper salts are employed in various industrial processes; the chloride is used as a disinfectant and in chemical operations; the sulphate is employed in the printing and dyeing of textiles, for preventing rot in timber, and as a fungicide.

The major sources of copper are native copper, chalcopyrite, chalcocite, erubescite and malachite, together with cupriferous pyrite, i.e. pyrite containing a few per cent of chalcopyrite. The following metals are frequently found associated with copper ores, and are recoverable at one or other stage of their treatment: gold, silver, platinum, palladium and bismuth. Copper ores usually carry a very small percentage of copper; thus some ores worked for copper with profit have only 0.5% of the metal, and perhaps the average copper content of all copper ore production is not more than 2%. Copper ores occur in a variety of forms: magmatic segregations, in veins and lodes, in contact metamorphic deposits, in bedded deposits, etc. Examples of these types are given under the descriptions of the copper minerals. The sulphides of copper which occur in depth in copper lodes are converted by oxidation and other chemical actions in the surface portion to the native metal, oxides and oxysalts.

Western world mine production of copper in 1985 was over 6.4 Mt. The main producers were Chile (1.3 Mt), the USA (1.1 Mt), Canada (0.8), Zaire (0.5 Mt) and Zambia (0.5 Mt), with smaller amounts (0.1–0.5 Mt) produced by Peru, the Phillipines, Australia, South Africa, Mexico, Papua New Guinea and Yugoslavia. Limited information is available on Eastern bloc countries, and the USSR is known to be a major producer with an output of over 0.6 Mt per annum.

The main countries carrying out copper refining are the USA (over 1 Mt), Japan and Chile, with the UK refining 0.126 Mt in 1985.

Tests. Copper oxides colour the flame emerald green when moistened with nitric acid, and copper chloride colours the flame an intense sky blue. The microcosmic salt bead is blue in the oxidizing flame, and opaque red in the reducing flame. The borax bead is somewhat similar. When heated with sodium carbonate and carbon on charcoal, copper compounds give a reddish mass, which readily blackens. Dilute solutions of copper minerals in acids become deep blue on addition of ammonia.

The copper minerals considered are:

element	native copper, Cu
oxides	cuprite, Cu_2O
	tenorite, CuO
sulphides	chalcopyrite, copper pyrites, $CuFeS_2$
	chalcocite, copper glance, Cu_2S
	covelline, CuS
	bornite, erubescite, Cu_5FeS_4
'grey coppers'	tetrahedrite, $(Cu,Fe)_{12}Sb_4S_{13}$
	tennantite, $(Cu,Fe)_{12}As_4S_{13}$
	famatinite, Cu_3SbS_4
	enargite, Cu_3AsS_4
	(bournonite, $CuPbSbS_3$)

sulphate	chalcanthite, $CuSO_4.5H_2O$
carbonates	malachite, $Cu_2CO_3(OH)_2$ azurite, $Cu_3(CO_3)_2(OH)_2$
silicates	chrysocolla, $CuSiO_3.2H_2O$ dioptase, $CuSiO_3.H_2O$
chloride	atacamite, $Cu_2(OH)_3Cl$

Silver minerals

Silver (Ag) occurs in nature in the free state, occasionally 99% pure, but generally containing copper, gold and other metals. It is a white metal which, next to gold, is the most malleable and ductile of all metals and the best conductor of heat and electricity. These properties of silver depend on its atomic structure, that of a face-centred cube, in which it is similar to copper and gold. The specific gravity of silver is 10.5, and its melting point is 960.5°C. It is unaltered by dry or moist air. Silver occurs also as sulphide, sulphosalts, arsenide, antimonide and chloride, and also associated with ores of lead, zinc, copper and other metals.

There are two main classes of silver ores. The first is the dry or siliceous ores, which are mined primarily for their silver content – the silver ores proper; but the greater part of the world production of silver is derived from the smelting of metalliferous ores, such as those of lead, copper and zinc, which contain a small percentage of silver.

Silver is recovered from silver ores proper by cyanidation or by cupellation, and from argentiferous lead ores by melting with a small percentage of zinc, the silver being more soluble in the molten zinc than is the lead. Silver and gold may be recovered together in the form of an alloy, which is afterwards refined or 'parted'. Silver containing gold is called 'dore silver'. Refined silver usually contains from 997.5 to 999.0 parts of silver per 1000 – pure silver being 1000 fine. On the London market the price is quoted for 999 fine; the standard alloy employed in Britain for coin, plate and jewellery is 925 fine and is generally called standard silver. The addition of a small amount of copper produces an alloy having a lower melting point, a greater hardness and affording a sharper casting than pure silver. Formerly, the major uses of silver were in coinage, plate and jewellery; but its industrial uses now far exceed these, particularly in the manufacture of electronic components, electrical machinery, mirrors, electroplate and batteries, in medicine, photography, glass-making, etc.

The Western world production of silver in 1985 was 9.95 million kg

(Mkg). The main producers are Mexico (2.15 Mkg), Peru (1.79 Mkg), the USA (1.19 Mkg), Canada (1.18 Mkg) and Australia (1.09 Mkg). In the Eastern bloc, the USSR (1.49 Mkg) and Poland (0.83 Mkg) are the chief producers. It is interesting to note that silver is derived mainly as a by-product of other metal refining, particularly copper, lead and zinc, but a certain amount comes from dry or siliceous ores.

Silver ores occur as veins, replacement deposits, contact metamorphic deposits or as alluvials. The most important primary ore is argentite, Ag_2S. The upper parts of silver deposits or lodes are weathered with the production of cerargyrite, AgCl, which is often accompanied by bromyrite, AgBr, and iodyrite, AgI; in several cases of such gossans, cerargyrite occurs above bromyrite, below which comes iodyrite. Below this halide zone there is, in many silver lodes, a zone of secondary enrichment in which native silver and rich secondary sulphides are developed; below this zone come the primary deposits in which the ore is usually much poorer.

Tests. Silver compounds, heated with sodium carbonate and charcoal, on charcoal, give a silver white bead; this bead is malleable, and does not tarnish. Hydrochloric acid, added to a solution of silver in nitric acid, produces a dense white precipitate of silver chloride which is soluble in ammonia; silver beads obtained by fusion with sodium carbonate may be tested in this way.

The silver minerals considered are:

element	native silver, Ag
sulphide	argentite, silver glance, Ag_2S
complex sulphides	stephanite, Ag_5SbS_4
	pyrargyrite, Ag_3SbS_3
	proustite, Ag_3AsS_3
	freieslebenite, $(Pb,Ag)_8Sb_5S_{12}$
	polybasite, $(Ag,Cu)_{16}(Sb,As)_2S_{11}$
telluride	hessite, Ag_2Te
chloride	cerargyrite, horn silver, AgCl

Gold minerals

Gold (Au) occurs very widely diffused in nature, chiefly in the free state, but invariably alloyed with some proportion of silver or copper, and occasionally with bismuth, mercury and other metals. Native gold has been known to contain as much as 99.8% gold, but as a rule ranges from 85% to 95%, the balance being usually silver for the most part. Gold, when pure, is the most malleable and ductile of all metals, these properties depending on its atomic structure, namely that of a face-centred cube (see Ch. 3). It becomes brittle, however, when it contains small amounts of bismuth, lead, arsenic, etc. It has a specific gravity of 19.3 and melts at about 1060°C.

Native gold is recovered from alluvial deposits by some form of water concentration, followed by amalgamation with mercury. That occurring in veins is milled and ground previous to amalgamation, in the case of 'free-milling' ores. When the ore is of a partly 'refractory' nature, or when the gold is very finely divided, cyanidation (i.e. solution of the gold in sodium cyanide) is employed. Often, treatment by mercury is followed by cyanidation of the tailings for the recovery of the unamalgamated fine and combined gold. Before the fine grinding of ores and treatment by cyanide that are the usual practice, it is sometimes necessary to roast ores to eliminate arsenic and antimony compounds, which decompose the cyanide and cause excessive consumption. Gold is also recovered by chlorination, and by smelting with lead ores.

In addition to the native metal, gold occurs in combination as tellurides, and possibly as selenides; large quantities of gold are obtained from sulphides, with which it is probably mechanically mixed. Gold ores can thus be classed into two groups: (1) free-milling ores, from which native gold is recoverable by crushing and amalgamation; and (2) refractory ores, tellurides and auriferous sulphides, which yield their gold by complex smelting processes.

Gold has a remarkable position in the world economy. Apart from its use for jewellery, it is employed for coinage, and for these purposes it is usually alloyed with silver or copper to withstand wear better. The purity or 'fineness' of gold is expressed in parts per 1000, the standard for coin in the British Commonwealth being 916.6 parts of gold to 83.4 of copper. In England the legal standard for jewellery is the carat of 22, 18, 15, 12 or 9 parts per 24. The fineness of gold in alloys therefore can be expressed either in carats or thousandths; thus pure or fine gold is 22 carats or 916.6 thousandths fine. For purposes of plate, jewellery, watch-cases, etc., the standard of 18 carats or 750 fine is legal, but the lower standards of 16 and 14 carats are also general. Less gold coin is minted at the present time, but gold is still required as a medium of exchange, a measure of value and a cover for paper currency.

The total Western world production in 1985 was estimated at 1212.8 tonnes, with South Africa contributing about half of this total (673.3 t).

Thereafter, the main producers are Canada (86.0 t), the USA (79.0 t), Brazil (63.3 t), Australia (57.0 t) and the Philippines (38.5 t), with Papua New Guinea, Columbia, Chile, Zimbabwe and Ghana also producing significant amounts. The USSR produced 270 t of gold in 1985. It is estimated that 40% of the US gold production is as a by-product of other metal production – particularly copper. During the past 50 years or so, the production from gold ores has fallen from about 70% and that from base-metal ores has risen from about 7%, the placer production remaining more or less constant.

Native gold occurs in veins of various types, and in alluvial deposits both modern and ancient; tellurides and auriferous sulphides also occur in veins. In the weathered parts of these vein deposits, gold may be concentrated mainly by removal of the useless associates. In gold-bearing quartz–pyrite veins, for example, the weathered portion may be made up of rusty quartz in which there are gold nuggets; in the gossans of telluride veins, gold appears as mustard-gold – spongy, filmy and finely divided free gold.

Tests. The physical properties of gold serve to distinguish the native metal; its yellow colour, malleability, fusibility, high specific gravity, and insolubility in any one acid are distinctive. All gold compounds yield a gold bead when heated on charcoal with sodium carbonate. Tellurides, as such, are detected by tests given in Appendix A.

The gold minerals dealt with are:

element	native gold, Au
element with other metallic elements	gold amalgam, Au with Hg, Ag
tellurides	sylvanite, $(Au,Ag)Te_2$
	calaverite, $(Au,Ag)Te_2$
	petzite, $(Ag,Au)_2Te$
	nagyagite, sulphotelluride of lead and gold

Group IIa   Calcium, strontium, barium and radium

Calcium minerals

Calcium (Ca) does not occur in the free state in nature but its compounds are extremely abundant. It may be produced by the electrolysis of fused calcium chloride, but modern production is by the reduction of lime by aluminium in retorts under a low pressure. The metal calcium is being increasingly used in alloys, and in many metallurgical and chemical operations.

Although not occurring native, calcium nevertheless enters into the composition of a very considerable portion of the Earth's crust, of which it forms about 3.5%. Whole formations, such as the Chalk and the Carboniferous Limestone, consist almost entirely of calcium carbonate, while thick and thin beds of limestone are more or less common throughout the entire series of stratified rocks. Calcium enters also into the composition of many rock-forming silicates, notably anorthite feldspar, $CaAl_2Si_2O_8$, pyroxenes and amphiboles, garnets, scapolite, epidotes, many zeolites and wollastonite, $CaSiO_3$; these are described with the other rock-forming silicates on later pages.

The non-silicate calcium minerals are of great economic value and their various uses are given in their descriptions below.

Tests. Some calcium minerals give a brick red flame coloration which is enhanced by moistening the substance with hydrochloric acid. Fused calcium compounds give an alkaline reaction with litmus. On the addition of sulphuric acid to solutions containing calcium salts, a white precipitate of calcium sulphate is formed.

The following are the more important non-silicate calcium minerals:

carbonates	calcite, $CaCO_3$ (trigonal)
	aragonite, $CaCO_3$ (orthorhombic)
	dolomite, $CaMg(CO_3)_2$
	gaylussite, $Na_2Ca(CO_3)_2.5H_2O$
	barytocalcite, $BaCa(CO_3)_2$
sulphates	anhydrite, $CaSO_4$
	gypsum, $CaSO_4.2H_2O$
	glauberite, $Na_2Ca(SO_4)_2$
	polyhalite, $K_2Ca_2Mg(SO_4)_4.2H_2O$
phosphate	apatite, $Ca_5(F,Cl)(PO_4)_3$
fluoride	fluorite, CaF_2
borates	ulexite, $NaCaB_5O_9.8H_2O$
	colemanite, $Ca_2B_6O_{11}.5H_2O$

Colemanite and ulexite are worked for their boron content.

Strontium minerals

Strontium (Sr) does not occur in a free state of nature, but may be prepared by the electrolysis of fused potassium and strontium chlorides. It much resembles calcium in its properties.

The major minerals of strontium are the sulphate and the carbonate, and these are the source of the strontium compounds used in industry. These strontium minerals occur as nodular deposits in sedimentary rocks or as veins, possibly of hydrothermal origin. Strontium compounds are used in the manufacture of pyrotechnics, such as flares and fuses, and fireworks in which red-coloured flames are required. Their former use in the purification of molasses has now dwindled, but they are employed in the manufacture of ceramics, plastics, paints, etc., and in some electrolytic refining processes.

Tests. Strontium compounds colour the blowpipe flame crimson. Fused strontium compounds give an alkaline reaction with litmus. With dilute sulphuric acid, solutions of strontium salts give a white precipitate of strontium sulphate.

The major minerals of strontium are:

sulphate	celestite, $SrSO_4$
carbonate	strontianite, $SrCO_3$

Barium minerals

The metal barium (Ba) is procured by the electrolysis of fused barium chloride, or by the reduction of barium oxides by aluminium in a vacuum furnace. It resembles calcium in its properties and is being used on an increasing scale in the production of certain alloys, for vacuum tube work, and as a hardener for lead.

Barium occurs in small amounts in many of the rock-forming silicates. The rare feldspar, celsian, is a barium aluminium silicate, $BaAl_2Si_2O8$, corresponding to anorthite, $CaAl_2Al_2Si_2O_8$. Economically, the most important barium minerals are the sulphate (barite) and carbonate (witherite), which are mined from vein deposits or from residual deposits resulting from the decay of rocks containing veins. Western world production of barite was 6.0 Mt in 1985 (a 25% reduction since 1981), and the major producers were China (1.10 Mt), the USA (0.81 Mt), India (0.42 Mt), Mexico (0.32 Mt), Morocco (0.32 Mt), Thailand (0.23 Mt) and Eire

(0.22 Mt), with Peru, West Germany, the UK, France, Turkey, Greece, North Korea and Brazil all having a production of between 0.1 and 0.15 Mt per annum. The Eastern bloc produces about 0.84 Mt per annum, with the USSR contributing about 0.5 Mt of that total.

The sulphate is used in the manufacture of white pigment, lithopone, and in various processes listed on p. 318; the carbonate is chiefly of value as a source of barium salts.

Tests. Barium compounds colour the blowpipe flame yellowish-green. Fused barium salts give an alkaline reaction with litmus. With dilute sulphuric acid, solutions of barium salts give a white precipitate of barium sulphate, $BaSO_4$. Barium minerals are usually whitish and have a high specific gravity.

The major minerals of barium are:

sulphate	barite, $BaSO_4$
carbonates	witherite, $BaCO_3$
	bromlite, $(Ba,Ca)CO_3$
	barytocalcite, $BaCa(CO_3)_2$

Group IIb   Beryllium, magnesium, zinc, cadmium and mercury

Beryllium minerals

The metal beryllium (Be) does not occur native, but can be obtained by the electrolysis of its fused compounds or by the reduction of the oxide or fluoride. Beryllium is a white metal with a specific gravity of about 1.85, and is thus much lighter than aluminium. The metal is employed in the production of special alloys, mainly with copper, but also with iron and nickel, and in nuclear reactors as a source of neutrons. One ore from which beryllium is obtained is beryl, $Be_3Al_2[Si_6O_{18}]$, which is available as a by-product in the mining of mica and feldspar deposits in pegmatites. another beryllium-bearing mineral is chrysoberyl, $BeAl_2O_4$. Both beryl and chrysoberyl are used as gemstones, but the important gemstone varieties are emerald and aquamarine.

The common beryllium ore is bertrandite $(Be_4Si_2O_7(OH)_2)$, containing about 15% of beryllium. The world (excluding China) produced 364 t in 1985, with the USA (236 t), the USSR (77 t) and Brazil (45 t) the main producers, and with Zimbabwe and Rwanda also producing small amounts. Production from China is increasing and may be significant in the next decade, and Brazil is planning to produce beryllium in the near future.

The beryllium minerals described are:

silicates	beryl, $Be_3Al_2[Si_6O_{18}]$
	bertrandite, $Be_{a4}Si_2O_7(OH)_2$
aluminate	chrysoberyl, $BeAl_2O_4$

The blowpipe tests for beryllium are not good, and the element is detected only by rather lengthy chemical tests.

Magnesium minerals

Magnesium (Mg) is not found free native, but is prepared by the electrolysis of a mixture of anhydrous magnesium chloride and potassium or sodium chloride. It is a silver white metal, easily tarnishing to the oxide magnesia, MgO. The metal is employed in the manufacture of light alloys and castings, especially for aircraft, and in various metallurgical processes. It is manufactured from magnesium chloride recovered from sea water, saline residues, or from the same compound produced from the carbonate, magnesite or from dolomite. The total world production in 1985 was 350 400 t, the main producers being the USA (150 000 t), the USSR (90 000 t) and Norway (60 000 t), with Canada, France, Italy, Japan, China and Yugoslavia also producing significant amounts.

Magnesium is estimated to constitute about 2.7% of the Earth's crust. It enters into the composition of a large number of rock-forming silicates, one major group of these being the ferromagnesian silicates, which include such common rock-forming minerals as biotite, pyroxene, amphibole, olivine, etc. The magnesium-bearing silicates are described with the other silicates in Chapter 9.

The chief magnesium mineral of economic importance is magnesite, $MgCO_3$, which is used for furnace linings; the largest deposit known is in China, with reserves estimated at 2000 Mt; dolomite, $MgCa(CO_3)_2$, is also of great industrial importance. The sulphates, epsomite, $MgSO_4.7H_2O$, and kieserite, $MgSO_4.H_2O$, are used in chemical manufacture, tanning, etc.; and the complex salts of magnesium, potassium and sometimes calcium, such as polyhalite, $K_2Ca_2Mg(SO_4)_4.2H_2O$, kainite, $MgSO_4.KCl.3H_2O$, and carnallite, $KMgCl_3.6H_2O$ occur as saline residues. Boracite, $Mg_3B_7O_{13}Cl$, is a source of boron compounds. Spinel, $MgO.Al_2O_3$, is used as a gemstone.

$ZnCO_3$, and hydrozincite, $Zn_5(OH)_6(CO_3)_2$, the hydrated silicate, hemimorphite, $Zn_4Si_2O_7(OH)_2.H_2O$, and sometimes the anhydrous silicate, willemite, Zn_2SiO_4. In such oxidized zones the hydrated sulphate, goslarite, $ZnSO_4.7H_2O$, often occurs as an efflorescence.

In most occurrences of zinc ore, the blende is accompanied by galena. There are several types of zinc deposit. In one very important type, illustrated by the great Tri-State field in the Mississippi Valley, galena and blende occur as metasomatic disseminations or gash, cavity or joint fillings in limestone; the ore is certainly epigenetic, but whether it was derived from below and transported by ascending solutions, or from above and carried down, is a matter for discussion. The important Broken Hill deposits occur in lodes along fault planes in a series of metamorphosed rocks, and are of hydrothermal origin. Other deposits of hydrothermal origin replace limestone and are exemplified by the Leadville field in Colorado. Other zinc–lead deposits are found as contact metamorphic deposits, but these are not very important. The ore of the famous Franklin Furnace deposit of New Jersey is franklinite, $(Fe,Zn,Mn)(Fe,Mn)_2O_4$, willemite and zincite, ZnO, and occurs as bands and lenses in crystalline limestone; this remarkable deposit is interpreted as of pyrometasomatic origin, but may possibly be a hydrothermal zinc deposit which has been subsequently contact metamorphosed. Finally, the decay of rocks such as limestones in which there are zinc–lead veins and deposits gives rise to residual deposits of these minerals.

Tests. Zinc minerals heated on charcoal give an encrustation which is yellow when hot, white when cold; this encrustation, moistened with cobalt nitrate and strongly reheated, assumes a fine green colour.

The nomenclature of some of the zinc minerals is rather confused; the sulphide is usually called zinc blende or blende in the UK, but is known as sphalerite in the USA; the anhydrous carbonate has been called calamine in the UK, but smithsonite in the USA; the hydrated silicate is hemimorphite (once called calamine in the USA). The zinc minerals considered here are:

element	native zinc (doubtful)
oxides	zincite, ZnO
	franklinite, $(Fe, Zn, Mn)(Fe,Mn)_2O_4$
sulphide	blende, sphalerite, ZnS
carbonate	smithsonite, $ZnCO_3$

basic carbonate	hydrozincite, $Zn_5(OH)_6(CO_3)_2$
silicates	willemite, Zn_2SiO_4
	hemimorphite, $Zn_4Si_2O_7(OH)_2.H_2O$
sulphate	goslarite, $ZnSO_4.7H_2O$

Cadmium minerals

Cadmium (Cd) is a bluish-white metal having a brilliant lustre and closely resembling zinc. It is very malleable and ductile, and has a specific gravity of about 8.6 and melts at 320°C. It is found in nature as the sulphide, greenockite, but this mineral is of rare occurrence. Cadmium also occurs in small quantities, probably as the sulphide also, in zinc ores, and the metal is obtained as a by-product in the distillation or electrolysis of zinc ores, in which it seldom occurs in greater amounts than 0.4%, and usually less.

The annual Western world production of cadmium in 1985 was 13 923 t (compared with 15 000 t in 1984), and the main producers are Japan, Canada, the USA, Belgium and West Germany. Australia and Mexico–Peru are also important producers of the metal. The USSR production of cadmium was estimated at 3000 t in 1983.

Cadmium and its compounds are used for a number of purposes. The metal is employed in several important alloys, fusible alloys used in fire extinguishers, bearing alloys in motor manufacture, etc.; it is also of importance in some process of electroplating and metal spraying, and in the manufacture of electric transmission wires. The cadmium salts are of importance as pigments giving, in combination with certain other materials such as selenium, brilliant reds and yellows to glass and other substances.

Tests. Cadmium minerals, when heated with sodium carbonate on charcoal, give a reddish-brown encrustation, which is yellow at some distance from the assay.

The only cadmium mineral considered is:

| sulphide | greenockite, CdS |

Mercury minerals

Mercury or quicksilver (Hg) exists native, but as such this is an unimportant source of the metal. It is a silver white metal, liquid at ordinary temperatures; it boils at 357°C, and has a specific gravity of 13.59. When

(1.18 Mt) and Turkey (0.72 Mt), produced over 80% of the total supply. Other important producers were the USSR and Argentina. The USA mines mostly kernite and colemanite deposits, whereas Turkey works colemanite and borax deposits.

Tests. Boron minerals give a rather nondescript yellow–green flame when heated before the blowpipe. Fusible borates, heated in the oxidizing flame or charcoal, moistened with cobalt nitrate and strongly reheated, give a blue glassy residue, but a similar residue is given by fusible silicates and phosphates. Boron minerals dissolved in dilute hydrochloric acid, after fusion with sodium carbonate if necessary, give a solution which has a characteristic effect on turmeric paper. This, moistened with the solution and dried at 100°C by placing it on a flask containing boiling water, assumes a reddish-brown colour which changes to inky black by moistening with ammonia.

The major borates are as follows:

boric acid	sassoline, H_3BO_3
hydrous sodium borates	borax, $Na_2B_4O_7 \cdot 10H_2O$
	kernite, $Na_2B_4O_7 \cdot 4H_2O$
hydrous calcium borate	colemanite, $Ca_2B_6O_{11} \cdot 5H_2O$
hydrous sodium calcium borate	ulexite, $NaCaB_5O_9 \cdot 8H_2O$
magnesium borate and chloride	boracite, $Mg_3B_7O_{13}Cl$

Aluminium minerals

Aluminium (Al) is not found in a free state, but in combination constitutes 8% of the Earth's crust, and is the most abundant of metals. It is an essential constituent of the clay minerals, and of a large number of important silicates such as the feldspars, micas, sillimanite, etc. The major industrial sources of aluminium and its compounds are bauxite (hydrated oxides) and to a less extent, cryolite, Na_3AlF_6, alunite, $KAl_3(SO_4)_2(OH)_6$, leucite, $KAlSi_2O_6$, and alum shales. Such industrial minerals as potter's clay, china clay or kaolin, fuller's earth, feldspar, garnet, mica etc., are aluminium silicates, while aluminium oxides occur as bauxite, corundum and emery.

The metal aluminium is produced in the electric furnace by the reduction of alumina obtained from bauxite. It is a silver-white ductile metal, capable of taking a high polish. Owing to its low specific gravity (2.58), it is

(1.18 Mt) and Turkey (0.72 Mt), produced over 80% of the total supply. Other important producers were the USSR and Argentina. The USA mines borax kernite and colemanite deposits, whereas Turkey works colemanite and borax deposits.

Tests. Boron minerals give a rather nondescript yellow–green flame when heated before the blowpipe. Fusible borates, heated in the oxidizing flame on charcoal, moistened with cobalt nitrate and strongly reheated, give a blue glassy residue, but a similar residue is given by fusible silicates and phosphates. Boron minerals dissolved in dilute hydrochloric acid, after fusion with sodium carbonate if necessary, give a solution which has a characteristic effect on turmeric paper. This, moistened with the solution and dried at 100°C by placing it on a flask containing boiling water, assumes a reddish-brown colour which changes to inky black by moistening with ammonia.

The major borates are as follows:

boric acid	sassoline, H_3BO_3
hydrous sodium borates	borax, $Na_2B_4O_7.10H_2O$
	kernite, $Na_2B_4O_7.4H_2O$
hydrous calcium borate	colemanite, $Ca_2B_6O_{11}.5H_2O$
hydrous sodium calcium borate	ulexite, $NaCaB_5O_9.8H_2O$
magnesium borate and chloride	boracite, $Mg_3B_7O_{13}Cl$

Aluminium minerals

Aluminium (Al) is not found in a free state, but in combination constitutes 8% of the Earth's crust, and is the most abundant of metals. It is an essential constituent of the clay minerals, and of a large number of important silicates such as the feldspars, micas, sillimanite, etc. The major industrial sources of aluminium and its compounds are bauxite (hydrated oxides) and to a less extent, cryolite, Na_3AlF_6, alunite, $KAl_3(SO_4)_2(OH)_6$, leucite, $KAlSi_2O_6$, and alum shales. Such industrial minerals as potter's clay, china clay or kaolin, fuller's earth, feldspar, garnet, mica etc., are aluminium silicates, while aluminium oxides occur as bauxite, corundum and emery.

The metal aluminium is produced in the electric furnace by the reduction of alumina obtained from bauxite. It is a silver-white durable metal, capable of taking a high polish. Owing to its low specific gravity (2.58), it is of great value in the manufacture of many articles where lightness is of importance. It melts at 658°C, and alloys with most metals and some non-metals, the light alloys of importance being those with zinc, copper or magnesium. Aluminium is also employed in the manufacture of household articles, of wrapping foil, cans, building-sheet, etc., in electrical equipment, in car and ship construction and in metallurgical and chemical processes. It is produced especially in localities where hydroelectric installations are practicable. The world production of aluminium metal in 1985 was 13.5 Mt. The main producers were the USA (3.5 Mt), the USSR (2.2 Mt), Canada (1.28 Mt), Australia, West Germany, Norway, Brazil, Venezuela, Spain, France, China, the UK (0.28 Mt), Yugoslavia and India.

The main bauxite producers were Australia (32.2 Mt) Guinea (14.7 Mt), Jamaica (6.1 Mt), Brazil (6.4 Mt) and Yugoslavia (3.5 Mt), with Surinam, Greece and Guyana making significant contributions (each more than 2.0 Mt). In the Eastern bloc, the USSR (4.6 Mt) and Hungary (2.8 Mt) were the largest annual producers of bauxite in 1985.

The energetic action of finely divided aluminium on a metallic oxide when heated together is utilized in the 'Thermit' process for the production of metallic chromium, manganese, molybdenum, tungsten, uranium, etc., and in the welding of rails, etc.; the aluminium combines directly with the oxygen of the oxide, and the heat evolved by this reaction is sufficient to promote the fusion of the reduced metal.

There are a number of industrially important silicates that contain aluminium, such as feldspar, muscovite, the clay group of minerals, the aluminosilicates, sillimanite, andalusite and kyanite, and garnet; most of these silicates are important rock-forming minerals and, as they are not worked for aluminium or its salts, their industrial value depends mainly upon their physical properties.

Tests. Most aluminium minerals, when finely powdered and heated before the blowpipe, moistened with cobalt nitrate solution and strongly reheated, give a mass showing a fine blue colour.

The non-silicate aluminium minerals described are:

oxides	corundum, Al_2O_3
	spinel, $MgAl_2O_4$
	chrysoberyl, $BeAl_2O_4$
hydroxides	diaspore, $HAlO_2$
	boehmite, $AlO(OH)$
	gibbsite, $Al(OH)_3$
	bauxite, mixture of Al hydroxides

oxides   rutile, TiO₂
                                       anatase, TiO₂
                                       brookite, TiO₂

oxide of titanium and iron   ilmenite, FeTiO₃

Zirconium minerals

Zirconium (Zr) does not occur free in nature, but it enters into the composition of a number of complex silicates, most of which are rare; however, the simple silicate, $ZrSiO_4$, is of widespread occurrence as the mineral zircon, and this mineral is the source of the metal and its compounds used in industry. The metal zirconium is allied to titanium in its properties, and may be prepared by reduction of zirconium oxide by magnesium; the oxide is produced by fusing zircon with acid potassium fluoride, treating with hydrochloric acid and precipitating the oxide with ammonia. The metal has a specific gravity of 6.06, a melting point of 1300°C, and is produced either as crystals or in powder, the later burning readily in air.

Zirconium metal is used in the construction of nuclear reactors, and in refractory and corrosion-resistant alloys; for chemical plants. Zirconium compounds are used as catalysts, as solid state devices and in rectifying applications. Zirconia, the oxide, is used as a refractory, in abrasives and enamels, etc. The source of the metal and its compounds is the silicate, zircon, which itself is employed as a refractory and as a gemstone. Zircon occurs as a constituent of acid igneous rocks, and pegmatite deposits of the mineral, or concentrations derived from the decay of such deposits, are worked.

The major producers of zircon sand $(ZrO_2.SiO_4)$ in 1985 were Australia (500 000 t), South Africa (130 000 t), the USA (60 000 t) and the USSR (99 000 t), out of a total world production of over 800 000 t.

Tests. The mineral zircon may be recognised by its physical properties. Zirconium minerals, fused with sodium carbonate and dissolved in hydrochloric acid, give a solution which turns turmeric paper an orange colour; the paper, dried by gently heating, assumes a yellow-red colour.

The only zirconium mineral dealt with is:

silicate   zircon, $ZrSiO_4$

oxides   rutile, TiO_2
                                            anatase, TiO_2
                                            brookite, TiO_2

oxide of titanium and iron   ilmenite, $FeTiO_3$

Zirconium minerals

Zirconium (Zr) does not occur free in nature, but it enters into the composition of a number of complex silicates, most of which are rare. However, the simple silicate, $ZrSiO_4$, is of widespread occurrence as the mineral zircon, and this mineral is the source of the metal and its compounds used in industry. The metal zirconium is allied to titanium in its properties, and may be prepared by reduction of zirconium oxide by magnesium; the oxide is produced by fusing zircon with acid potassium fluoride, treating with hydrochloric acid and precipitating the oxide with ammonia. The metal has a specific gravity of 4.08, a melting point of 1300°C, and is produced either as crystals or in powder, the latter burning readily in air.

Zirconium metal is used in the construction of nuclear reactors, and in refractory and corrosion-resistant alloys for chemical plants. Zirconium compounds are used as catalysts, as solid state devices and in reinforcing applications. Zirconia, the oxide, is used as a refractory, in abrasives and enamels, etc. The source of the metal and its compounds is the silicate, zircon, which itself is employed as a refractory and as a gemstone. Zircon occurs as a constituent of acid igneous rocks, and pegmatitic deposits of the mineral, or concentrations derived from the decay of such deposits, are worked.

The major producers of zircon sand ($ZrO_2.SiO_2$) in 1985 were Australia (500 000 t), South Africa (140 000 t), the USA (80 000 t) and the USSR (90 000 t), out of a total world production of over 800 000 t.

Tests. The mineral zircon may be recognized by its physical properties. Zirconium minerals, fused with sodium carbonate and dissolved in hydrochloric acid, give a solution which turns turmeric paper an orange colour; the paper, dried by gently heating, assumes a yellow–red colour.

The only zirconium mineral dealt with is:

silicate   zircon, $ZrSiO_4$

The rare element **hafnium** (Hf) replaces zirconium in up to about 1% of zircon. Hafnium is used mainly in control rods in naval nuclear reactors and also in ceramics, refractories, alloys and in hafnium–columbium carbide cutting tools. About 50 t of hafnium was produced in the USA in 1985.

Rare earth minerals

The rare earth elements (REEs) include cerium, lanthanum, erbium, yttrium, europium and gadolinium. The most important are **cerium** (Ce), **yttrium** (Y), **gadolinium** (Gd), **samarium** (Sm) and **neodymium** (Nd). **Thorium** is associated with this group being common in monazite, a cerium-bearing mineral.

The uses of the REE oxides vary from petroleum catalysts and metallurgical uses to the manufacture of ceramics and glass. Cerium, yttrium and europium oxides are used in cathode ray and colour television tubes, and as a coating on camera lenses. Neodymium is used in high strength neodymium–iron–boron permanent magnets.

Many minerals contain REEs, and in particular orthite or allanite (a cerium-bearing member of the epidote group of rock-forming minerals), monazite (an RE-bearing phosphate containing thorium), bastnaesite (a hydrated RE carbonate), and xenotime (yttrium phosphate).

Rare earth oxides are normally obtained from bastnaesite and monazite. Yttrium is produced from monazite or xenotime. Most REEs in the USA and China are obtained from bastnaesite deposits, and in Australia from monazite deposits.

World production of rare earth ores in 1985 was estimated at 47 000 t, with the principal mineral concentrate producers being the USA (25 310 t bastnaesite), China (12 000 t bastnaesite), Australia (14 000 t monazite), Brazil (6000 t monazite) and India 4000 t monazite). Canada is about to start mining xenotime (for yttrium).

Tests. Satisfactory tests for cerium and other REEs are complicated chemical ones, and are beyond the scope of this book. Rare earth metals can be detected by spectroscopic methods. Under the microscope, the identity of grains of monazite can be established by use of the spectroscopic eyepiece, whereby characteristic absorption bands are observed.

The cerium minerals considered are:

(37 000 Afm[3]), Norway (26 400 Mm[3]) and Rumania (22 000 Mm[3]), with Mexico also producing significant amounts.

Carbonic acid is a combination of carbon, oxygen and hydrogen, and has the chemical formula H_2CO_3. The salts of this acid are called carbonates, and are very common minerals, existing in the Earth's crust in enormous quantities. For example, the common limestones are composed mainly of calcium carbonate, $CaCO_3$; the rock dolomite is a carbonate of calcium and magnesium, $CaMg(CO_3)_2$; the important iron ore, siderite, is iron carbonate, $FeCO_3$; and a host of other economically important carbonates are known. The carbonates are described in Chapter 8.

Silicon minerals

Silicon (Si) does not occur in a free state in nature, but its compounds are extraordinarily abundant. It constitutes about 28% of the Earth's crust. The oxide, quartz, and the great group of the silicates are the most important rock-forming minerals. Silica is used in glass-making particularly. Silicon is used in the manufacture of silicon chips and semi-conductors, as a de-oxidizer in steel-making, as ferrosilicon alloy, and in aluminium casting and other metallurgical processes. Its organic compounds, especially the silicones, are exceedingly important.

Silica, SiO_2, is the only oxide of silicon. It occurs in the form of quartz, chalcedony, agate, flint, etc. Sand is usually made up mostly of small grains of quartz (more rarely of flint) and consolidated sands provide the important sedimentary rock, sandstone. Opal is a hydrated form of silica. Silicon is obtained from silica-rich sands, sandstones or quartzites, which are commonplace (to a limited degree of purity) in almost every country.

The large number of minerals known as the silicates were formerly considered as salts of various theoretical silicic acids, but are now classified on the basis of the structural arrangements of their constituent atoms. These structures have been elucidated by the methods of X-ray analysis, and are described in Chapter 9, where all silicate minerals are described.

Tin minerals

Tin (Sn) is said to have been found native but, if it does occur so, is of very rare occurrence. It is chiefly found in the form of the oxide, cassiterite or tinstone, which is the main source of the metal, only a small amount being obtained from tin sulphides occurring with cassiterite in Bolivia. Tin is a bright, white metal, malleable and ductile. It has a specific gravity of 7.3 and melts at 232°C. A bar of the metal emits a crackling sound when bent.

(37 000 Mm^3), Norway (26 400 Mm^3) and Rumania (22 000 Mm^3), with Mexico also producing significant amounts.

Carbonic acid is a combination of carbon, oxygen and hydrogen, and has the chemical formula H_2CO_3. The salts of this acid are called carbonates, and are very common minerals, existing in the Earth's crust in enormous quantities. For example, the common limestones are composed mainly of calcium carbonate, $CaCO_3$; the rock dolomite is a carbonate of calcium and magnesium, $CaMg(CO_3)_2$; the important iron ore, siderite, is iron carbonate, $FeCO_3$; and a host of other economically important carbonates are known. The carbonates are described in Chapter 8.

Silicon minerals

Silicon (Si) does not occur in a free state in nature, but its compounds are extraordinarily abundant. It constitutes about 28% of the Earth's crust. The oxide, quartz, and the great group of the silicates are the most important rock-forming minerals. Silica is used in glass-making particularly. Silicon is used in the manufacture of silicon chips and semiconductors, as a de-oxidizer in steel-making, as ferrosilicon alloy, and in aluminium casting and other metallurgical processes. Its organic compounds, especially the silicones, are exceedingly important.

Silica, SiO_2, is the only oxide of silicon. It occurs in the form of quartz, chalcedony, agate, flint, etc. Sand is usually made up mostly of small grains of quartz (more rarely of flint), and consolidated sands provide the important sedimentary rock, sandstone. Opal is a hydrated form of silica.

Silicon is obtained from silica-rich sands, sandstones or quartzites, which are commonplace (to a limited degree of purity) in almost every country.

The large number of minerals known as the silicates were formerly considered as salts of various theoretical silicic acids, but are now classified on the basis of the structural arrangements of their constituent atoms. These structures have been elucidated by the methods of X-ray analysis, and are described in Chapter 9, where all silicate minerals are described.

Tin minerals

Tin (Sn) is said to have been found native but, if it does occur so, is of very rare occurrence. It is chiefly found in the form of the oxide, cassiterite or tinstone, which is the main source of the metal, only a small amount being obtained from tin sulphides occurring with cassiterite in Bolivia. Tin is a bright, white metal, malleable and ductile. It has a specific gravity of 7.3 and melts at 232°C. A bar of the metal emits a crackling sound when bent.

Cassiterite is obtained commercially from both lodes and alluvial (placer) deposits. In the former it may be associated with arsenic, copper and iron minerals, wolfram, etc.; in alluvial deposits it is often associated with ilmenite or titaniferous iron ore, monazite, zircon, topaz, tourmaline, etc. The proportion of tin in ores is usually expressed as black tin, that is, cassiterite containing about 70% of the metal per tonne of ore. In the case of alluvial deposits, less than 1 kg per tonne has been profitably worked. The alluvial or mine deposits are concentrated up to as near 70% of metallic tin as is practicable, by means of shaking tables and other mechanical contrivances; when the cassiterite is associated with ilmenite, wolfram or other magnetic minerals, electromagnetic separators are employed. The dressed product or concentrate is reduced in reverberatory furnaces, and the metal further purified by electrolytic processes.

The chief use of tin is in the manufacture of tin-plate, which is sheet-iron coated with a very thin coating of tin, and tin-plate being employed for the production of cans, etc. Another very important use is for the manufacture of a number of important alloys, such as pewter, the various solders, bearing metals, type-metal, bronze, gun-metal, bell-metal, fusible metal, etc. Salts of tin are employed in calico printing, dyeing, silk-making, in the ceramic industry, etc.

Western world mine production of tin-in-concentrates was 160 295 t in 1985. The major producing countries (79% of output) were Malaysia (36 880 t), Brazil (27 750 t), Indonesia (22 000 t), Bolivia (16 950 t), Thailand (16 590 t), Brazil (13 100 t), Australia (6840 t) and the UK (5150). Each of the other tin-mining countries, Zaire and Nigeria, produced less than 2500 t of tin-in-concentrates.

The USSR and China were estimated to produce around 23 000 t and 14 700 t respectively of tin ore in 1985. Tin ore is one of the few important industrial ores not produced in significant amount in the USA. The recovery of scrap tin – secondary tin – was important in industry, especially in the USA, where 22% of the annual consumption of the metal (11 000 t) was supplied from this source; the tin coating on cans is now too thin to repay recovery. Tin-smelting plants are concentrated in Malaysia, Indonesia, Thailand, Bolivia, the USSR, China, the UK, Brazil, Australia and the USA which, between them, account for 90% of annual world output.

Tests. The following are the chief tests for tin. When heated on charcoal with sodium carbonate and charcoal, tin compounds are reduced to the metal, which is soft and malleable. The encrustation given by heating tin compounds alone on charcoal, when moistened with cobalt nitrate and strongly reheated, assumes a blue–green colour. The tin bead, when treated with warm nitric acid, becomes coated with a white covering of hydroxide.

As already noted, the major mineral of tin is the oxide, cassiterite. In addition to this compound, tin occurs in a few other minerals, mostly complex sulphides, only one of which is sufficiently important to be described here; namely, stannine or tin pyrites, Cu_2SnFeS_4. The important tin minerals are therefore:

oxide   cassiterite, tinstone, SnO_2
sulphide   stannine, tin pyrites, Cu_2SnFeS_4

Lead minerals

Lead (Pb) is known native, but is of exceedingly rare occurrence. Lead is a bluish-gray metal, the freshly cut surface of which shows a bright metallic lustre which, however, quickly oxidizes on exposure to air. It is soft, may be scratched with the fingernail, and makes a black streak on paper. The specific gravity of the metal is 11.34. It fuses at 327°C, and crystallizes when cooled slowly. It has little tenacity, and cannot be drawn into wire but is, however, readily rolled or pressed into thin sheets, or extruded when in a semi-molten condition through dies to form piping. These properties are a consequence of its close-packed cubic structure; they are materially affected by the presence of small quantities of impurities.

Lead is easily reduced from its compounds. It is readily soluble in nitric acid, but is little affected by hydrochloric or sulphuric acid. It forms a number of compounds of great commercial importance, as noted below.

The principal ores of lead are the sulphide, galena, PbS, and the sulphate, anglesite, $PbSO_4$, and carbonate, cerussite, $PbCO_3$. For the production of lead, the ore is first partly roasted or calcined, and then smelted in reverberatory or blast furnaces. Most lead ores contain silver, and this metal is obtained from the lead by cupellation, repeated melting and crystallization, alloying with zinc, or by electrolytic processes.

Sphalerite or blende is frequently associated with galena, and the presence of zinc causes difficulties in smelting. Mechanical separation (dressing) by jigs, flotation processes, etc., of the two minerals is resorted to, and, with the improvement of such processes, low-grade mixed ores are now being worked at a profit. When antimony is associated with the galena, the ore may be smelted direct for the production of antimonial lead.

The uses of lead and its compounds are manifold. The metal is employed in the construction of accumulators, as is the oxide, but this use has been decreasing in recent years. Maintenance-free batteries based on lead–tin–calcium alloys are now dominating the US market, and making rapid progress in other markets. The use of lead additives to petrol (gasoline) is now declining throughout the world as the campaign for lead-free air gathers momentum, and eventually it is considered that lead-free petrol will be used universally (see also under rhenium). Other uses for lead include lead sheeting and piping, cable-covers, ammunition, foil, etc. It is a constituent of many valuable alloys, such as pewter, solder, babbitt-metal, type-metal, bronzes, anti-friction metal and fusible metal. Lead compounds, such as the oxides, red lead and litharge, and the basic carbonate, white lead, are employed extensively as pigments. The oxide is used in glass-making, as a flux, and in the rubber industry. The nitrate is employed in calico dyeing and printing processes, the arsenate is used as an insecticide, and the acetate is employed in medicine.

In 1985, the smelter production of lead was 4.11 Mt, the dominant producers being the USA (1.00 Mt), Japan (0.51 Mt), the USSR (0.50 Mt), West Germany (0.36 Mt), the UK (0.31 Mt), Canada (0.24), Australia (0.22 Mt) and France (0.22 Mt). The principal countries mining lead in 1983 were Australia, the USA, the USSR, Canada, Peru, Mexico, South Africa, Morocco and Yugoslavia, each with a production of over 100 000 t of ore (metal content).

The major primary ore is galena, PbS; deposits of galena oxidize in their upper parts into oxysalts, of which the most important, economically, are cerussite, $PbCO_3$, and anglesite, $PbSO_4$. Lead ores occur in a number of forms, not all of economic importance; the primary modes of occurrence are as lodes or veins, as metasomatic replacements and contact metamorphic deposits, or as disseminations.

Tests. The chief tests for lead are as follows: lead compounds colour the blowpipe flame a pale sky blue; this is a poor colour and of little value and, further, lead compounds attack the platinum wire. When lead minerals are heated alone on charcoal, they give a sulphur yellow encrustation. When heated with potassium iodide and sulphur, they give a brilliant yellow encrustation – this is a good test. Roasted with sodium carbonate and charcoal, on charcoal, lead minerals are reduced to metallic lead, which shows a lead grey bead, bright while hot but dull when cold; the bead is malleable and marks paper.

The minerals of lead considered here are:

sulphide	galena, PbS
oxide	minium Pb_3O_4
carbonate	cerussite, $PbCo_3$
chorocarbonate	phosgenite, $PbCO_3.PbCl_2$
sulphatocarbonate	leadhillite, $PbSO_4.2PbCO_3.Pb(OH)_2$
sulphate	anglesite, $PbSO_4$
basic sulphates	plumbojarosite, $PbFe_6(OH)_{12}(SO_4)_4$ linarite, $(PbCu)SO_4.(Pb,Cu)(OH)_2$
chlorophosphate	pyromorphite, $3Pb_3P_2O_8.PbCl_2$ or $(PbCl)Pb_4(PO_4)_3$
chloro-arsenate	mimetite, $3Pb_3As_2O_8.PbCl_2$ or $(PbCl)$ $Pb_4(AsO_4)_3$
chlorovanadate	vanadinite, $3Pb_3V_2O_8.PbCl_2$ or $(PbCl)Pb_4(VO_4)_3$
chromate	crocoisite, $PbCrO_4$
molybdate	wulfenite, $PbMoO_4$

Note also jamesonite, $Pb_4FeSb_6O_{14}$, bournonite, $CuPbSbS_3$, freiesleberite, $(Pb,Ag)_8Sb_5S_{12}$, and nagyagite, a sulphotelluride of lead and gold.

Group Va   Vanadium, niobium and tantalum

Vanadium minerals

Vanadium (V) does not occur free in nature. It is a whitish silvery metal, melting at about 1720°C. It has a great affinity for oxygen, a property underlying its use in metallurgy. Vanadium ores are treated in various ways, by smelting in the electric furnace, and reduction by the Thermit process to produce ferrovanadium, with some 30% vanadium.

Vanadium is used mainly in the manufacture of special steels, such as high-speed tool steels, the vanadium acting as a scavenger for oxygen, and also imparting special properties of toughness, etc., to the steel. In addition, other alloys are becoming important. Vanadium salts are used for various processes connected with chemical manufacture, printing of fabrics, dyeing, ceramics, etc. Vanadium is obtained from magnetite deposits, and is given as the equivalent of V_2O_5 (vanadium pentoxide) obtained.

Western world production total in 1985 is not known with certainty, but is estimated at about 35 000 t. The major producers were South Africa (20 000 t equiv. V_2O_5), China (7000 t) and the USA (2700 t), with Japan and Mexico also producing significant quantities (probably more than 1000 t per annum); New Zealand will produce ~ 1000 t per annum from 1986 onwards, and Australia, which has a production capacity of 1800 per annum, has temporarily ceased production; USSR production is not known, but about 12 000 t of vanadium was exported in 1985.

Tests. Vanadium compounds give characteristic reactions in the beads. In the oxidizing flame, the borax head is yellow when hot, and yellow–green to colourless when cold; in the reducing flame, it is dirty green when hot, and clear green when cold. In the oxidizing flame, the microcosmic salt bead is yellow to amber coloured, and in the reducing flame, it is green.

Vanadium minerals are not abundant, but include:

sulphide	patronite ?VS_4
silicate	roscoelite (a vanadium-bearing mica)
vanadate	carnotite, $K_2(UO_2)_2(VO_4)_2.3H_2O$
chlorovanadate	vanadinite, $(PbCl)Pb_4(VO_4)_3$

Niobium and tantalum minerals

Niobium (Nb) and Tantalum (Ta) are rare metallic elements possessing high tensile strengths. They are both resistant to corrosion. Niobium is used in the production of high-strength, low-alloy (HSLA) steels. The use of tantalum has decreased over the years but it is still used in special steels, particularly those used for dental and surgical instruments. The main use of tantalum now is in capacitors produced for the aerospace industry, computers, and the telecommunications industries. It is employed also in electrodes, and a compound, tantalum carbide, one of the hardest materials known, is used in tools. The only sources of any commercial importance are the minerals tantalite and pyrochlore, but most tantalum is produced as a by-product of tin smelting. Tantalite occurs in association with wolfram and cassiterite in granitic pegmatites, and the several small production units operate on deposits of this nature, or on alluvial deposits derived from similar occurrences. A part of the tantalum in tantalite is almost invariably replaced by the closely allied metal, niobium, and when the tantalum is subordinate in amount, the mineral is known as columbite or niobite.

In 1985 world production of niobium was between 15 000 t and 17 500 t (Nb_2O_5, from 50 000 t of concentrates), of which 82% comes from Brazil (mainly from the Araxa Mine) and 17% from Canada, with the remainder from Nigeria and Zaire.

In 1985 world production of tantalum was about 840 t of equivalent tantalum pentoxide (Ta_2O_5), which represents a fall from 1979 and 1980. Most tantalum is recovered from slags, with the major producers being Thailand, Australia, Malaysia and Brazil; the main Canadian mine in Manitoba was closed during 1983

The tantalum–niobium minerals considered here are:

oxide of tantalum–niobium, iron and manganese	tantalite-columbite, $(Fe,Mn)(Ta,Nb)_2O_6$
oxide of tantalum–niobium, sodium and calcium	pyrochlore, $(Na,Ca)(Ta, Nb)_2O_6(O,OH,F)$

Group Vb   Nitrogen, phosphorus, arsenic, antimony and bismuth

Nitrogen minerals

The gas nitrogen (N) makes up some 78% by volume of the atmosphere. It occurs in combination in two principal types of minerals, the nitrates and the ammonium minerals.

The nitrates are salts of nitric acid, HNO_3. These salts are mostly very soluble in water, so that their occurrence as minerals is restricted. The two major mineral nitrates are:

sodium nitrate	soda nitre, Chile saltpetre, $NaNO_3$
potassium nitrate	nitre, saltpetre, KNO_3

The ammonium radicle, NH_4, occurs as the cation portion of several mineral salts, such as:

ammonium chloride	sal ammoniac, NH_4Cl
ammonium sulphate	mascagnite, $(NH_4)_2SO_4$

Several other, still rarer, ammonium salts are know as minerals; for example, taylorite, $(NH_4)_2SO_4.5K_2SO_4$, an ammonium alum, $(NH_4)Al(SO_4)_2.12H_2O$, and ammonioborite, $(NH_4)_2B_{10}O_{16}.5H_2O$, but only sal ammoniac and mascagnite are described. Chile produced 327 000 t of sodium nitrate and 113 000 t of potash nitrate in 1985.

Tests. All the ammonium salts are more or less soluble in water, and are easily and entirely volatized before the blowpipe; this character suffices to distinguish them from other minerals. They also give the characteristic ammonia odour when heated with quicklime, or when ground up with lime and moistened at the same time with water.

Phosphorus minerals

Phosphorus (P) forms an acid, phosphoric acid, H_3PO_4, and salts of this acid are fairly common as minerals. Usually, however, mineral phosphates are more complex, being phosphates of two or more metals, basic phosphates of various types, or compounds into which other radicles enter. Phosphate rock is the term used to describe minerals containing phosphate, and world production is given in terms of 'phosphate rock'. Over 80% of the demand for phosphate rock comes from the fertilizer industries, where it is the raw material for phosphate fertilizers ('superphosphate', etc.). In 1985, world production of phosphate rock was 153.62 Mt, of which the major producers were the USA (51.1 Mt), the USSR (31.3 Mt), Morocco (20.7 Mt) and China (14.8 Mt). Other important producers (>1 Mt per annum) were Jordan, Tunisia, Brazil, Israel, South Africa, Togo, Senegal, Nauru and Christmas Island in the Pacific, and Syria.

The uses of the various economically important phosphates are detailed under their descriptions.

Tests. Tests for phosphates are as follows: In the flame test, many phosphates give a blue–green colour, which is increased in many cases if the mineral is moistened with strong sulphuric acid. When phosphates are fused with sodium carbonate on charcoal, and the fused mass removed and transferred to a closed tube with a little powdered magnesium and ignited, the phosphate is reduced to phosphide which, when, moistened, gives the well known disagreeable smell of phosphoretted hydrogen, PH_3.

The main mineral phosphates are as follows:

amblygonite	lithium aluminium phosphate, $Li(F,OH)AlPO_4$
apatite	calcium fluorophosphate and chlorophosphate, $Ca_5(F,Cl)(PO_4)_3$
autunite	hydrous copper phosphate, $4CuO.P_2O_5.H_2O$
monazite	phosphate of the cerium metals, $(Ce,La,Yt,Th)PO_4$

phosphochalcite	hydrous copper phosphate, $6CuO.P_2O_5.3H_2O$
pyromorphite	chlorophosphate of lead, $Pb_5Cl(PO_4)_3$
torbernite	hydrous phosphate of copper and uranium, $Cu(UO_2)_2(PO_4)_2.12H_2O$
turquoise	basic hydrous aluminium copper phosphate, $CuAl_6(PO_4)_4(OH)_8.4H_2O$
vivianite	hydrous iron phosphate, $Fe_3(PO_4)_2.8H_2O$
wavellite	hydrous aluminium phosphate, $Al_6(PO_4)_4(OH)_6.9H_2O$

Arsenic minerals

Arsenic (As) is found native, usually associated with other metals, but never in sufficient quantity to repay working; in combination it is very widely distributed, and occurs in many sulphide ores.

Arsenic is an extremely brittle, steel grey metal of a brilliant lustre, and having a specific gravity of 5.7. It is obtained from its ores by heating in retorts, but most of the production is as a by-product in the smelting of arsenical lead, silver, cobalt or copper ores.

The metal arsenic is employed in small quantity in the manufacture of lead shot and certain alloys. The most important industrial compound of arsenic is white arsenic, arsenious oxide, As_2O_3, which is obtained in the form of flue-dust or 'soot' in the smelting of arsenical ores proper and of the numerous arsenical ores of other metals. The greater proportion of the arsenic used in industry depends for its employment on the poisonous properties of arsenic compounds, the manufacture of insecticides, weed-killers, sheep-dips, etc., absorbing some 70% of the annual output. Other uses are as a decolorizer of glass, in paint manufacture, textile printing, etc. It is probable that the world consumption of arsenic annually is over 50 000 tonnes of white arsenic, the major producers being the US, Sweden, Mexico, France, Belgium and Japan, but information on arsenic production by individual countries is not easily available, except for the USSR which produced 8100 t in 1985; it is an instructive comment on the arsenic situation to realize that one Swedish gold–copper mine could supply the whole world demand.

The most important minerals of arsenic from the economic viewpoint are mispickel or arsenopyrite, FeAsS, with the sulphides orpiment As_2S_3, and realgar, As_2S_2, of less account. Native arsenic is fairly widespread, but not commercially important. Arsenic enters into the composition of a number of complex sulphides, from some of which white arsenic is obtained as a by-product; examples of such minerals are enargite, Cu_3AsS_4, tennantite, $(Cu,Fe)_{12}As_4S_{13}$, proustite, Ag_3AsS_3, and rarer sulphides of arsenic, copper and lead. As indicated below, these minerals are described under their more important metallic component. In addition to the iron sulpharsenide, mispickel, another sulpharsenide, that of cobalt, cobaltite, $CoAsS$, is important, but it is described on p. 249 as an ore of cobalt. Another group of arsenic minerals are the arsenides of cobalt and nickel described with the minerals of these two metals; these arsenides are smaltite, $(Co,Ni)As_{3-n}$, kupfernickel or niccolite, $NiAs$, and chloanthite, $NiAs_2$. Certain arsenates, such as mimetite, $(PbCl)Pb_4(AsO_4)_3$, and olivenite, $Cu_3As_2O_8.Cu(OH)_2$, occur in the oxidized portion of arsenical lead and copper veins. Other arsenates such as erythrite, $Co_3As_2O_8.8H_2O$, and annabergite, $Ni_3As_2O_8.8H_2O$, characterize weathered cobalt and nickel arsenide ores respectively. Finally, the oxide, arsenolite, As_2O_3, is known in small quantity as a decomposition product of other arsenic ores.

The primary arsenic minerals occur in lodes or veins more or less directly connected with igneous intrusions; the arsenates, arsenolite, realgar and orpiment are characteristic of the oxidized portion of such deposits.

Tests. The following are the main tests for arsenic. Arsenic compounds, when heated on charcoal, give a white encrustation far from the assay, and at the same time, fumes having a garlic odour are emitted. Heating in the open tube, arsenic compounds give a white sublimate, which is volatile on heating. Heated in the closed tube, some arsenic compounds give a shining black sublimate, the arsenic mirror; most arsenates give a similar mirror when heated with charcoal or sodium carbonate in the closed tube.

From the foregoing, the following list of arsenic minerals may be compiled:

element	native arsenic, As
oxide	arsenolite, white arsenic, As_2O_3
sulphides	orpiment, As_2S_3
sulpharsenides	mispickel, arsenopyrite, FeAsS cobaltite, CoAsS
arsenides	kupfernickel, niccolite, NiAs chloanthite, $NiAs_2$ smaltite, $(Co,Ni)As_{3-n}$

arsenates	mimetite, $(PbCl)Pb_4(AsO_4)_3$
	olivenite, $Cu_3As_2O_8.Cu(OH)_2$
	erythrite, $Co_3As_2O_8.8H_2O$
	annabergite $Ni_3As_2O_8.8H_2O$
complex sulphides	enargite, Cu_3AsS_4
	tennantite, $(Cu,Fe)_{12}As_4S_{13}$
	proustite, Ag_3AsS_3

Antimony minerals

Antimony (Sb) in a free state is of extremely rare occurrence. The chief source of the metal is the sulphide, stibnite or antimonite, Sb_2S_3, which is widely distributed, but found in workable quantities in comparatively few localities. Metallic antimony is a tin-white, very brittle metal, with a crystalline structure. It has a specific gravity of 6.7 and melts at 630°C.

For the production of the metal, the sulphide is freed from its gangue by liquidation and reduced in reverberatory furnaces, or the crude ore is volatilized in a blast furnace, and the condensed fumes reduced in reverberatory furnaces. On the market, the liquidated sulphide is called 'crude antimony', while the metal is called 'regulus of antimony'. The first quality of refined antimony is known as 'star antimony', owing to the fern-like markings on its surface. Antimonial lead ores, free from gold and silver, are commonly smelted direct for 'hard' or 'antimonial' lead.

Antimony is used as a hardening agent in lead for battery grids, and as a vulcanizing agent for rubber tyres. Antimony is extensively used in fire retardants (flameproofing) in modern building projects. It is used in the production of some alloys such as type-metal and anti-friction metals. Its compounds are used for a variety of purposes, such as for pigments, in medicine, as a mordant, in the manufacture of opaque enamel ware, and in glass and pottery manufacture.

Total world production of antimony in 1985 was 52 980 t of metal (from ores and concentrates), and the major producers were China (12 700 t), the USSR (10 800 t), Bolivia (9000 t) and South Africa (8160 t), together accounting for about 77% of the total supply. Other important producers include Mexico (2000 t), Turkey, Yugoslavia, Thailand and Australia (each 900 t per annum), with Morocco, Czechoslovakia, the USA and Peru contributing significant amounts.

The antimony minerals are as follows. The element occurs as native antimony in small amounts, but the most important mineral is the sulphide, antimonite or stibnite, Sb_2S_3. A series of oxygen compounds occur as weathered products of the sulphide ores, and among these are the oxides senarmontite, Sb_2O_3 (cubic), valentinite, Sb_2O_3 (orthorhombic), cervantite $Sb_2O_3.Sb_2O_5$ and the oxy-sulphide, kermesite, $2Sb_2S_3.Sb_2O_3$. In addition to these purely antimony minerals, the element also enters into the composition of a large number of complex sulphides, of which the following are considered in this book: jamesonite, $Pb_4FeSb_6S_{14}$; a group of silver antimony sulphides, such as stephanite, Ag_5SbS_4, pyragyrite, Ag_3SbS_3, polybasite, $(Ag,Cu)_{16}(Sb,As)_2S_{11}$, freieslebenite, $(Pb,Ag)_8Sb_5S_{12}$, etc.; and a group of copper antimony sulphides, such as bournonite, $CuPbSbS_3$, tetrahedrite, $(Cu,Fe)_{12}Sb_4S_{13}$, and famatinite, Cu_3SbS_4.

Antimony ores occur both in deposits associated with volcanic rocks, and also more deep-seated veins formed under moderate to high temperatures and pressures. Thus stibnite often occurs with mercury ores, but is more common in veins with a gangue of quartz and only a small proportion of other sulphides. Certain of the replacement galena deposits show antimony minerals such as jamesonite and stibnite. The surface oxidation of these primary antimony ores leads to the formation of the oxides and oxy-sulphide mentioned above.

Tests. The following are important tests for antimony. When antimony compounds are heated on charcoal, a dense white sublimate is formed as an encrustation near the assay. The nearness to the assay, and the absence of any characteristic fumes distinguish the reaction of antimony from that given by arsenic. Heated in the open tube, antimony compounds give a white sublimate of oxide of antimony, which appears as a ring near the assay; heated with sodium carbonate in the closed tube, antimony compounds give a red–brown sublimate, which is black when hot.

The following minerals of antimony are described in this book:

element	native antimony, Sb
oxides	senarmontite, Sb_2O_3 (cubic)
	valentinite, Sb_2O_3 (orthorhombic)
	cervantite, $Sb_2O_3.Sb_2O_5$
sulphide	antimonite, stibnite, Sb_2S_3
oxysulphide	kermesite, $2Sb_2S_3.Sb_2O_3$

complex sulphides	jamesonite, $Pb_2FeSb_6S_{14}$
	stephanite, Ag_5SbS_4
	pyrargyrite, Ag_3SbS_3
	polybasite, $(Ag,Cu)_{16}(Sb,As)_2S_{11}$
	freieslebenite, $(Pb,Ag)_8Sb_5S_{12}$
	bournonite, $CuPbSbS_3$
	tetrahedrite, $(Cu,Fe)_{12}Sb_4S_{13}$
	famatinite, Cu_3SbS_4

Bismuth minerals

Bismuth (Bi) occurs in a free state in nature, often associated with silver, gold, copper, lead and other minerals. It is a greyish-white metal, with a slightly reddish tinge, lustrous and very brittle, having a specific gravity of 9.8 and melting at 271°C. Bismuth was obtained by smelting the dressed ores in small reverberatory furnaces or crucibles. Bolivia was the only major source of native ore, but has been forced to discontinue production because the selling price was less than the cost of production. World production in 1985, which was probably between 2000 and 3000 t, came entirely from by-product sources, particularly from the treatment of lead ores but also from copper, tungsten, tin and zinc ores. The main suppliers in 1983 were Peru (737 t), the USA (500 t), Mexico (395 t), Belgium and Japan, but only a few output figures are available. Other sources include Canada (222 t), the UK, the People's Republic of Korea (100 t), China, Bulgaria and West Germany. Bolivia supplied to 1979 from primary ore, but the present world price is not high enough to re-open the mine there.

Bismuth salts are used in medicine, and to a limited extent in pigments, glass, etc.; with tin, lead, mercury, etc., bismuth forms a series of alloys with low melting points, the fusible metals, which are important in certain industrial processes (e.g. casting) and are also employed in certain appliances such as automatic sprinklers and similar apparatus.

As already noted, bismuth occurs as the native element which, with the sulphide, bismuthinite, Bi_2S_3, and various complex lead–bismuth sulphides not dealt with in this book, are of primary origin. The notable oxidized minerals are the oxide, bismuth ochre or bismite, Bi_2O_3, the basic carbonate, bismutite, $Bi_2Co_5.H_2O$ and a telluride, tetradymite, $Bi_2(Te,S)_3$.

Bismuth ores occur in three main associations; (1) with tin and copper minerals, as in the Bolivian deposits; (2) with cobalt, as at Schneeberg, Saxony; and (3) with gold, as in the Australian deposits.

Tests. Bismuth compounds react as follows. When heated on charcoal with sodium carbonate and charcoal, they give a brittle, metallic bead, which volatizes on heating to give a yellow encrustation. Heated with potassium iodide and sulphur, bismuth compounds give an encrustation which is yellow near the essay, and scarlet in the outer parts. Solutions of bismuth salts become milky on the addition of water, owing to the formation of insoluble basic compounds, which are redissolved on the addition of an acid.

The bismuth minerals dealt with in this book are:

element	native bismuth, Bi
oxide	bismuth ochre, bismite, Bi_2O_3
carbonate	bismutite, $Bi_2CO_5.H_2O$
sulphide	bismuthinite, Bi_2S_3
telluride	tetradymite, $Bi_2(Te,S)_3$

Group VIa   Chromium, molybdenum, tungsten and uranium

Chromium minerals

Chromium (Cr), never found in nature except in combinations, is produced by the reduction of its ore by carbon in the electric furnace, or by the Thermit process. It is a brilliant white metal, having a specific gravity of about 6.5, and melting at about 1800°C. It possesses the property of imparting to iron and steel a high degree of hardness and tenacity, and for that reason has, in recent years, become of great industrial importance. For this purpose, an alloy of iron and chromium (ferrochrome, produced in the electric furnace) is commonly used; it is cheaper to make, melts at a lower temperature, and is consequently better under control than the pure metal. Stainless steel contains as much as 18% of chromium. The compounds of chromium are also of considerable industrial importance. Chromite, an oxide of iron and chromium, is used very extensively as a refractory material for furnace linings. Other salts, artificially prepared, are used as pigments, and in various industries, such as chromium plating, dyeing, tanning, photography, etc.

The only source of chromium is chromite, chrome iron ore, $FeCr_2O_4$. In 1985, total world production was 9.73 Mt, and the major Western producing countries were South Africa (3.34 Mt), Finland (0.51 Mt), Turkey (0.50 Mt), Zimbabwe (0.48 Mt), India (0.44 Mt), Brazil (0.31 Mt) and the Philippines (0.28 Mt). The USSR produced 2.94 Mt of high-grade chromite in 1985, and Albania was estimated to produce 1.0 Mt, also in 1985. A new source of chromite was recently discovered in Tibet, and this is expected to give China self-sufficiency in chrome production in the very near future. Chromium also occurs in various rock-forming minerals such as the chrome–spinel, picotite, the chrome garnet, uvarovite and crocoisite lead chromate, $PbCrO_4$.

Tests. The best tests for chromium are provided by the beads; chromium compounds produce a fine green colour in both borax and microcosmic salt beads.

Although chromium is present in several silicate materials, chrome diopside, chrome zoisite and uvarovite (garnet), there is only one important ore mineral:

Oxide   Chromite, Fe Cr_2O_4

Molybdenum minerals

Molybdenum (Mo) does not occur in a free state in nature but may be prepared from its sulphide, the mineral molybdenite, directly in the electric furnace, by the reduction of its oxide by means of carbon, or by the Thermit process. The metal has a specific gravity of about 10.2, melts at about 2620°C, and is white or greyish, and brittle. Molybdic acid forms salts known as molybdates, examples of which occur as minerals; thus wulfenite is lead molybdate, and molybdite is possibly a hydrated iron molybdate. Molybdic acid is employed only in the laboratory. Ammonium molybdate is a special reagent for the detection of phosphoric acid, a small quantity added to an acid solution containing phosphates producing a yellow precipitate after some time. The principal use of molybdenum is in the manufacture of special steels, and for this purpose ferromolybdenum alloy is frequently used in place of the metal. Molybdenum alloys are gaining importance, and the metal is also employed in certain electrical apparatus and in catalysts for the petrochemical industries.

There are two ores of molybdenum, the more important one being the sulphide, molybdenite, MoS_2, and the less important one being lead molybdate, wulfenite, $PbMoO_4$. Molybdenite occurs in deposits associated with acid igneous rocks, while wulfenite is found in the oxidized portions of lead- and molybdenum-bearing deposits. Molybdenite is a widely distributed ore, but frequently occurs in small veins or scattered in tiny flakes through the rocks, so that a concentration by table-dressing or oil flotation is necessary.

In 1985, Western world production was 178 million pounds (M lb) (approx. 89 800 t), and the major suppliers were the USA (47 200 t), Chile (18 100 t) and Canada (6800 t), which together account for 80% of the total supply. The USSR produced about 11 300 t of molybdenum in 1985.

Tests. The major tests for molybdenum are the following: when heated on charcoal in the oxidizing flame, molybdenum compounds give a white encrustation; this becomes blue where touched by the reducing flame. In the oxidizing flame molybdenum compounds colour the microcosmic salt bead yellow when hot, and colourless when cold; the bead is green in the reducing flame.

The two main minerals of molybdenum have been mentioned, namely the sulphide, molybdenite, and the lead molybdate, wulfenite. It was considered until recently that the oxide occurred as the mineral molybdite, but it has been shown that this mineral is probably a hydrated iron molybdate, possibly $Fe_2O_3.3MoO_3.8H_2O$. The molybdenum minerals considered in this book are therefore:

sulphide	molybdenum, MoS_2
molybdates	wulfenite, $PbMoO_4$
	molybdite, $Fe_2O_3.3MoO_3.8H_2O$

Tungsten minerals

Tungsten (W) is not found native, but is produced in the form of a greyish-black powder, with a specific gravity of about 19. The metal is obtained by reduction of its ores by carbon, or by the Thermit process.

The main tungsten minerals are tungstates of iron, manganese and calcium, the tungsten ores being wolfram, $(Fe,Mn)WO_4$, and scheelite, $CaWO_4$. Wolfram is usually associated with cassiterite, and for a long time the separation of these minerals was a matter of some dificulty. It is effected by electromagnetic separation. Scheelite is less often mixed with cassiterite and, in this case, the separation is performed by roasting the crushed ore with sodium carbonate, by which operation sodium tungstate is formed, and the tin ores can be removed. In the case of several tin-mining companies, the presence of wolfram enables profits to be made which, but for the 'mixed minerals', would otherwise be impossible.

Tungsten ores, chiefly wolfram, appear on the market in the form of concentrates, varying between 60% and 70% tungstic acid, WO_3, and are purchased on the basis of their tungstic acid content.

World production of tungsten ores was 45 100 t in 1985, with the major producers being China (13 500 t), the USSR (9100 t), Canada (3000 t), South Korea (2600 t), Australia (2000 t), Bolivia (1700 t), Portugal (1400 t), Austria (1400 t), Brazil (1200 t), USA (1100 t) and Burma (1100 t).

The main use of tungsten, either as metallic tungsten or in the form of ferrotungsten alloys produced in the electric furnace, is in cemented carbides such as the tungsten-carbide–cobalt cutting tool, and hard-facing alloys. Other important uses are in tool steels, lamp filaments and electrical contacts. Tungsten salts are used as mordants and for fireproofing purposes.

Two oxides of tungsten are known, WO_2 and WO_3; the latter is tungstic acid, and from it are formed various salts called tungstates, several of which occur as minerals, as in wolfram and scheelite, which have already been mentioned. Tungstic acid, WO_3, occurs as the mineral tungstite or tungstic ochre, and is formed as an alteration product of the tungstate minerals. The tungstate minerals occur for the most part as primary minerals associated with cassiterite and in close connection with acid igneous rocks; the calcium tungstate, scheelite, is found in pyrometasomatic deposits at the contact of granitic rocks and limestones.

Tests. The tests for tungsten are as follows. Tungsten salts colour the microcosmic salt bead blue in the reducing flame. Tungsten minerals, fused with sodium carbonate, and then dissolved in hydrochloric acid, give, on the addition of a small piece of tin or zinc, a blue solution when heated.

The tungsten minerals considered are:

oxide	tungstite, tungstic ochre, WO_3
tungstates	wolfram, $(Fe, Mn)WO_4$
	hubnerite, $MnWO_4$
	ferberite, $FeWO_4$
	scheelite, $CaWO_4$

Uranium (and radium) minerals

Uranium (U), which is not found native, is a hard, white metal, having a specific gravity of 18.7, and melting at a white heat. It is a constituent of a number of rare minerals, the most important of which are pitchblende and carnotite, together with tobernite, autunite and various hydrous derivatives. Metallic uranium is prepared by reduction of its oxide with carbon in the electric furnace. Pitchblende and carnotite are worked also for their content of radium (Ra), which is contained in all uranium minerals.

Uranium forms two oxides, and from these a complex group of uranium salts is derived. The uses of the oxides as such is limited; they are employed in glass-staining, for glazes, in dyeing and in photography. The metal has been employed in the production of certain special steels, but is not used for this purpose to any great extent, since steels of similar properties can be produced with less expensive components. There was little market for uranium until the exploitation of atomic fission in 1945 and the development of nuclear reactors; the demand for radium is on the increase and the minerals are valued according to their percentage of U_3O_8, which always carries a certain proportion of radium.

Radium (Ra) results from the disintegration of uranium, and uranium minerals contain 320 mg of radium per tonne of uranium. The main sources of radiur are pitchblende, carnotite and various decomposition products of pitchblende. Radium is used in the treatment of cancer, in certain X-ray apparatus, and for luminous paint.

Western world production of uranium (as U) in 1985 was 35 500 t, the major producers of which were Canada (12 800 t), the USA (6000 t), South Africa (5000 t), Australia (5000 t) and Namibia (3600 t), together accounting for 68% of the Western world supply. No information is available on the USSR (or any Eastern bloc country) production, but China was estimated to have produced 5000 t of uranium in 1983.

Tests. In the microcosmic bead, uranium compounds give a light moss-green colour, both hot and cold, in the oxidizing flame.

The major uranium minerals are the following:

oxide	pitchblende, uraninite, UO_2
hydrous phosphates	torbernite, copper uranite, $Cu(UO_2)_2(PO_4)_2.12H_2O$ autunite, lime uranite, $Ca(UO_2)_2(PO_4)_2.10–12H_2O$
hydrous vanadate	carnotite, $K_2(UO_2)_2(VO_4)_2.3H_2O$
hydrated sulphates	zippeite, uraconite and others of uncertain composition

Group VIb   Sulphur, selenium and tellurium

Sulphur minerals

Sulphur (S) occurs native in orthorhombic and monoclinic forms, the orthorhombic variety being the low-temperature common type, and the high-temperature monoclinic varieties being rare.

Native sulphur and metallic sulphides, mainly iron pyrites, FeS_2, form practically the sole source of the sulphuric acid of commerce, and may be regarded as the most important minerals in connection with the chemical industry.

Native sulphur (called brimstone in the industry) is purified from the associated gangue by melting in ovens, etc., or by distilling in closed vessels, with the production of cast-stick sulphur, or of flowers of sulphur condensed in flues. Sulphur can also be recovered from sour natural gas, oil refinery gases, coal and tar sands, but is only significant at present from oil and gas and from tar sands. In America, sulphur, over 99% pure, is obtained in great quantity from deep-seated deposits in the Gulf states; the sulphur is extracted by the Frasch process, in which a double tube is driven into the sulphur beds, and superheated steam or hot air is forced down, with the result that the sulphur melts and is forced up the inner tube and into vats.

Frasch sulphur is also obtained from Mexico, Iraq, Poland and the USSR. The mode of occurrence and origin of sulphur deposits is discussed under the description of that mineral.

For the manufacture of sulphuric acid, iron pyrites, FeS_2, which theoretically contains 53.46% of sulphur and which is commonly sold on a guarantee of 45–50%, is more used than any other mineral, with the exception of sulphur itself. The oxide of iron which is formed during the roasting of pyrites is saleable for its iron content; and if the mineral contains copper, gold or silver, even in small quantities, it is paid for at a higher rate.

The value of sulphides when smelting oxidized ores is well known, and the calorific value of burning sulphur is utilized in pyritic smelting. Enormous quantities of sulphur dioxide are present in furnace gases, and many large smelters now recover it as sulphuric acid. The common association of arsenic with sulphur in mineral sulphides, and especially in pyrites, necessitates special care in the manufacture of sulphuric acid; but native sulphur, although seldom containing arsenic, is more liable to be contaminated with selenium, which is also objectionable.

Sulphur is used in the manufacture of fertilizers, which is by far its greatest use. Other uses are in the fibres, pigments, metallurgical, plastics and oil industries.

World production of sulphur in 1985 was 59.24 Mt, of which 38.96 Mt was from native sulphur, 9.15 Mt was from pyrite, and the remainder (11.13 Mt) from oil and gas refining. The major producers were the USA (5.0 Mt Frasch + 5.2 Mt recovered + 1.2 Mt others = 11.4 Mt in total), Canada (9.02 Mt), the USSR (6.60 Mt), Saudi Arabia (~2 Mt) and Mexico (2.0 Mt), accounting for over 50% of the total.

With hydrogen, sulphur forms an acid, sulphuretted hydrogen, H_2S. This gas, being readily absorbed in water, is found in certain mineral springs, as at Harrogate, England. A sulphide is formed by the replacement of the hydrogen of sulphuretted hydrogen by a metal; mineral sulphides are abundant and exceedingly important minerals. Examples of common mineral sulphides are galena, PbS, blende, ZnS, cinnabar, HgS, and pyrite or iron pyrites, FeS_2. These and other sulphides are described in Chapter 8, section 3.

With oxygen and hydrogen, sulphur forms many compounds, only one of which is important as an acid in mineralogy; this is sulphuric acid, H_2SO_4. Sulphates are formed by the substitution of a metal for the hydrogen of this acid, and are a very important group of minerals; examples of mineral sulphates are gypsum, $CaSO_4.2H_2O$, anglesite, $PbSO_4$, and barite, $BaSO_4$. The mineral sulphates are described under the headings of their metallic constituents.

Tests. Sulphur may be recognized when present in a mineral by the silver coin test. In this, the powdered mineral is first fused with sodium carbonate on charcoal; the fused mass is then placed on a silver coin and a black stain is produced when moistened. On roasting in the open tube or on charcoal, sulphides give a sharp pungent odour of sulphur dioxide, SO_2. On the addition of barium chloride to the solution, sulphates give a dense white precipitate of barium sulphate; also, after reduction by heating to the sulphide, they give the silver coin test as above.

Minerals containing sulphur may be found in the sulphide and sulphate groups of minerals in Chapter 8.

Selenium minerals

Selenium (Se) belongs to the same structural group as sulphur and tellurium. It occurs in native sulphur and in all pyritic ores, although often in negligible traces. Its mineral compounds are salts of the acid H_2Se, which is analogous with sulphuretted hydrogen, H_2S; thus, clausthalite is lead selenide, PbSe, berzelianite is copper selenide, Cu_2Se, tiemannite is mercury selenide, HgSe, and naumannite is silver selenide, Ag_2Se. The principle sources of selenium are the deposits in sulphuric acid chambers, and the anode mud or slime obtained in the electrolytic refining of copper.

Selenium is used in a number of diverse industrial applications, of which photocopiers and electronics, glass production and pigments are the most important. Photoreceptors based on high-purity selenium continue to dominate the plain-paper copier market. The action of light in discharging a selenium surface in a localized manner is used to reproduce images as electrostatic charge patterns. Complex selenium–tellurium alloys have been developed, offering a wide range of spectral responses and electrical characteristics, and are used in X-ray imaging equipment. Selenium is used in the selenium rectifier, and also in glassware to counteract the green (iron impurity) tinge. Selenium pigments range in colour from orange to deep red, and are used mainly by the plastics industry. In addition to these uses, selenium is used in special steels and as an alloy addition in the grids of lead-acid batteries. Selenium is recognized as an essential trace nutrient in animal diets.

World production in 1985 was estimated at 1700 t. The major producers were Japan (512 t), Canada (305 t), the USA (185 t) and Yugoslavia. The USSR is self-sufficient in selenium.

Tests. The main test for selenium compounds depends on the production, when they are heated on charcoal before the blowpipe, of a curious smell described as that of decaying horse-radish – a smell that can be readily recognized when once it has been encountered.

Selenium-bearing varieties of sulphur and tellurium are called selensulphur and selentellurium respectively. The description of the selenides mentioned above is beyond the scope of this book.

Tellurium minerals

Tellurium (Te) occurs free in nature in small quantities, in sulphur and pyrites. However, it is mostly found combined with metals as tellurides, such metals being bismuth, lead and, most importantly, gold and silver.

Tellurium is obtained with selenium in the anode slime from electrolytic copper refineries. When pure, tellurium has a greyish-white colour and a metallic lustre. It has a specific gravity of 6.3, melts at 450°C, and boils at 1400°C.

Tellurium is added to copper to improve machinability without reducing conductivity. Other important uses are in secondary vulcanizing agents, and as a constituent of various catalysts. It is also used in photocopiers (see under Selenium in the previous section). A new use of tellurium is as a data-storage medium in optical disks in computers, and selenium and tellurium are increasingly being used in the field of solar energy in various photoelectric and photothermal devices.

The Western world production of tellurium in 1985 was 215 t, with the major producers being the USA (106 t), Belgium (90 t), Japan (50 t), Peru (20 t) and Canada (11 t). No information is available on USSR and Eastern bloc production.

Tests. The best test for tellurium is to heat the compound in the closed tube with strong sulphuric acid, which then assumes a brilliant reddish-violet colour.

The main tellurium minerals are the native metal, the oxide tellurite, TeO_2, and the tellurides. The gold tellurides are important ores of gold, and there is also a silver telluride, hessite, and a bismuth telluride, tetradymite, $Bi_2(Te,S)_3$. The tellurium minerals considered in this book are therefore:

element	native telluride, Te
oxide	tellurite, TeO_2
tellurides	tetradymite, $Bi_2(Te,S)_3$
	hessite, Ag_2Te
	sylvanite, $(Ag,Au)_2Te$
	calaverite, gold and silver telluride, $(Ag,Au)_2Te$
	petzite, $(Ag,Au)_2Te$
sulphotelluride	nagyagite, $Pb_5Au(Te,Sb)_4S_{5-8}$

The gold-bearing tellurides are extremely important gold ores; they occur in veins and replacement deposits. In the upper parts of the veins, the tellurides are decomposed; some tellurium oxide is formed but most is removed in solution.

Group VIIa   Manganese and rhenium

Manganese minerals

Manganese (Mn) does not occur in an uncombined state in nature but may, like chromium, be produced in the electric furnace and by the Thermit process. It is a light, pinky-grey metal, melting at about 1260°C, and having a specific gravity of about 7.4.

The main application of manganese is in the manufacture of steel and manganese is added to the furnace, either as ferromanganese, containing 15–80% manganese, or as silicomanganese, containing 10–20% silicon, in which it acts as a deoxidizing agent.

The chief sources of manganese and its salts are the oxide minerals. Pyrolusite, MnO_2, is also used, as such, for a number of purposes such as the decolorization of glass, as a dryer in the manufacture of paint and varnish, and in dry batteries and, importantly, for the manufacture of chlorine, bromine and oxygen, and of permanganates and other manganese compounds. The permanganates of sodium and potassium are used in disinfectants.

For chemical uses, a high percentage, 80% of manganese in the form of peroxide is demanded, MnO_2 being taken as the basis price; lime should be present only in quantities of less than 2%. For metallurgical purposes, 50% MnO_2 is a common basis. Ores of iron containing manganese are smelted direct for the production of manganese pig-iron, but such ores should be regarded as iron ores, and the manganese would not be paid for except at the same rate as iron.

World production of manganese ore in 1985 was over 22 Mt, the major producers being the USSR (10.0 Mt), South Africa (3.2 Mt), Gabon (2.2 Mt), Brazil (2.0 Mt), Australia (1.75 Mt), China (1.7 Mt) and India (1.4 Mt), which together account for over 96% of the total.

Manganese is very widely distributed and, to a greater or lesser degree, replaces two sets of elements: first, the alkaline earths, calcium, barium and magnesium; and second, aluminium and iron. Purely manganese minerals of the greatest importance are the oxides; others of less importance are the carbonate, silicate and sulphide.

Manganese minerals occur in varied ways. Dialogite and rhodonite occur as veinstones in some silver lodes, the gossans of which carry the oxide minerals such as pyrolusite and psilomelane. Dialogite also occurs as a metasomatic replacement of limestone. The most important and interesting deposits of manganese, however, are oxides of sedimentary or residual origin. These oxides, pyrolusite, psilomelane, polianite, wad, braunite and manganite, occur in two main types of deposit. In the sedimentary deposits, the manganese has been precipitated in beds or layers of nodules together with iron compounds with which it is invariably associated, a process at the present time being carried on in moderately deep water. Many of the workable manganese bodies have been formed by the uplifting of these deep water deposits. Another deposit of this type is formed by precipitation of manganese oxides in lakes, etc., by the action of minute plants, giving rise to the bog-manganese deposits, as in Sweden, Spain and the US.

The second type of manganese deposit is formed by the alteration of rocks containing manganese-bearing minerals, mainly silicates. By the weathering of such rocks, the manganese oxides aggregate together as nodules and layers in the residual clay which forms on the outcrop of the weathered rock, and form a residual or lateritic deposit, such as occurs and is worked in India, Brazil, Gabon, Ghana and Arkansas, USA. It will be seen that all deposits of manganese oxides have been formed by the breaking-up of the manganese-bearing minerals of igneous and metamorphic rocks.

Workable deposits of hausmannite, braunite and franklinite, as in Sweden, Piedmont and at Franklin Furnace, USA, are formed by the metamorphism of sedimentary or residual manganese deposits.

Tests. Manganese minerals give distinctive bead reactions. The borax and microcosmic salt beads are reddish-violet in the oxidizing flame, and colourless in the reducing flame. The sodium carbonate bead is bluish-green in the oxidizing flame.

The manganese minerals dealt with are:

oxides	hausmannite, Mn_3O_4
	braunite, Mn_2O_3
	manganite, $MnO(OH)$
	pyrolusite, MnO_2
	polianite, MnO_2
	psilomelane, a hydrated oxide with Ba and K
	wad, like psilomelane
	asbolan, cobaltiferous wad
	franklinite, $(Fe,Zn,Mn)(Fe,Mn)_2O_4$
carbonate	dialogite, rhodochrosite, $MnCO_3$
silicate	rhodonite, $MnSiO_3$
sulphide	alabandite, MnS

In addition to these specifically manganese minerals, the element enters into many silicates, of which the manganese garnet, spessartite, is one of the most important.

The very rare element **rhenium** (Re) occurs in molybdenite obtained from porphyry copper deposits. Rhenium has a melting point of 3450°C and a specific gravity of 10.4. Rhenium is used mainly in platinum–rhenium catalysts to produce lead-free high-octane petrol, and also in the electrical industry. World production of rhenium in 1985 was approximately 18 t, the main producers being Chile (6 t of the mineral ammonium perrhenate), the USSR (4.5 t), West Germany, Canada and Finland. U.S. figures are withheld, but are believed to be about 3.9 t per annum.

Group VIIb   Fluorine, chlorine, bromine and iodine

Halogen (fluorine, chlorine, bromine, iodine) minerals

The four elements, fluorine (F), chlorine (Cl), bromine (Br) and iodine (I), constitute a well marked group in the Periodic classification (see Table 1.8), and show a number of similarities and progressive variations in their properties. All unite readily with the metals to form salts such as fluorides, chlorides, etc., many of which are of great importance in economic mineralogy.

The most important occurrence of fluorine is in salts of hydrofluoric acid, HF; the most important of these fluorides being fluorite, CaF_2, and cryolite, Na_3AlF_6, both of which are of considerable industrial importance. Fluorine also enters into the composition of some silicates, such as topaz and amblygonite, and the important mineral apatite is calcium fluoride and phosphate. Fluorine is detected in fluorides by heating with strong sulphuric acid, when hydrofluoric acid, which etches glass, is liberated.

Certain chlorides, salts of hydrochloric acid, HCl, are abundant and important minerals, notable examples described in this book being the following: rock salt, NaCl; sylvine, KCl; carnallite, $KMgCl_3$; cerargyrite, AgCl; and calomel, HgCl. Oxy-chlorides occur as minerals, the most important example being atacamite, $Cu_2(OH)_3Cl$. Chlorophosphates, chloro-arsenates and chlorovanadates are represented by the minerals of the pyromorphite set, e.g. pyromorphite, $(PbCl)Pb_4(PO_4)_3$. Finally, the sodium chloride molecule enters into the composition of certain silicates as, for example, sodalite and marialite–scapolite. Chlorine is detected by the precipitation of silver chloride on the addition of a silver nitrate solution of the mineral in nitric acid, and by the copper oxide–microcosmic salt test.

Fluorides (fluorite) are widely used in the steel, aluminium and chemicals industry. World production in 1985 was 4.73 Mt, the main producers being Mexico (0.73 Mt), Mongolia (0.70 Mt), the USSR (0.56 Mt), South

Africa (0.35 Mt), Spain (0.30 Mt), France (0.26 Mt), the UK (0.15 Mt) and Italy (0.15 Mt).

Rock salt (see also under Sodium, this chapter) is used for the production of chlorine, caustic soda and soda ash, products which are involved in a large number of end uses. The total production in 1985 was 160 Mt, the major producers being the USA (36.0 Mt), the USSR (17.0 Mt), China (14.5 Mt), Canada (10.1 Mt), West Germany (7.5 Mt), Australia (5.44 Mt) and Mexico (5.9 Mt). The UK produces some 1.6 Mt of rock salt per year.

Bromine is not an abundant mineral constituent. It occurs as bromides, salts of hydrobromic acid, HBr, the most important of which are silver bromides, such as bromyrite, AgBr, and embolite, Ag(Cl,Br). When heated with potassium bisulphate and pyrolusite, red vapours of bromine are liberated from bromine compounds. The commercial production of bromine compounds comes from sea water, salt brines, or from some saline residues. Bromine is used in medicine, photography and for petrol additives. In 1985 production was about 160 000 t in the USA, and 125 000 t in Israel.

Iodine is rare in nature. The chief source is the Chile nitrate deposits, where certain calcium iodates occur in small amounts. The possible production exceeds the demand. Iodine is also produced from seaweed and kelp, and from the brines from oil wells. Iodine is used mainly in catalysts, animal feed and pharmaceuticals. Data are usually withheld, but it is believed that the USA requires 3000–4000 t per annum, of which 2300 t are imported. Chile produced 3040 t of iodine in 1985.

Group VIIIa   Iron, cobalt and nickel

Iron minerals

Next to aluminium, iron (Fe) is the most widely distributed and abundant metal, consisting about 4.6% of the Earth's crust. It is found native in meteoritic masses and in eruptive rocks, mostly associated with allied metals such as nickel and cobalt. In addition to the essentially iron minerals, iron enters into the composition of a great number of rock-forming silicates. Iron and steel form the foundation of modern industry, and are used in enormous quantities.

World production of iron ore in 1985 was 896.4 Mt. This represents almost a 50% increase over the 20–year period since the last revision of this book; in 1966, annual production was 600 Mt. The main producers in 1985 were as follows:

USSR	248 Mt	Sweden	20.6 Mt
China	130 Mt	Liberia	16.1 Mt
Brazil	120 Mt	Venezuela	14.8 Mt
Australia	95.3 Mt	France	14.7 Mt
USA	48.8 Mt	Mexico	9.5 Mt
India	42.5 Mt	Mauritania	9.2 Mt
Canada	39.0 Mt	North Korea	8.0 Mt
South Africa	24.4 Mt	Spain	6.7 Mt

Purely as a matter of comparison (and interest), the UK produced 400 000 t of iron ore in 1985.

In the same year, steel production was 716.2 Mt, the main producers being as follows:

USSR	155.2 Mt	Poland	16.1 Mt
Japan	105.2 Mt	UK	15.7 Mt
USA	80.4 Mt	Czechoslovakia	15.2 Mt
China	46.5 Mt	Canada	14.7 Mt
West Germany	40.5 Mt	Rumania	14.4 Mt
Italy	23.7 Mt	Spain	14.2 Mt
Brazil	20.5 Mt	South Korea	13.5 Mt
France	18.8 Mt		

Metallic iron is unaffected by dry air, but oxidizes to 'rust' under the influence of moist air. Cast iron, wrought iron and steel are the main forms in which iron appears in commerce. Their different properties are primarily due to the presence of varying amounts of carbon. Steel is again divided into several classes or grades, each named after its particular properties (mild, hard, etc.), the use to which it is put (tool-steel, etc.), or the metal with which it is alloyed (manganese, chrome, nickel, or tungsten steel). These special steels have, in practice, been found to be especially valuable for different purposes, such as armour plate, guns, high-speed cutting tools, rails, springs, etc. Nickel steel is of particular importance, its tensile strength and elasticity being enormously greater than those of ordinary steel.

Pig-iron, from which all the various grades of iron and steel are obtained, is produced in the blast furnace by the reduction of iron ore by coke. Increasing percentages of the raw ores are now improved by sintering or agglomeration into pellets before dispatch to the smelters.

The main ores of iron are: the oxides magnetite, Fe_3O_4, containing 72.4% Fe; hematite, Fe_2O_3, containing 70% Fe; goethite, $FeO(OH)$, containing 68.5% Fe; limonite; and the carbonate, siderite, chalybite or spathic iron ore, containing 48.3% Fe. Less important ores are the sulphides, pyrite, FeS_2, pyrrhotite, $Fe_{1-n}S$, and the complex oxide, franklinite, $(Fe,Zn,Mn)(Fe,Mn)_2O_4$. Finally, certain important iron ores are composed to some extent of hydrous iron silicates, such as chamosite, thuringite, greenalite and glauconite. Magnetite of exceptional purity occurs in large quantities in Sweden, and is the source of the noted 'Swedish iron'. Undesirable impurities in iron ores are arsenic, sulphur and phosphorus, except in the case of the manufacture of basic or non-Bessemer steels. Iron ores containing 30% Fe and upwards are profitably smelted, but the value of iron ores (in common with most other ores, of course) depends on their situation, and also on the composition of their gangue in addition to their iron content. For example, an iron ore containing 30% Fe, and a gangue of silica, alumina and lime in such proportions as to make it self-fluxing, may be more valuable than a richer ore containing impurities which it would be necessary to remove. Most iron ores in Europe require some mechanical cleaning, and in some cases electromagnetic separation is employed. Roasting or calcining is also frequently resorted to for the purpose of removing water, carbonic acid and sulphur.

Considerable quantities of the natural oxides and silicates are mined and prepared for the market, for use in the manufacture of paints, and as linoleum fillers, etc. For example, ochres are hydrated ferric oxides; sienna and umber are silicates of aluminium containing iron and manganese; red and brown ochres are the natural hydrated or anhydrous oxides, or they may be produced by the calcination of carbonates, whereby a wide range of shades is obtained.

Iron ores occur in a number of types of deposit. Magmatic segregations, often followed by injection into the surrounding rocks, are exemplified by the great magnetite deposits of Kiruna in northern Sweden. Pyrometasomatic deposits are widespread, as in the Urals and the western USA. Hydrothermal disseminations, veins and replacements are especially common. Important iron ores occur as syngenetic sedimentary beds as, for example, in the Jurassic rocks of the English Midlands. Ancient sedimentary deposits are present in enormous bulk in the Precambrian rocks of Canada, the Lake Superior district, Brazil, India, South Africa, West Africa and Australia; very rich iron ores are formed when these chemical precipitates undergo leaching and removal of their primary silica. It has been estimated that the reserves of such enriched ores exceed those of all other types combined. Residual deposits of various kinds, especially the limonitic ores found in the gossans of sulphide deposits, are another important type of occurrence.

Tests. The following tests are useful in the detection of iron in compounds. With borax, in the oxidizing flame, iron compounds give a bead which is yellow when hot, and colourless when cold or, if more material is added to the bead, brownish-red when hot and yellow when cold. In the reducing flame, the usual colour is bottle green of various shades. The microcosmic salt bead is similar to that of borax in the oxidizing flame, but in the reducing flame it is brownish-red hot and passes, on cooling, to yellow–green and finally to colourless.

The important iron minerals are the following. The native element has been recorded in several localities. The chief oxides are magnetite, Fe_3O_4, and hematite, Fe_2O_3; ilmenite, $FeTiO_3$, chromite, $FeCr_2O_4$, and franklinite $(Fe,Zn,Mn)(Fe,Mn)_2O_4$, are described with titanium, chromium and zinc respectively; hercynite, $FeAl_2O_4$, is a spinel. The chief hydrated oxides are goethite, $FeO(OH)$, limonite, $FeO(OH).nH_2O$, and turgite, $Fe_2O_3.nH_2O$. The carbonate is siderite or chalybite, $FeCO_3$. Sulphides are represented by pyrite, FeS_2, and pyrrhotite, $Fe_{1-n}S$, and complex sulphides by chalcopyrite, $CuFeS_2$, and arsenopyrite, $FeAsS$. Copperas is a hydrated sulphate of iron, $FeSO_4.7H_2O$; vivianite is a hydrated phosphate, $Fe_3P_2O_8.8H_2O$. Hydrated silicates of various types are ilvaite, $CaFe^{2+}Fe^{3+}$-$O(Si_2O_7)(OH)$, and chamosite, thuringite, greenalite and glauconite. In addition to its occurrence in such silicates, iron also enters in considerable amounts into many of the rock-forming silicates, such as the pyroxenes, hypersthene and hedenbergite, the amphiboles, anthophyllite, glaucophane and riebeckite, the iron garnets, the biotitic micas, and other silicates such as staurolite and chloritoid. Finally, columbite is an iron tantalate, and wolfram an iron tungstate. The following iron minerals are dealt with in this book:

element	native iron, Fe
oxides	magnetite, Fe_3O_4 hematite, Fe_2O_3
hydrated oxides	goethite, $FeO(OH)$ limonite, $FeO(OH).nH_2O$ turgite, $Fe_2O_3.nH_2O$
carbonate	siderite, chalybite, $FeCO_3$
sulphides	pyrite, FeS_2 marcasite, FeS_2 pyrrhotite, $Fe_{1-n}S$

hydrated sulphate	copperas, $FeSO_4.7H_2O$
hydrated phosphate	vivianite, $Fe_3P_2O_8.8H_2O$
hydrated silicate	ilvaite, $CaFe^{2+}Fe^{3+}O[Si_2O_7](OH)$

Cobalt minerals

Cobalt (Co) is a malleable metal, closely resembling nickel in appearance. It has a high melting point, approaching 1500°C. It can be produced by the reduction of its oxides by carbon. Cobalt and its compounds are used in three main ways: it is employed in the production of a valuable series of alloys used in rustless and high-speed steels and certain non-ferrous alloys; it is used to a certain extent in electroplating, and finally, its compounds are extensively used in the manufacture of pigments, especially blues, employed in the glass, enamel and pottery industries.

In 1985, Western world production was 21 840 t, the main producers being Zaire (10 677 t), Zambia (4290 t), Canada (3707 t), Finland (1425 t) and Japan (1243 t). Other important, although small, producers were France and China. In the Eastern bloc, the USSR produced 2700 t and Cuba produced a small amount. In Zambia, cobalt is a by-product in copper smelting. In Canada, cobalt is associated with arsenic, nickel and silver, and its profitable exploitation often depends on the price of silver. Cobalt deposits of economic value occur as cobalt veins carrying smaltite and cobaltite, as cobaltiferous pyrrhotite, or as asbolite which results from the weathering of cobaltiferous basic and ultrabasic rocks (asbolite being analogous to the garnierite deposits of nickel).

Tests. Cobalt minerals colour both borax and microcosmic salt beads a rich blue. The residue obtained by heating cobalt minerals with sodium carbonate and charcoal is feebly magnetic. Cobalt minerals weather on their exterior to pinkish cobalt 'blooms' – cobalt indicators.

The primary cobalt sulphides, sulpharsenides and arsenides weather in the oxidation zone into oxides and hydrated oxy-salts. The chief cobalt minerals are therefore:

arsenide	smaltite, $(Co,Ni)As_{3-n}$
sulphide	linnaeite, Co_3S_4
sulpharsenide	cobaltite, $CoAsS$
'bloom', hydrated arsenate	erythrite, $Co_3(AsO_4)_2.8H_2O$
'oxide'	asbolite (wad); an oxide of manganese, sometimes containing up to 40% Co

Nickel minerals

Nickel (Ni), which never occurs native, is a white malleable metal, unaffected by moist or dry air, and capable of taking a high polish. It is obtained by the reduction of its oxide, or by the 'Mond' process, which consists of the formation of a volatile nickel carbonyl produced by passing carbon monoxide over heated nickel oxide, and the dissociation of this compound at a higher temperature into nickel and carbon monoxide, which can be used again. The ore is usually first treated for the production of matte and, additionally, the copper–nickel ores of Canada are smelted for the direct production of 'model metal', an alloy of nickel and copper, whose applications are of great industrial importance.

Nickel is used in the coinage of a large number of countries. It is used to some extent in electroplating (nickel plating); it is employed in the construction of certain storage batteries, and several of its salts are used in chemical industry. The main use of nickel, however, is in its applications in the form of alloys with other metals; for example, German silver (an alloy of copper, nickel and zinc in varying proportions), white metal, nickel bronzes, etc. The manufacture of nickel–steel alloys containing from 2.5% to as much as 79% of nickel absorbs the largest proportion of the nickel produced. The properties of the alloys vary remarkably with the amount of nickel but, in general, nickel steel has a greater hardness and tensile strength than carbon steel, and is used for a great number of purposes – for armour plate, aircraft construction, motor cars, etc.

Nickel deposits may be divided into three types, similar to the cobalt deposits; that is, veins, nickeliferous pyrrhotite and nickeliferous serpentines (garnierite). The most important source of nickel is the mineral pentlandite, which is commonly associated with pyrrhotite, as at Sudbury, Ontario, Canada. With the pyrrhotite are associated arsenic, copper and cobalt, and often a considerable amount of silver, and a minute proportion of platinum. Some of these metals now constitute an important part of the metal output of the Sudbury field.

The Western world production of nickel (metal in ore) in 1985 was 510 000 t, the major producers being Canada (176 000 t), New Caledonia (44 760 t), Australia (43 000 t), the Dominican Republic (25 400 t), South Africa (20 000 t), Botswana (18 000 t), Finland (15 800 t), Columbia (12 000 t) and Zimbabwe (10 000 t), with the Philippines and Brazil also producing significant amounts. The USA and Norway produce 54 000 t and 37 500 t respectively, but in the USA only 4000 t of the total is from local ores, and in Norway 11 000 t of the total is from local ores. Japan is also a main producer of nickel (metal), at 23 000 t per annum, all of which is produced from imported materials and not from mining operations there. In the Eastern bloc, the USSR is an important producer, with 170 000 t output in 1983. Poland produces a small amount, and Cuba is an important producer with 35 000 t nickel metal being produced in 1985, from a nickel sulphide deposit which is the second largest in the world.

Tests. Blowpipe reactions for nickel are poor. In the borax bead, nickel compounds give a reddish-brown colour in the oxidizing flame, which changes to an opaque grey in the reducing flame; in the microcosmic salt bead, the colours are reddish-browns. Nickel compounds give a feebly magnetic residue when heated on charcoal with sodium carbonate. Nickel ores oxidize on the surface to a green colour, the nickel blooms, due to the formation of oxy-salts of the metal.

The primary nickel minerals are sulphides and arsenides, and these are oxidized in the upper parts of the deposits into the nickel blooms, of a green colour. The minerals of nickel dealt with are:

arsenides	kupfernickel, niccolite, $NiAs$ chloanthite, $(Ni,Co)As_{3-n}$
antimonide	breithauptite, $NiSb$
sulphides	millerite, NiS pentlandite, $(Fe,Ni)S$
'blooms'	emerald nickel, zaratite, $NiCO_3.2Ni(OH)_2.4H_2O$ nickel vitriol, morenosite, $NiSO_4.7H_2O$ nickel bloom, anabergite, $Ni_3(AsO)_2.8H_2O$
silicates	garnierite and genthite, hydrated nickeliferous magnesium silicates

Group VIIIb Ruthenium, rhodium, palladium, osmium, iridium and platinum

Platinum group minerals

The members of the platinum group of metals, platinum, palladium, osmium, iridium, rhodium and ruthenium, occur together in nature as native metals or alloys. The most abundant of these metals is platinum, the others occurring in small quantities with it. The metals are used in jewellery, in the electrical trades, in electroplating, in chemical industries, in dentistry, for certain photographic purposes and especially as catalysts in the chemical and petroleum industries. An important use is in the electronics industry.

Platinum (Pt) occurs native, and in that form constitutes the most important source of the metal. It is a greyish-white lustrous metal, having a specific gravity of 21.46 and melting at 1760°C. It is malleable and ductile, and may be welded at a bright red heat. Its resistance to acids, and to chemical influence generally, renders it of particular use in the laboratory, and in the electrical and other industries. Platinum is mainly used as a catalyst in automobile exhaust converters and petrol refining. It is used in the electronics industry, and also as a catalytic agent in the manufacture of chemicals by the contact process, and in dentistry and jewellery. Platinum is refined and separated from associated metals by a somewhat complicated series of operations.

In addition to the native metal, an important source of platinum is sperrylite, $PtAs_2$, as this occurs in the important platinum deposits of Sudbury, Ontario. Platinum occurs in a number of forms. It is found disseminated as small original grains in basic and ultrabasic igneous rocks such as olivine–gabbros and peridotites, as in the Urals; it occurs also in similar rocks in the Bushveld norite complex in South Africa, and in chromite-rich layers in the same complex. In the pyrrhotite deposits of Sudbury, Canada, which are possibly magmatic segregations, sperrylite and the platinum group metals afford a large share of the world output. Platinum occurs in quartz veins in the Transvaal, and a small amount is present in many copper deposits. From all these types of occurrence, placer or alluvial deposits are formed and, up to recently, this type of deposit supplied the greater part of the output. The chief producers of platinum group metals in 1985 were the USSR (117 000 kg, of which 8400 kg were sold to the West), South Africa (70 400 kg) and Canada (4700 kg).

World production in 1985 was approximately the sum of these outputs at 200 000 kg or 200 t, but data are difficult to obtain. The US consumption is estimated at 100 t per annum, of which 8 t is produced in the USA. Since world demand can often exceed supply in any one year, platinum can be withdrawn from US stockpiles (85 000 t Pd; 37 000 t Pt; 2700 t Ir) to satisfy any larger demand.

The main platinum minerals considered are:

| element | native platinum, Pt |
| arsenide | sperrylite, $PtAs_2$ |

Rhodium (Rh) is used in the glass industry, and **palladium** (Pd) in the electronics industry.

8

The non-silicate minerals

ABBREVIATIONS

Physical properties

CS	crystal system	L	lustre
F&H	form and habit	TR	transparency
TW	twinning (if present)	T	tenacity
COL	colour	HD	hardness (e.g. 2.5–3.0)
S	streak (if present)	SG	specific gravity (e.g. 3.55–3.72)
CL	cleavage (if present)	DF	distinguishing features (when
F	fracture (if present)		needed)

Other properties are rare and are written out in full (e.g. **FEEL**).

Optical properties

n_α, n_β, n_e or n_o, n_e or n are the mineral RIs
δ birefringence
$2V = X^o - Y^o$ + ve OAP is parallel to, e.g., (010)

COL	colour	B	birefringence
P	pleochroism (if	IF	interference figure
	present)	E	extinction angle (if present)
H	habit	TW	twinning (if present)
CL	cleavage	Z	zoning (if present)
R	relief	DF	distinguishing features (when
A	alteration (if present)		needed)

The order within each section is as follows:

> MINERAL NAME, with alternatives
> C composition of mineral
> Physical properties (see above)
> Optical properties (see above)
> Varieties
> Tests
> Occurrence
> Uses (or Production and uses)

The above is the scheme for common minerals; for rare minerals the various subdivisions are condensed or not included.

8.1 Native elements

8.1.1 Metals

NATIVE GOLD

C Au; silver has been known to amount to over 26% in argentiferous gold or electrum, and Cu, Fe, Pd and Rh may also be present

Physical properties

CS Cubic

F&H Crystals rare as cube {100} or octahedron {111} or rhombic dodecahedron {110}, but usually found as grains or scales in alluvial deposits associated with heavy obdurate minerals such as garnet, zircon, etc. Gold also occurs as rounded masses called nuggets, in alluvial deposits or embedded in quartz veins. Occasionally, the metal occurs in strings, threads, etc., and the following names indicate the varied forms assumed; grain gold, thread gold, wire gold, foil gold, moss gold, tree gold, mustard gold, sponge gold.

COL Shades of yellow; some specimens from Kashmir possess a coppery or bronze yellow colour. When much silver is present the metal may appear almost silver white. In thin transparent leaves gold appears green (as on the visors of space suits).

S Gold and shining

F Hackly

L Metallic

TR Opaque

T Very ductile, malleable and sectile, being easily cut with a knife

HD 2.5–3.0

SG 12.0–20.0 (the variation is due to the metals with which the gold specimen may be alloyed)

Varieties Forty-three per cent of rhodium (Rh) has been reported in a variety called **rhodium gold** from Mexico; **maldonite**, a black Australian variety containing Bi, has a composition Au_2Bi.

Tests The colour, combined with the malleability, weight and sectility of gold, distinguish it from other minerals. Iron pyrite, which has sometimes been mistaken for gold, and is known as fool's gold, cannot be cut with a knife, while chalcopyrite crumbles beneath the blade and gives a greenish-black streak. Heated on charcoal, all gold compounds give a yellow malleable globule of gold.

Occurrence Gold is found in hydrothermal deposits often associated with igneous rocks, and in auriferous quartz veins, where it seems to be present throughout the entire temperature range of mineralization. Other important sources of gold are the deposits of placer or alluvial type. These deposits are derived from the weathering and disintegration of the primary gold-bearing rock. Gold is found in the residual or lateritic deposits formed on the outcrop of weathered gold-bearing rock – **eluvial placers**. Great placer deposits usually occur in the valleys and may be of recent or ancient date, and may be at the surface or at a great depth (deep lead). Concentration of gold may arise by many processes of sedimentation, such as wind action, wave action (beach placers) and especially by river action. Examples of shallow placers are those of the Urals and Siberia, India, China, Ghana, Alaska (Klondyke), British Columbia, western USA (California, Montana), South America (Equador, Chile, Bolivia) and Australia (Ballarat, River Torrens). Alluvial gold is found in small quantities in the deposits of several British rivers, such as in Cornwall, in North Wales and in Scotland, at Leadhills and the River Helmsdale in Sutherland. Various ancient placers of little importance in gold production have been described, such as in the Permo-Carboniferous rocks of New South Wales, and in the basal Cambrian of the Black Hills, South Dakota. Possibly the greatest of all placers, whether ancient or modern, is that of the **Banket** or gold-bearing conglomerates of the Rand, South Africa, which supply about two thirds of the world's gold. These gold-bearing conglomerates are Precambrian in age, are composed of small quartz pebbles in a metamorphosed sandy matrix, and are found at several horizons in the Witwatersrand system. The origin of this gold is probably an ancient deltaic placer deposit in which the detrital gold has undergone solution, redeposition and recrystallization, with the gold originating from even older gold–arsenopyrite veins along the boundary between the Onvervacht and Fig Tree Shale formations in the Barberton district of East Transvaal. Alluvial or placer gold has been derived from primary gold deposits, usually true veins (the reefs of the gold miner), such veins usually consisting of gold, quartz and pyrite. Many of them occur in Tertiary volcanic rocks, chiefly andesites and rhyolites as in certain goldfields of New Zealand, Mexico, Transylvania and the western USA. Deeperseated, gold–quartz veins occur in California and the western USA. Ballarat and Bendigo in Victoria, New South Wales, Queensland, Nova Scotia, Alaska, Brazil, Austria, the Urals, etc. In the great Porcupine goldfield of Ontario, free gold is accompanied by tellurides. Argentite is associated with quartz and gold in the Comstock Lode and at Tonopah (Nevada) and elsewhere. Small gold deposits occur as replacements in limestone where gold-bearing jaspery rocks are found, and another minor type of deposit is of pyrometasomatic origin. Small but exceedingly rich deposits are found associated with alunite at Goldfield, Nevada. In the UK, gold was worked in quartz veins in the Menevian slates of Middle Cambrian age in North Wales.

GOLD AMALGAM

An amalgam of Au, Hg and Ag, the gold averaging about 40%. It is sometimes found crystallized, but usually in small white or yellowish-white grains which crumble easily. It is usually associated with platinum and has been reported from California, Colombia, the Urals and elsewhere.

NATIVE SILVER

C   Ag; small amounts of other metals, such as Cu, Au, Hg, Pt and Bi are associated with silver

Physical properties

CS Cubic	**S** Silver white, and shining
F&H Rarely found as distorted crystals, but mostly found filiform, arborescent or massive	**F** Hackly
	T Sectile, malleable and ductile
	HD 2.5–3.0
COL Silver white, tarnishing readily	**SG** 10.1–11.1

Tests   Silver is soluble in nitric acid; a clean piece of copper immersed in the solution becomes coated with silver and, when a pinch of common salt or a drop of hydrochloric acid is added to the solution, a white precipitate of silver chloride occurs which is soluble in ammonia. Before the blowpipe, silver fuses readily on charcoal to a silver globule which crystallizes on cooling.

Occurrence   Silver is found associated with cobalt–nickel–iron arsenides in basic igneous rocks, as at Annaberg (Saxony) and Cobalt (Ontario). Silver occurs in the oxidized zones of galena-bearing veins. Many veins recorded as silver veins are, in fact, argentiferous galena veins. Such deposits are found at Kongsberg (Norway) and Andreasberg in the Harz, where its formation is connected with zeolitization; that is, the deposition of zeolites by heated waters passing through the rocks. Native silver is also associated with native copper in the Lake Superior region. Native silver occurs in the upper parts of silver sulphide lodes below the chloride capping, and is often concentrated to form rich deposits, as in Mexico, the Comstock Lode (Nevada), Broken Hill (NSW), Peru, and many other silver-mining districts.

NATIVE COPPER

C   Cu, sometimes with some silver and bismuth

Physical properties

CS Cubic	**COL** Copper red
F&H Cubic crystals frequently twinned on {111}, often massive. It sometimes occurs in thin sheets or plates, infilling narrow fissures. Arborescent habits are also common, and occasionally confused threads of copper are found.	**S** Metallic and shining
	CL None
	F Hackly
	T Ductile and malleable
	HD 2.5–3.0
	SG 8.9

Tests Before the blowpipe, copper fuses easily and becomes coated with black oxide of copper. It dissolves in nitric acid, and gives a blue solution with addition of ammonia.

Occurrence Copper is often found with cuprite and copper–iron sulphide deposits associated with basic extrusives. It occurs as a hydrothermal and metasomatic deposit filling cracks or amygdales, and forming partial replacements in basic lava flows. The most important deposits of native copper are those of the Lake Superior region, where the copper occurs in a series of ancient lava flows and conglomerates. The native copper is found in the amygdaloidal volcanic rock clasts in the Calumet conglomerate, and in metasomatic veins. Native copper occurs associated with a basic intrusive rock at Monte Catini, Italy. Native copper also occurs in the 'oxidation zone' of copper lodes, the zone of weathering, and rich but mostly small deposits of this type have been worked.

NATIVE MERCURY, Quicksilver

C Hg; some silver may be present

Physical properties

CS Trigonal, if frozen at −39°C	**COL** Tin white
F&H Occurs as small fluid globules disseminated through the matrix in which it occurs	**L** Metallic
	TR Opaque
	SG 13.59

Tests Mercury dissolves readily in nitric acid. Heated before the blowpipe, mercury volatilizes with little or no residue, but should any residue be left, the presence of silver may be shown by fusion with sodium carbonate on charcoal and the production of a silver bead.

Occurrence Native mercury occurs as fluid globules scattered through cinnabar, HgS, as at Almaden (Spain), Idria (Italy), etc., and is sometimes found in some quantity, filling cavities. It is a rare mineral, and is of secondary origin. It is deposited with cinnabar from the waters of certain hot springs.

NATIVE AMALGAM, silver amalgam
C   Hg and Ag, in varying proportions

Physical properties

CS Cubic	**L** Metallic
F&H Rhombic dodecahedron {110} common; also massive	**TR** Opaque
COL Silver white	**T** Brittle and grates under the knife when cut
S Silver white	**HD** 3.0–3.5
F Conchoidal or uneven	**SG** 10.5–14.0

Variety **Arquerite** is a soft, ductile silver-rich variety, found in the mines of Arqueros, Coquimbo, Chile.

Tests Heated in the closed tube, mercury sublimes and condenses on the cold portion of the tube, leaving a residue of silver. Heated before the blowpipe, mercury volatilizes, and a globule of silver is left. When rubbed on copper, amalgam imparts a silvery lustre. Amalgam is soluble in nitric acid.

Occurrence Native amalgam occurs as scattered grains along with cinnabar, as at Almaden, Spain, or in the oxidation zone of silver deposits where it is associated with cerargyrite.

NATIVE PLATINUM
C   Pt, alloyed with Fe, Ir, Os, Au, Rh, Pd and Cu; in 21 analyses cited by Dana, the Pt ranges from 45% to 86%

Physical properties

CS Cubic	**F** Hackly
F&H Crystals rare; usually found in grains and irregularly shaped lumps – a nugget found in the Urals weighed about 10 kg	**L** Metallic
	TR Opaque
	T Ductile
COL White or steel grey	**HD** 4.0–4.5
S White or steel grey	**SG** 21.46 (for pure Pt)

Tests Native platinum sometimes exhibits magnetic polarity, some Russian specimens being said to attract iron filings more powerfully than an ordinary magnet. The high specific gravity, infusibility and insolubility of platinum serve to distinguish it from other minerals.

Occurrence The occurrence of platinum has been summarized in the introduction to the platinum minerals in Chapter 7.

Other platinum group elements

PALLADIUM

Characters and occurrence
HD 4.5–5.0   **SG** 11.3–12.0   **MELTING POINT** 1546°C
Palladium (Pd) occurs native in crude platinum, and in small quantities in cupriferous pyrites, especially those containing nickel and pyrrhotite. It is a silver white metal, as hard, but not so ductile, as platinum. It oxidizes more readily than Pt.

Palladium is much used in dental alloys, as a catalyser, for coating the surfaces of silver reflectors used in searchlights, etc., in the construction of delicate graduated scales, and in the manufacture of Pd–Au (Palau) crucibles, used in analyses because of the lack of reaction.

The major source of supply to the Western world is the copper–nickel ore of Sudbury, from whose matte it is recovered. The metal is also found with platinum in Brazil, the Urals and elsewhere, and usually contains iridium as well as platinum.

OSMIUM and IRIDIUM

Osmium (Os) occurs native in crude platinum and in **osmiridium**, an alloy with iridium described below. Osmium is a bluish-grey metal with an SG of 22.48. It is the heaviest of metals, and fuses at 2200°C. No reliable statistics are available as to production. It is of little commercial importance, and the supply is in excess of the demand.

Iridium (Ir) occurs native in crude platinum, and is alloyed with osmium as osmiridium (iridosmine). It is a steel white metal with an SG of 22.4, and a melting point of 2290°C. Its chief source is from crude platinum, but about 150 kg of osmiridium are produced annually from gold ores in South Africa. The consumption of osmium is increasing, its main application being the dental, electrical and jewellery trades.

OSMIRIDIUM, iridosmine
C   An alloy of Ir and Os in variable proportions

Physical properites

CS	Trigonal	**L**	Metallic
F&H	Chiefly in small, flattened grains	**HD**	6.0–7.0
COL	Tin white to steel grey	**SG**	19.3–21.12

Occurrence Both osmiridium and iridium are found in the gold washings of the Urals, Bingera (NSW), Brazil, and in Canada. A considerable but varying production comes from placer deposits in Tasmania, Sudbury (Ontario), recoveries from platinum refineries, and the South African gold ores.

RHODIUM and RUTHENIUM

Rhodium (Rh) occurs in native platinum and in the pyrrhotitic ores of Sudbury, Ontario. It is a white metal, ductile and malleable at a red heat. It has a specific gravity of 12.1 and melts at about 2000°C. It can occur up to 2% in crude platinum ore. It has few applications, but is used in the manufacture of thermal couples and for crucibles.

    Ruthenium (Ru) occurs native in crude platinum, and is recovered in the refining. It is a white, hard and brittle metal with an SG of 12.2. It has very few industrial applications.

NATIVE IRON

C   Fe, usually alloyed with nickel, cobalt or some other metal, or mixed with other iron compounds

Physical properties

CS	Cubic	**CL**	{100} perfect
F&H	Crystals occur as octahedra, but usually found massive, and with a somewhat granular structure	**F**	Hackly
		T	Malleable and ductile
		HD	4.0–5.0
COL	Iron grey	**SG**	7.3–7.8
S	Iron grey		

Test Strongly magnetic.

Occurrence Native iron of terrestrial origin has been reported as occurring in Brazil, Auvergne, Bohemia, and in grains disseminated in basalt from Giant's Causeway (Ireland) and elsewhere. The largest masses known to be of terrestrial origin occur in Disko Island and elsewhere on the coast of West Greenland, where masses ranging of up to 25 t in weight weather out from a basalt. Native iron also occurs as grains in some placer deposits, as at Gorge River, New Zealand, and in Piedmont. Native iron may be formed by the alteration of iron minerals, as in eastern Canada.

Meteoritic iron is found in meteorites, masses which have fallen from the outer atmosphere on to the surface of the Earth. Meteorites may consist either entirely of iron, or partly of olivine and other silicates. Meteoritic iron, which is known as **kamacite**, or α-iron, in the low-temperature form and **taenite**, or γ-iron, in the high-temperature form, is usually alloyed with nickel and small quantities of cobalt, manganese, tin, chromium, sulphur, carbon, chlorine, copper and phosphorous. It rusts much less readily than terrestrial iron, on account of the nickel which it contains. The minerals olivine, enstatite, augite and anorthite have been found in meteorites. In them, too, has been detected a phosphide of nickel and iron, called **schriebersite**, and also a sulphide of iron, known as **troilite**, having the formula FeS.

NATIVE ZINC

Native zinc is said to have been found in basalt near Melbourne, Australia, and has also been reported as occurring in the auriferous sands of the Nittamitta River in the same district, associated with topaz and corundum. However, the existence of native zinc is still doubted. It has, however, been artificially crystallized in hexagonal prisms, with low pyramidal terminations.

8.1.2 Semi-metals and non-metals

NATIVE ARSENIC

C   As, often with Sb and traces of other metals

Physical properties

CS Trigonal	**S** Tin white
F&H Often occurs granular, massive or reniform, but sometimes columnar or stalactitic	**F** Uneven and granular
	F Nearly metallic
	T Brittle
COL Tin white on recent fracture surfaces, but quickly tarnishing to a dark grey	**HD** 3.5
	SG 5.7

Tests Arsenic compounds, when heated on charcoal, give a white encrustation from the assay, with the fumes having a light odour of garlic. Heated in an open tube, arsenic compounds give a white sublimate which is volatile on heating. Heated in a closed tube, some arsenic compounds give a shining black sublimate, the arsenic mirror, and most arsenates give a similar mirror when heated in a closed tube with charcoal or sodium carbonate.

Occurrence Native arsenic occurs principally as a minor constituent of certain lead, silver, nickel or cobalt ores, as at Freiberg in Saxony, Joachimsthal in Bohemia, etc.

NATIVE ANTIMONY

C Sb, sometimes with traces of Ag, Fe and As

Physical properties

CS Trigonal	F Uneven
F&H Usually occurs massive, granular or lamellar	L Metallic
	TR Opaque
COL Tin white	T Very brittle
S Tin white	HD 3.0–3.5
CL {0001} perfect	SG 6.6–6.7

Tests Heated before the blowpipe, antimony fuses easily, giving off white fumes of antimonious oxide which condense and form a white encrustation on the charcoal near the assay.

Occurrence In veins, associated with stibnite or ores of silver or arsenic.

NATIVE BISMUTH

C Bi, sometimes with traces of S, As and Te

Physical properties

CS Trigonal	S Silver white
F&H Crystals are rhombohedra resembling cubes in form. Usually found massive, foliaceous, or granular, and also in reticulated or plumose habits.	CL {0001} perfect, {10$\bar{1}$1} less good
	L Metallic; tarnishes easily
	TR Opaque
	T Brittle when cold; somewhat malleable when heated; also sectile
COL Silver white, with a faint tinge of red	HD 2.0–2.5
	SG 9.7–9.8

Tests Heated on charcoal, native bismuth fuses and volatilizes, forming an orange–red encrustation. After fusion in a ladle, bismuth crystallizes readily. It dissolves in nitric acid, the solution becoming milky when water is added. When heated on charcoal with potassium iodide and sulphur, bismuth gives a brilliantly coloured encrustation, yellow near the assay, scarlet further away.

Occurrence Occurs in veins associated with ores of tin, silver, cobalt and nickel, and also in association with minerals such as pyrites, chalcopyrite, quartz, etc., as in Bolivia, Schneeberg in Saxony, Australia, etc.

NATIVE TELLURIUM
C   Te, sometimes with a little gold and iron

Physical properties

CS	Trigonal	**S**	Tin white
F&H	Crystallizes in hexagonal prisms with the basal edges modified; usually found massive and granular	**L**	Metallic
		T	Brittle
		HD	2.0–2.5
COL	Tin white	**SG**	6.1–6.3

Tests   Heated in the open tube, it forms a white sublimate of tellurous acid, which fuses to small transparent colourless drops before the blow-pipe. When heated with strong sulphuric acid it gives a reddish-violet solution.

Occurrence   In the Maria Loretto Mine, near Zalanthna, Transylvania, where it was formerly worked for the small quantity of gold (less than 3%) that it contained. It has also been reported from Western Australia and Colorado.

NATIVE SULPHUR
C   S, but often contaminated with other substances such as clay, bitumen, etc.

Physical properties

CS	Orthorhombic	**CL**	Imperfect {110} and {111} cleavages present
F&H	Crystals bounded by acute pyramids; also found massive and in encrustations	**L**	Resinous
		TR	Transparent to substranslucent
COL	Sulphur yellow, often with a reddish or greenish tinge	**HD**	1.5–2.5
		SG	2.07
S	Sulphur yellow		

Varieties   Crude and impure forms of sulphur are purified by heating, the pure material being collected on cool surfaces. This fine sulphur is usually melted and cast into sticks, and is then known as **brimstone** or roll sulphur.

Tests   Sulphur may be recognized by its burning with a blue flame, during which suffocating fumes of sulphur dioxide are formed. Sulphur is insoluble in water, not acted upon by acids but is dissolved by carbon disulphide. Its colour and low specific gravity are distinctive.

Occurrence   Sulphur is found in the craters and crevices of extinct volcanoes, and has been deposited by gases of volcanic origin, as in Japan, where it is mined from an old crater lake. Another mode of origin is by the action of hot springs, by which the sulphur is deposited with tufa, etc., as in Wyoming, California and Utah. The most important occurrences of sulphur, from the commercial standpoint, are those in which the element is bedded or layered with gypsum, and in this case a common association of minerals is sulphur, gypsum, aragonite, celestite, and often crude oil. Such occurrences are located on salt domes in the Gulf states of the USA, and examples of these are provided by the famous Sicily, Texas and Louisiana sulphur deposits. Opinion is divided as to whether these bedded deposits have been formed by true sedimentation, or by the alteration by various processes of gypsum.

Uses Sulphur is used for the manufacture of sulphuric acid, in making matches, gunpowder, fireworks and insecticides, for vulcanizing India rubber, and in bleaching processes involving the use of sulphur dioxide.

DIAMOND
C C

Physical properties

C Cubic

F&H Octahedral crystals common, often with curved faces

TW Commonly twinned on {111}

COL White or colourless, sometimes yellow, red, red, green or, very rarely, blue or black. Those most free from colour are termed diamonds of the first water, and are the most valuable.

CL {111} perfect

F Conchoidal

L Brilliantly adamantine

TR Transparent to translucent when dark coloured

HD 10 (highest on Moh's scale)

SG 3.52

Optical properties

$n = 2.417$

Isotropic

COL Colourless

F&H Usually occurs as small octahedral crystals

CL Perfect octahedral {111}

R Extremely high, and this coupled with high dispersion gives the diamond a sparkle or 'fire'

Varieties **Bort**, or bortz, and carbonado (a black diamond) are compact varieties of diamond occurring in granular or rounded aggregates, and although of no value as gems, are used extensively for abrasive and cutting purposes in gem-cutting, for the cutting edges of diamond drills, and for dressing emery wheels. Much diamond not of the first water is also employed for these purposes, shaped diamonds being used as turning tools, etc.

Occurrence Diamond occurs in two types of deposits; either in igneous rocks of ultrabasic composition, especially kimberlites, or in alluvial deposits derived from these primary sources. In the Kimberley diamond fields, diamond occurs in an ultrabasic igneous breccia called **kimberlite** (blue ground) which forms pipes in black shale, and which is derived from an upper mantle source from depths of at least 100–150 km. Recent work has shown that diamonds are, in general, very old indeed (*c.* 3500 Ma) and have not crystallized from the kimberlite magma in which they occur but have been picked up from very deep Archaean crustal layers within which the diamonds had originally formed. Diamonds have been found in similar igneous rocks elsewhere in South Africa, Sierra Leone, Tanzania, the USSR, Brazil, and Arkansas in the USA, and have been reported in dolerite in New South Wales. Diamond occurs in alluvial deposits associated with other minerals of high specific gravity and extreme obduracy, as in South, West and South-West Africa, and in Brazil, India, etc.

Uses The uses of diamond and its varities for gems and abrasives are given in Chapter 7 (see under Carbon).

GRAPHITE, plumbago, black lead
C   C, sometimes contaminated with small amounts of silica, iron oxides, clay, etc.

Physical properties

CS Hexagonal

F&H Crystals uncommon, usually occurring in scales, laminae, or columnar masses; sometimes granular but rarely earthy

COL Iron grey to dark steel grey

S Black and shining

CL Perfect basal {0001}, parallel to the large surface of the scales

L Metallic

T Thin laminae are flexible, and the mineral is sectile

FEEL Feels cold like metal when handled, because it is a good conductor of heat

DF Graphite resembles molybdenite in most of its physical properties, but is distinguished by its jet black streak, whereas the streak of molybdenite is greenish black

HD 1.0–2.0

SG 2.0–2.3

Occurrence Graphite has been considered to occur as a primary constituent of igneous rocks, but it is probable that, in these cases, the mineral has been derived from the adjacent country rock. The main occurrences are of three types: (1) veins of a true fissure character as in Ceylon, Irkutsk, and at Borrowdale in the Lake District; (2) bedded masses of a lenticular and patchy nature in gneiss, crystaline limestone, etc., as in eastern Canada; and (3) disseminations through the country rock, often near veins or associated with the contact of an igneous rock, as in eastern USA, , etc. Graphite is also common in metasediments where it forms organic material and, if abundant, a graphite schist results. This graphite is indicative of reducing conditions, and pyrite is also present.

Uses The main uses of graphite are for facings in foundry moulds, for paint, and for crucibles. Additional important uses are as a lubricant, in commutators, as a stove polish, for lead pencils, in electroplating (where it provides a conducting surface on non-metallic substances which are to be plated), and as electrodes for the electric furnace. Different grades of the material are suitable for different purposes; low-grade graphite, with perhaps no more than 40% of the element, is useful for paint manufacture; the flaked graphite is almost essential for crucibles and the Ceylon graphite of this quality is eminently suitable; and very pure artificial graphite prepared in the electric furnace is preferred for electrodes and many other purposes. Often, the physical properties and the freedom from grit are more important than the amount of carbon actually present.

8.2 Halides

8.2.1 Simple halides

Formula *AX*

HALITE, ROCK SALT, common salt

C $NaCl$; $CaSO_4$, $CaCl_2$ and $MgCl_2$ are usually present as impurities, sometimes with $MgSO_4$. The presence of magnesium compounds causes the mineral to become wet and lumpy.

Physical properties

CS Cubic

F&H Crystals form cubes, rarely octahedra. The cube faces are often hollow, giving hopper crystals. It also occurs massive and granular, rarely fibrous

COL Colourless or white if pure; often yellowish, blue, purple tints, red

CL Cubic {100} perfect

F Conchoidal

L Vitreous or greasy

TR Transparent to translucent

T Brittle

TASTE Saline

SOLUBILITY Soluble in water

HD 2.0–2.5

SG 2.2

Optical properties
$n = 1.544$
Isotropic
Halite is colourless, with a perfect cubic leavage, and an RI almost exactly equal to 1.54.

Tests Taste quite distinctive. Halite colours the fla
sodium. It crackles and decrepitates when heated,
chlorine flame with copper oxide in the microcosm
halite gives a white precipitate of silver chloride on a
solution.

Occurrence Deposits of rock salt occur as extensive geo
are the result of the evaporation of enclosed or partly enclo
sea water. During this concentration and evaporation, the salts
out in a definite order. In the great German Permian deposits at Stassfurt
the order is:

Top

6	carnallite	potassium magnesium chloride
5	kieserite	magnesium sulphate
4	polyhalite	calcium magnesium potassium sulphate
3	rock salt	sodium chloride
2	gypsum and anhydrite	calcium sulphates
1	dolomite and calcite	calcium and magnesium carbonates

Bottom

Hence there is a fairly regular sequence in the deposits from bottom to top in the order named, the least soluble minerals, the calcium and magnesium carbonates, being at the bottom, and the most soluble minerals, polyhalite, kieserite and carnallite (the **bitterns**) being at the top. In other similar occurrences the process has not been carried so far, and the bitterns are usually absent from the succession. Salt beds occur at various geological horizons: for example, in the Silurian and Carboniferous of Michigan, New York State and Ontario; in the Permian of Stassfurt, Germany; in the Trias of Cheshire (England), Lorraine (France), Wurtemberg (Germany) and Salzburg (Austria); and in the Tertiary of Wieliczka, Poland. The salt is extracted by ordinary mining by shafts and galleries, or by pumping the brine from the salt bed to the surface, and there recovering the salt by evaporation. Salt is present in the waters of the ocean, and vast inland saltwater lakes exist, such as the Dead Sea, the Great Salt Lake of Utah etc. Since sea water contains only about 3.5% of total dissolved material in it, the formation of natural beds of salt thousands of feet in thickness requires explanation. It is suggested that thick salt beds (**sabkha**-type evaporite successions) could be laid down by the continual replenishment of a lagoon (such as the Gulf of Karabugas in the Caspian) with supplies of salt water, and the consequent enrichment in salt of the waters of the lagoon by continued evaporation. Beds of salt differ from other rocks in their reaction to pressure; salt flows while other rocks fracture or fold when subjected to crustal movements. Salt glaciers have been described from Iran, and 'intrusive' plugs of salt (the salt domes) are of great economic importance in the gulf state of the USA, since they have provided the proper conditions for the accumulation of vast reservoirs of crude oil.

Uses Rock salt is used for culinary and preserving purposes, and in a great number of chemical manufacturing processes, such as the manufacture of sodium carbonate for glass-making, soap-making, etc.

SYLVITE, sylvine
C   KCl

Physical properties

CS Cubic	**TR** Transparent to translucent
F&H Cube modified by octahedron; it also occurs massive, and granular	**TASTE** Saline, more bitter than that of halite
COL Colourless or white	**HD** 2.0
CL Cubic {100} perfect	**SG** 2.0
L vitreous	

Optical properties
$n = 1.490$ Isotropic
It is colourless, with low (negative) relief (<1.54).

Tests Sylvite is soluble in water. It shows the lilac flame of potassium, and gives a blue flame in the copper oxide–microcosmic salt bead test.

Occurrence Occurs as a saline residue in the Stassfurt and other salt deposits, associated with rock salt and carnallite. It also occurs around the fumaroles of Vesuvius.

SAL AMMONIAC
C   NH₄Cl

Physical properties

CS Cubic

F&H Rarely occurs as octahedra; usually efflorescent and encrusting

COL White if pure, but usually yellowish or grey

L Vitreous on fresh surfaces; externally dull

TR Translucent to opaque

TASTE Pungent, cool and saline

HD 1.0–2.0

SG 1.5–1.9

Optical properties In thin section, sal ammoniac is colourless with moderate relief ($n = 1.639$) and is isotropic.

Tests When ground in a mortar with soda-lime, gives an ammoniacal odour.

Occurrence Occurs as a white efflorescence in volcanic districts. All sal ammoniac of commerce is an artificial product.

Uses Sal ammoniac is used in medicine, dyeing, soldering and various metallurgical processes, as a chemical reagent, and also in electric batteries.

CERARGYRITE, kerargyrite, horn silver

C AgCl, with 75.3% silver

Physical properties

CS Cubic

F&H Rarely occurs as cubes, but usually massive and wax-like; frequently in encrustations

COL Shades of grey, greenish or bluish; colourless when pure

S' Shining

F Subconchoidal

L Resinous to adamantine

TR Transparent to subtransparent or opaque

T Sectile; cuts like wax

HD 2.5

SG 5.55

Optical properties In thin section, cerargyrite is colourless with high relief ($n = 2.071$), and is isotropic.

Tests Cerargyrite is soluble in ammonia, but not in nitric acid. It fuses in the candle flame, and heated on charcoal yields a globule of metallic silver. When placed in the microcosmic salt bead, to which copper oxide has been added, it gives an intense azure-blue colour to the oxidizing flame, indicating chlorine. A plate of iron rubbed with the mineral becomes silvered.

Occurrence   Occurs in the **gossan** or upper parts of silver veins, associated with other silver halides such as bromyrite, $AgBr$, iodyrite, AgI, and embolite, $Ag (Cl, Br)$, and is formed by the action of descending waters containing chlorides and other halides on the oxidized primary ores. Cerargyrite often forms extremely rich but small silver deposits. Some localities are Freiberg (Saxony), Andreasberg (Harz), Broken Hill (NSW), Atacama (Chile), Comstock Lode, and Tonopah (Nevada).

CALOMEL, horn quicksilver
C Hg_2Cl_2

Characters and occurrence
HD 1.0–2.0   **SG** 6.48
Forms small crystals, whitish, greyish or brownish in colour, and is found associated with cinnabar at Idria, Italy, and Almaden, Spain, and elsewhere.

Formula AX_2

FLUORITE, FLUORSPAR, Blue John, Derbyshire spar
C CaF_2

Physical properties

CS Cubic	**S** White
F&H Cubes very common, rarely octahedra or tetrahexahedra {210}; fluorite also occurs in compact or granular habits	**CL** Octahedral {111} perfect **F** Conchoidal to uneven **L** Vitreous **TR** Transparent to translucent
TW Interpenetrant twinned cubes common	**T** Brittle **HD** 4.0 (a mineral on Mohs' scale)
COL Variable, including colourless, white, green, purple, yellow or blue	**SG** 3.0–3.25

Optical properties
$n = 1.433$–1.435 Isotropic
Colourless, with excellent octahedral cleavage, and moderate (negative) relief (RI much less than 1.54).

Variety   **Blue John** is a purple or blue variety from Derbyshire used for vases, etc.

Tests   Fluorite gives the reddish flame of calcium. When it is heated with sulphuric acid, greasy bubbles of hydrofluoric acid gas are formed, which cause a white film of silica to be deposited on a drop of water held on a glass rod at the mouth of the tube. Heated with potassium bisulphate in a closed tube, it gives hydrofluoric acid; this attacks the glass to form silicon fluoride, which is decomposed in the presence of water to give a white ring of silica on the tube.

Occurrence Fluorite is a late-stage crystallizing mineral in acid igneous rocks. It crystallizes at low temperatures, *c* 510°C in pegmatites and alkaline igneous rocks. In late-stage pneumatolytic deposits, fluorite occurs with cassiterite, topaz, apatite and lepidolite, whereas in hydrothermal veins and replacement deposits fluorite is associated with sulphides (galena, blende, etc.), calcite, quartz and barite. Fluorite forms the cementing material in some sandstones, as in the Elgin Trias of Scotland.

Uses The finest grade of fluorite is used in enamelling iron for baths, etc., in the manufacture of opaque and opalescent glasses, and for the production of hydrofluoric acid. The inferior grades are used as a flux in steel-making and for foundry work. Transparent fluorite is being used in increasing quantities for the construction of camera lenses.

ATACAMITE, remolinite
C $Cu_2(OH)_3Cl$, with 59.4% copper

Physical properties

CS Orthorhombic

F&H Rarely prismatic; usually massive and lamellar

COL Bright deep green to blackish green

S Apple green

CL {010} perfect; {101} distinct

L Adamantine to vitreous

T Translucent or subtranslucent

HD 3.0–3.5

SG 3.78–3.87

Optical properties In thin section, atacamite is pleochroic in greens with a high relief ($n_{max} = 1.880$, $n_{min} = 1.861$); $2V \approx 75°$ −ve

Tests Heated in the closed tube, atacamite gives off water and forms a grey sublimate. It is easily soluble in acids. On charcoal, it fuses in time to metallic copper, colouring the flame azure blue (chloride), and forming a brownish and greyish-white deposit on the charcoal, which volatilizes in the reducing flame, again giving an azure-blue coloration.

Occurrence Atacamite occurs in the zone of weathering of copper lodes, especially when this weathering has been effected under desert conditions.

It is found at Botallack Mine (St Just, Cornwall), Los Remolinos and the Atacama Desert (South America), and at Linares (Spain) and Burra Mine (South Australia).

8.2.2 Multiple halides

Formula $A_m B_n X_p$ with $(m + n) : p$ greater than $1 : 2$

CARNALLITE
C $KMgCl_3.6H_2O$; KCl and NaCl may be present as impurities

Physical properties

CS Orthorhombic	**L** Greasy or vitreous
F&H Crystals rare, usually occurring massive and granular	**TR** Transparent to translucent
	TASTE Bitter
COL White, but pink or reddish if iron oxide present	**SOLUBILITY** Soluble in water
	HD 2.5
F Conchoidal	**SG** 1.60

Optical properties
$n_{min} = 1.465$ $n_{max} = 1.497$ $\delta = 0.030$
Colourless in thin section, with moderate (negative) relief; $2V \approx 70°$ +ve.

Tests Carnallite gives off water on heating in the closed tube. It shows the lilac flame of potassium; and a blue flame of chloride in the copper oxide–microcosmic salt bead test. If the residue from roasting on charcoal is heated with cobalt nitrate, a pink mass due to magnesium is formed.

Occurrence Carnallite occurs as a saline residue at Stassfurt, and represents the final stage in the drying-up of the salt lake.

Uses In the natural state, carnallite is used as a fertilizer. It is also important as a source of potassium salts.

CRYOLITE
C Na_3AlF_6

Physical properties

CS Monoclinic	**F** Uneven
F&H Crystals rare, usually occurring massive, with a lamellar structure	**L** Vitreous
	TR Subtransparent to substranslucent
COL Colourless, snow white, reddish, brownish, sometimes black	**T** Brittle
	HD 2.5
CL Good {001} and {110} partings; less good on {101}	**SG** 2.97

Optical properties
$n_\alpha = n_\beta = n_\gamma = 1.338$
Cryolite is colourless in thin section, monoclinic (pseudo-cubic) with a high negative relief (RIs < 1.54); $2V \approx 43°$ +ve.

Tests Heated alone before the blowpipe, cryolite fuses easily, colouring the flame intense yellow from sodium. When the residue, from heating on charcoal, is moistened with cobalt nitrate and strongly reheated, a blue mass from aluminium results. If cryolite is heated with sulphuric acid, greasy bubbles of hydrofluoric acid are evolved, indicating a fluoride.

Occurrence Cryolite occurs in a pegmatite vein in granite at Ivigtut in West Greenland, associated with galena, sphalerite, siderite, fluorite, etc. This deposit was worked, and the cryolite proven, to a depth of 40 m.

Uses Cryolite is employed in the manufacture of aluminium, for making sodium and aluminium salts, and in the manufacture of a white porcellanous glass. In industry, natural cryolite has largely been replaced by synthetic cryolite.

8.3 Sulphides

Sulphides, selenides, tellurides, arsenides, antimonides, oxysulphides, and multiple sulphides, or sulfosalts, are included in this division. The arrangement is according to the decreasing $A : X$ ratio in sulphides and the $(A + B) : X$ ratio in the multiple sulphides, where A represents the metallic elements, B the semi-metals (such as As in multiple sulphides), and X the non-metals (especially S).

8.3.1 Simple sulphides

Formula $A_m X_p$, with $m : p$ greater than 3 : 1

TETRADYMITE, telluric bismuth
C $Bi_2Te_2S_3$; sulphur and selenium may sometimes be present

Physical properties

CS Trigonal
F&H Crystals tabular; also massive, granular and foliaceous
COL Pale steel grey
S Grey, marks paper

CL Basal {0001} perfect
L Metallic or splendent
T Laminae are flexible
HD 1.5–2.0
SG 7.2–7.6

Tests Heated in the open tube, tetradymite gives a white sublimate of tellurous acid. Heated on charcoal, it gives off white fumes, and forms a coating of tellurous acid and orange bismuth oxide, eventually fusing and volatilizing completely.

Occurrence Associated with gold tellurides (see p. 251).

NAGYAGITE, black tellurium
C $Pb_5Au(Te,Sb)_4S_{5-8}$, with 6–12% gold

Physical properties

CS Orthorhombic
F&H Usually massive or foliaceous
COL Dark, lead grey
S Dark, lead grey
CL Perfect {010} prismatic

L Metallic
TR Opaque
T Sectile; thin laminae are flexible
HD 1.0–1.5
SG 6.8–7.2

Tests Heated in open tube, nagyagite gives a sublimate of lead antimonate and lead tellurate, and with antimony and tellurium oxides in the higher parts of the tube. The antimony oxide volati''zes when reheated, and the tellurium oxide, at a high temperature, fuses to colourless transparent drops.

Occurrence See under Petzite (p. 234).

Formula A_2X

ARGENTITE, silver glance
C Ag_2S, with 87.1% silver

Physical properties

CS Cubic
F&H Forms {100} and {111} occur in distorted crystals; it also occurs reticulated, arborescent and, usually massive
COL Black lead grey
S Black lead grey, shining

F Small, subconchoidal or uneven
L Metallic
TR Opaque
T Sectile
HD 2.0–2.5
SG 7.19–7.36

Tests Heated in open tube, argentite gives off sulphurous fumes. Heated on charcoal in the oxidizing flame, it fuses with intumescence, gives off sulphurous fumes, and yields metallic silver. It is soluble in dilute nitric acid.

Occurrence Argentite is the most common primary ore of silver. It occurs in small quantities in the sedimentary Kupferschiefer of Mansfeld, Germany. Its main occurrence is in various types of veins such as argentite veins, in which it is accompanied by stephanite and polybasite, and in propylitized volcanic rocks which are important for silver production in Mexico. Argentite, in association with galena and blende, occurs in veins in the San Juan mining district of Colorado. Gold-bearing argentite–quartz veins are common as in Tonapah (Nevada) and the famous Comstock Lode of Nevada. Replacement veins carrying argentite are typified by that of Portland Canal, British Columbia. Argentite accompanies native silver in the cobalt–nickel veins of Cobalt (Ontario) and Annaberg (Saxony), and in the silver stringers of Kongsberg (Norway).

HESSITE
C Ag_2Te

Physical properties

CS	Cubic	**L**	Metallic
F&H	Massive	**HD**	2.5
COL	Lead grey	**SG**	8.4

Varieties Hessite often contains some gold, and so grades into petzite.

Tests Powdered hessite, heated with strong sulphuric acid, gives a reddish-violet solution, indicating tellurium. Heated on charcoal, it gives a silver bead.

Occurrence Hessite occurs with other tellurides in the Kalgoorlie goldfield, Western Australia, in the gold veins of the Porcupine mining area of Ontario, and in various veins in Chile, Mexico and California.

PETZITE
C $(Ag,Au)_2Te$

Physical properties

F&H	Massive or granular	**T**	Brittle, slightly sectile
COL	Steel grey to iron black	**HD**	2.5–3.0
F	Uneven	**SG**	8.7–9.0

Occurrence   Petzite (and nagyagite) are found in veins and replacement deposits in association with free gold, pyrite and other sulphides. Localities include Kalgoorlie (Western Australia), Cripple Creek and Boulder (Colorado).

CHALCOCITE, copper glance, redruthite, vitreous copper ore
C Cu_2S, with 79.8% copper; traces in iron may be present

Physical properties

CS Orthorhombic	**S** Blackish lead grey, sometimes
F&H Crystals prismatic with basal	shining
faces; twinning frequently giving a	**CL** Prismatic poor
stellate grouping of three individuals;	**F** Conchoidal
usually massive, with a granular or	**L** Metallic
compact structure	**HD** 2.5–3.0
COL Blackish lead grey, often with a	**SG** 5.5–5.8
bluish or greenish tarnish	

Tests   Heated on charcoal, chalcocite boils and ultimately fuses to a globule of copper. It is soluble in hot nitric acid, leading a precipitate of sulphur. The typical reactions for copper are seen.

Occurrence   Chalcocite is a very valuable copper ore. It is formed by the alteration of primary copper sulphides in the zone of secondary enrichment, often through the agency of meteoric waters. It occurs in low-temperature hydrothermal veins associated with other copper minerals, particularly covellite, as in Cornwall, Siberia, Kongsberg (Norway), Monte Catini (Italy), Mexico, Peru, Chile, Butte (Montana), and Zambia. Chalcocite and **digenite** (a closely related mineral with formula Cu_9S_5) occur in cupriferous, red bed type sedimentary rocks.

Formula A_3X_2

BORNITE, erubescite, variegated copper ore
C Cu_5FeS_4

Physical properties

CS Tetragonal (pseudo-cubic)	**S** Pale greyish black, and slightly
F&H Crystals as cubes and octahedra;	shining
usually massive	**F** Conchoidal or uneven
COL Coppery red or golden brown; it	**L** Metallic
tarnishes quickly on exposure to air	**T** Brittle
and becomes iridescent (peacock ore)	**HD** 3.0
	SG 4.9–5.4

Tests Heated in open tube, bornite yields sulphur dioxide fumes, but gives no sublimate. It fuses in the reducing flame to a brittle magnetic globule. It is soluble in nitric acid, leaving a deposit of sulphur. Bornite is characterized by its tarnish, and by the red colour of fresh surfaces.

Occurrence Bornite is a valuable ore of copper. It occurs as a primary deposit in many copper lodes, and as a constituent of the zone of secondary enrichment of these. Bornite is usually associated with other copper and iron sulphide minerals in the 'secondary environment'. It is nearly always hypogene in origin, but may result from unmixing of high-temperature Cu + Fe + S solid solutions on cooling. It is found in some of the Cornish mines, where it was known as horse-flesh ore. In many occurrences, bornite is closely associated with igneous magmas, appearing either as magmatic segregations (as in Namaqualand), or as rather later deposits connected with pegmatites or end-stage consolidation. Bornite forms veins with quartz, or with quartz and chalcopyrite. It is the dominant constituent of the bedded copper deposit of the Kupferschiefer of Mansfeld, Germany.

Formula *AX*

GALENA, lead, glance, blue lead
C PbS

Silver sulphide is almost always present, and galena is one of the most important sources of silver. When sufficient silver is present to be worth extracting, the ore is called **argentiferous galena**, and zinc, cadmium, iron, copper, antimony and gold have also been detected in analyses of this mineral. No external characters serve to distinguish even the highly argentiferous ores from ordinary galena, and the composition can only be determined by analysis.

Physical properites

CS Cubic

F&H Cubes common, often modified by octahedra and other forms; it also occurs massive, and coarsely or finely granular

COL Lead grey

S Lead grey

CL Perfect {100} cubic cleavage, specimens cleaving easily when struck into small cubes

F Flat, even or subconchoidal

L Metallic on fresh surfaces, but dull when exposed due to tarnish

HD 2.5

SG 7.4–7.6

Tests Heated in the open tube, galena gives off sulphurous fumes. When heated on charcoal, it emits sulphurous fumes, forms a yellow encrustation of lead oxide and fuses to a malleable metallic globule, which marks paper. If heated with potassium iodide and sulphur on charcoal, it forms a brilliant yellow encrustation. Galena is decomposed by hydrochloric acid, with evolution of sulphuretted hydrogen; on cooling the solution, white crystals of lead chloride are deposited, which are soluble on heating.

Occurrence Galena often occurs associated with sphalerite (see p. 237). Metasomatic disseminations are exemplified by the lead–zinc deposits of the important Tri-State field in the Mississippi Valley. Galena is common in hydrothermal veins and as replacement deposits in many rock types, but especially in limestones as at Leadville, Colorado. The famous Broken Hill lode in Australia is of similar hydrothermal origin and provides argentiferous galena, but an unusual feature is that garnet is a gangue mineral. In several other types of lead lodes, the galena is associated with silver minerals. Pyrometasomatic or contact metamorphic deposits of galena are relatively unimportant, but one such commercial deposit in which the ore occurs in a limestone at its contact with granite porphyry is at the Magdalena mines, New Mexico. Galena is also common in Proterozoic stratiform massive sulphide deposits. Galena of sedimentary origin is not important, and one example of such deposits is in Permo-Triassic rocks at Aix-la-Chapelle, where the sandstone contains a small proportion of galena and cerussite, most likely leached from lead ores out cropping within the denudation area. In the UK, small deposits of galena of metasomatic or hydrothermal origin occur in Derbyshire, Flint, Cumberland, Cardigan, the Isle of Man and Cornwall, and a larger deposit has been proven in the Strontian district, west of Fort William in the Scottish Highlands.

Uses Galena is the most important ore of lead, almost all commercial lead being derived from this source.

ALABANDITE
C MnS

Physical properties

CS Cubic	**L** Submetallic
F&H Granular and massive	**TR** Opaque
Col Iron black	**HD** 3.5–4.0
S Greenish black	**SG** 3.95
CL Cubic {100} perfect	

Tests Heated before the blowpipe, alabandite fuses easily. A purple bead is given with borax in the oxidizing flame, especially with the roasted mineral.

Occurrence As a primary mineral, alabandite occurs in veins associated with sulphides and rhodonite.

SPHALERITE, BLENDE, black jack
C ZnS; iron replaces zinc, and cadmium may also be present, but always less than 5%

Physical properties

CS Cubic

F&H Tetrahedra or dodecahedra common as crystals; it also occurs massive and compact; occasionally botryoidal or fibrous

TW Twinning occurs on {111}, the forms present being difficult to determine

COL Black or brown usual, occasionally yellow or white, and rarely colourless

S White to reddish brown

CL {110} perfect

F Conchoidal

L Resinous to adamantine

TR Transparent to translucent or opaque

T Brittle

HD 3.5–4.0

SG 3.9–4.2

Optical properties

n 2.37–2.50
(increasing with Fe^{2+} replacing Zn)

Isotropic

COL Colourless or pale brown

H Small crystals often with curved faces (see above)

CL {110} perfect (6 directions of cleavage)

R Extremely high

A Alteration is common, to smithsonite, etc.

Tests Heated alone before the blowpipe, sphalerite is either infusible or very difficult to fuse. With sodium carbonate on charcoal in the reducing flame, it colours the flame strongly green. On charcoal, when roasted in the oxidizing flame, and then intensely heated in the reducing flame, sphalerite yields an encrustation of zinc oxide, which is yellow when hot and white when cold, and this encrustation assumes a green colour when heated with cobalt nitrate solution. Some varieties of sphalerite, heated on charcoal with sodium carbonate, first give a reddish-brown coating of cadmium oxide. Sphalerite is soluble in hydrochloric acid, with evolution of sulphuretted hydrogen, and the solution gives a white precipitate of zinc sulphide on the addition of ammonium sulphide.

Occurrence Sphalerite is common in stratabound veins and massive sulphide deposits, where it is associated with galena. It also occurs with galena, pyrite and chalocopyrite in calcareous nodules or veinlets, which may be of diagenetic origin. Iron-rich sphalerite often occurs with pyrrhotite. Sphalerite deposits are found in the Tri-State field of the USA, Broken Hill in NSW, Westphalia, Colorado, and in the UK in Derbyshire, Cumberland and Cornwall. Sphalerite has been reported from contact metamorphic deposits in New Mexico. Sphalerite is the most important ore of zinc.

CHALCOPYRITE, COPPER PYRITES
C $CuFeS_2$, with 34.5% copper

Physical properties
CS Tetragonal

F&H Twinned crystals resembling tetrahedra common, usually wedge-shaped; frequently massive

TW Twinning occurs on a number of planes {012}, {110}, etc. and gives lamellar twins, deformation twins and also interpenetrant twins

COL Brass yellow, often with an iridescent tarnish

S Greenish black, very slightly shining

CL Poor {011} and {111}

F Conchoidal or uneven

L Metallic

TR Opaque

HD 3.5–4.0

SG 4.1–4.3

Tests Chalcopyrite decrepitates when heated in the closed tube and gives a sublimate of sulphur. Before the blowpipe on charcoal, it fuses to a metallic magnetic globule, and gives off sulphurous fumes. With fluxes chalcopyrite gives reactions for both copper and iron. It may be distinguished from pyrite by its inferior hardness; chalocpyrite crumbling when cut with a knife, and pyrite resisting the attempt to cut it. Furthermore, pyrite emits sparks when struck with steel and chalcopyrite does not. Pyrite has a black streak, whereas that of chalcopyrite is greenish-black. Chalcopyrite may be distinguished from gold by its brittle nature and its non-malleability, gold being soft, malleable and easily cut with a knife. Chalcopyrite is soluble in nitric acid, whereas gold is not.

Occurrence Chalcopyrite is the principal commercial source of copper, and occurs in a number of ways. It is the major copper mineral in porphyry copper deposits, and occurs with bornite in the stratiform sulphide deposits of the Copper belt in Zambia, and also in the famous Kupferschiefer of Permian age at Mansfeld, Germany, as grains in a shale. Chalcopyrite appears to be relatively mobile and may replace other minerals, particularly pyrite. Magmatic segregations of chalcopyrite are known, but are not important. Pneumatolytic veins with chalcopyrite occur in Cornwall, Norway, Oregon, South Australia and Chile. An important mode of occurrence is as hydrothermal or metasomatic veins, as in California,

Montana, Arizona, Alaska and Canada. Pyrometasomatic deposits are important; in these the chalcopyrite occurs with other sulphides and skarn minerals at or near the contact between bodies of intrusive granodioritic rock and limestone; examples include Clifton-Morenci, Bisbee (Arizona), Bingham (Utah), Alaska, Canada, Australia, Japan and Korea.

STANNITE, stannine, tin pyrites, bell metal ore
C Cu_2SnFeS_4, with 27.5% tin; zinc is usually present in varying amounts

Physical properties

CS Tetragonal (pseudo-cubic)	**S** Blackish
F&H Crystals rare; usually occurs massive, granular or disseminated	**F** Uneven
	L Metallic
COL Steel grey if pure; otherwise iron black or bronze, often with a blue tarnish. It has a yellowish colour if present in an admixture with chalcopyrite.	**TR** Opaque
	T Brittle
	HD 4.0
	SG 4.4

Tests Heated in the open tube, stannite gives off fumes of sulphur dioxide, and also forms a sublimate of tin oxide close to the assay. Heated on charcoal, it fuses after long roasting to a brittle metallic globule which, heated in the oxidizing flame, gives off sulphur, and coats the support with white tin oxide. In borax, the roasted mineral gives reactions for iron and copper.

Occurrence Stannite occurs associated with cassiterite, chalcopyrite, sphalerite, and galena, as in Cornwall, Saxony, etc. It occurs in the tin lodes of Boliva, associated with cassiterite, silver minerals, and sulphides of copper, antimony, lead, zinc, bismuth, etc.

GREENOCKITE, xanthochroite
C   CdS

Physical properties

CS Hexagonal	**S** Orange–yellow to red
F&H Rarely as squat hexagonal crystals; often as a coating on zinc ores	**L** Resinous to adamantine
	TR Transparent to subtransparent
	HD 3.0–3.5
COL Honey-coloured, lemon or orange–yellow	**SG** 4.8–4.9

Optical properties
$n_o = 2.506$ $n_e = 2.529$ $\delta = 0.023$ Uniaxial +ve

A very rare mineral, pale yellow in thin section and weakly pleochroic. The extreme relief (RIs ≈ 2.5) is important.

Tests Heated in the closed tube, greenockite turns carmine red, and reverts to its original colour on cooling. Heated in the open tube, it gives off sulphurous fumes. Heated with sodium carbonate on charcoal, greenockite gives a reddish-brown encrustation. Such an encrustation is yielded before the zinc encrustation, so that careful observation of the behaviour of samples of sphalerite may indicate the presence of cadmium in them. Greenockite dissolves in hydrochloric acid, with the evolution of sulphuretted hydrogen.

Occurrence Almost invariably occurs associated with zinc ores, on which it forms a coating. It has also been reported in cavities in basic rocks associated with zeolites. Localities are Bishopton (Scotland), Pribram (Czechoslovakia) and Freidensville (Pennsylvania).

PYRRHOTITE, pyrrhotine, magnetic pyrites
C $Fe_{1-n}S$, $n = 0$–0.2; up to 5% nickel may be present and pyrrhotite, if rich in nickel, can be valuable as a source of that metal

Physical properties

CS Hexagonal	**F** Uneven or imperfect conchoidal
F&H Hexagonal tabular prisms occasionally occur; generally massive	**L** Metallic
	TR Opaque
COL Reddish, brownish, bronze or copper coloured; always tarnished on exposure to air	**T** Brittle
	MAGNETISM Magnetic
	HD 3.5–4.5
S Dark greyish black	**SG** 4.40–4.65
CL Poor {0001} basal	

Tests Pyrrhotite is soluble in hydrochloric acid, with evolution of sulphuretted hydrogen. Heated before the blowpipe, it fuses in the reducing flame to a black magnetic globule, but in the oxidizing flame it is converted into a globule of red iron oxide. Pyrrhotite is distinguished from pyrites by its lesser hardness and by its colour, from copper pyrites by its colour, and from niccolite by its SG and blowpipe reactions.

Occurrence Pyrrhotite indicates a relatively low sulphur availability. It is common in igneous rocks, metamorphic rocks and stratiform copper–iron sulphide deposits. In veins it is usually taken to indicate precipitation from relatively high-temperature hydrothermal solutions.

The most important occurrence of pyrrhotite is at Sudbury, Canada, where the mineral is accompanied by the nickel-bearing sulphide pentlandite, and the deposits constitute the world's largest known source of nickel. Other less important localities of pyrrhotite occur in Morocco, Finland and Norway, and it also occurs at Botallack (Cornwall), Beer Alston (Devonshire), Dolgelly (Wales), Kongsberg (Norway), Andreasberg in the Harz and elsewhere.

Uses The most valuable ore of nickel.

NICCOLITE, kupfernickel, nickeline, copper nickel, arsenical nickel
C   NiAs; Sb, Co, Fe and S may be present as impurities

Physical properties

CS Hexagonal	**L** Metallic
F&H Massive	**TR** Opaque
COL Pale copper red, sometimes with	**T** Brittle
a tarnish	**HD** 5.0–5.5
S Pale brownish black	**SG** 7.3–7.6
F Uneven	

Tests Heated before the blowpipe, niccolite gives off arsenic fumes, and fuses to a globule, and this globule, if subsequently heated in the borax bead, gives reactions for nickel; reddish-brown in the oxidizing flame and opaque grey in the reducing flame. Cobalt and iron flame reactions also occur. It is soluble in nitric acid, giving an apple-green solution.

Occurrence Niccolite occurs in Ni–Co–Ag–As–Au deposits, which are probably low-temperature hydrothermal veins and replacements. These deposits are associated with basic igneous rocks and organic-rich sedimentary rocks. Important localities are Cobalt (Ontario), and many mines in Cornwall, Saxony, the Harz, etc.

BREITHAUPTITE, antimonial nickel
C   NiSb, usually with varying amounts of PbS

Physical properites

CS Hexagonal	**L** Metallic, bright
F&H Crystals rare, usually found	**HD** 5.5
massive	**SG** 7.5
COL Light copper red on fresh	
surfaces	

Tests Heated on charcoal breithauptite gives a white coating of antimony oxide. After roasting it gives nickel reactions in the borax bead.

Occurrence Breithauptite occurs with niccolite as at Andreasberg in the Harz. It also occurs at Cobalt, Ontario.

MILLERITE, nickel pyrites, capillary pyrites
C   NiS, often with traces of cobalt, copper and iron

Physical properties

CS Trigonal	**S** Greenish black
F&H Occurs in capillary crystals of extreme delicacy, hence the name capillary or hair pyrites. It sometimes occurs in columnar tufted coatings; rarely as rhombohedra.	**CL** Perfect {10$\bar{1}$1} rhombohedral
	L Metallic
	HD 3.0–3.5
	SG 5.3–5.6
COL Brass yellow to bronze yellow, often tarnished	

Tests Heated in the open tube, millerite gives off sulphurous fumes. After roasting, millerite is heated before the blowpipe with borax and with microscosmic salt. In the oxidizing flame it gives a violet bead; in the reducing flame, it gives a grey bead, owing to the reduction to metallic nickel, the impurities also frequently giving reactions in the beads. Heated on charcoal with sodium carbonate and charcoal, it gives a metallic magnetic mass of nickel.

Occurrence Millerite occurs as nodules in clay ironstone, as in South Wales. It is found in veins associated with other nickel and cobalt minerals, as at Cobalt (Ontario), Cornwall, Saxony, and especially at the Gap Mine, Lancaster County, Pennsylvania, USA.

PENTLANDITE
C   (Fe,Ni)$_9$S$_8$

Physical properties

CS Cubic	**L** Metallic
F&H Massive or granular	**T** Brittle
COL Bronze yellow	**HD** 3.5–4.0
S Black	**SG** 5.0
F Uneven	

Tests Heated in the open tube, pentlandite gives off sulphurous fumes. After roasting, it gives nickel reactions in the borax bead. Before the blowpipe, it yields a magnetic mass. Pentlandite is soluble in nitric acid, the solution giving a reddish-brown precipitate of ferric hydroxide on the addition of ammonia.

Occurrence Pentlandite occurs associated with pyrrhotite and other copper, nickel and iron sulphides in the Sudbury, Ontario, nickel deposit. It occurs in some basic igneous rocks, particularly norites, and in some massive sulphide deposits.

Use One of the chief ores of nickel.

COVELLITE, covelline
C CuS, with 66.4% copper

Physical properties

CS	Hexagonal	CL	Basal {0001} perfect
F&H	Platy hexagonal crystals; usually massive	TR	Opaque
		HD	1.5–2.0
COL	Indigo blue	SG	4.6

Occurrence Covellite occurs at many localities in the zone of secondary enrichment of copper lodes etc., replacing copper–iron sulphides. This zone, which is situated between the gossan and the unaltered zone, contains products of the chemical reactions between the original vein material and descending solutions, and is often the richest part of the lode.

CINNABAR
C HgS, with impurities of clay, bitumen, etc.

Physical properties

CS	Trigonal	CL	Prismatic {10$\bar{1}$0} perfect
F&H	Rhombohedra or trigonal prisms common, the crystals often being tabular; usually massive, granular or encrusting	F	Subconchoidal to uneven
		L	Adamantine (of crystals), dull if massive
		TR	Subtransparent to opaque
COL	Cochineal red, sometimes brownish or dark coloured	T	Sectile
		HD	2.0–2.5
S	Scarlet	SG	8.09

Optical properties Cinnabar is deep red in thin section, uniaxial +ve and with truly extreme relief (RIs 2.905–3.256).

Variety **Hepatic cinnabar** is a compact variety with a liver-brown colour, and a brownish streak.

Tests Heated in the open tube, cinnabar yields a sublimate of metallic mercury and also a black one of mercury sulphide and fumes of sulphur dioxide. Heated in the closed tube, cinnabar gives a black sublimate, which becomes red if detached and rubbed on a streak plate. When heated with sodium carbonate and charcoal in the closed tube, it gives metallic mercury. Heated on charcoal with potassium iodide and sulphur, cinnabar gives greenish fumes and a slight greenish encrustation.

Occurrence Cinnabar occurs as disseminations, impregnations and stockworks in a variety of rocks, but in many cases under circumstances that indicate that it is the result of volcanic activity. It occurs in low-temperature hydrothermal veins, impregnations and replacement deposits, often associated with recent volcanics. It often replaces quartz and sulphides and is associated with native mercury, mercurian tetrahedrite, stibnite, pyrite and marcasite. The most important locality is Almaden in Spain, where the mineral occurs as impregnations of small veins in quartzite; other important localities are Idria in Italy, and the Western states of the USA, especially California.

Uses Cinnabar supplies practically all the mercury of commerce. The paint, vermilion, which has the same composition, is prepared from this ore.

REALGAR
C As_2S_2, with 70.1% arsenic

Physical properties

CS Monoclinic	**F** Conchoidal or uneven
F&H Prismatic crystals rare, usually occurring massive or granular	**L** Resinous
	TR Transparent to translucent
COL Red or orange	**HD** 1.5–2.0
S Red or orange (same as colour)	**SG** 3.56

Optical properties In thin section, realgar is pale yellowish in colour with extreme relief (RIs = 2.5–2.7), biaxial +ve, with $2V \approx 40°$. It alters to orpiment.

Tests Heated in closed tube, realgar gives a reddish-yellow sublimate. When heated on charcoal, it emits sulphurous and garlic fumes and forms a white sublimate far from the assay. Heated in the open tube, white volatile sublimate and garlic fumes are formed.

Occurrence Realgar occurs associated with orpiment, to which mineral it changes on exposure. Realgar occurs as a deposit in hot springs, and as a volcanic sublimate. It is common in veins, where it may occur as nests or nodules in clay, or associated with cinnabar, as in Tuscany, Galicia and Spain.

Formula A_3X_4

LINNAEITE
C Co_3S_4

Characters and occurrence
CS   Cubic
Linnaeite is a steel grey mineral, tarnishing coppery red, and occurring in octahedral crystals or massive forms, in sulphide veins, associated with chalcopyrite, pyrite, etc. When heated in the closed tube, linnaeite gives sulphur after a time, and the roasted material gives the blue colour of cobalt in the borax bead.

Formula A_2X_3

ORPIMENT
C As_2S_3, with 61.0% arsenic

Physical properties
CS   Monoclinic
F&H   Crystals rare, usually occurring foliaceous or massive
COL   Lemon yellow
S   Yellow
CL   Prismatic {010} perfect in crystal forms

L   Pearly to brilliant on cleavage faces; elsewhere resinous or dull
TR   Subtransparent to subtranslucent
T   Sectile; laminae are flexible but not elastic
HD   1.5–2.0
SG   3.4–3.5

Optical properties   In thin section, orpiment is yellow to greenish yellow with extreme relief ($RI \approx 3.0$) and strong dispersion. $2V$ large (70–80°) but sign indeterminable.

Tests   Heated in the closed tube, orpiment gives a reddish-yellow sublimate. Heated on charcoal, it emits sulphurous and garlic fumes and forms a white sublimate far from the assay. Heated in the open tube, a white volatile sublimate and garlic fumes form.

Occurrence   Orpiment occurs in the oxidized portions of arsenic veins. It is associated with antimony ores in veins as at Kapnik in Transylvania, Kurdistan, etc. It also occurs as a deposit from some hot springs, as at Steamboat Springs, Nevada, and as a sublimate from volcanoes at Naples.

STIBNITE, ANTIMONITE, antimony glance, grey antimony
C Sb_2S_3, with 71.7% antimony

Physical properties

CS Orthorhombic	**L** Metallic, but liable to tarnish and
F&H Common as elongated, bladed	sometimes iridescent on the surface
crystals straited longitudinally, occur-	**T** Sectile or brittle; thin laminae are
ring in masses of radiating crystals	slightly flexible
COL Lead grey	**HD** 2.0
S Lead grey	**SG** 4.5–4.6
F Subchoncoidal	

Tests   Heated in the open tube, stibnite gives off antimonous and sulphurous fumes, the former condensing as a white non-volatile sublimate, while the latter may be recognized by the odour. Heated on charcoal, it fuses easily and gives a white encrustation near the assay. Stibnite fuses easily in the flame of a candle.

Occurrence   The chief mode of occurrence of stibnite is in low-temperature hydrothermal veins where it is associated with quartz. It is associated with complex antimony- and arsenic-bearing sulphides, pyrite, gold and mercury. It occurs with quartz, dolomite, calcite and barite in veins, as in Cornwall, Westphalia, Saxony, etc. The great Chinese deposits of Hunan occur in brecciated sandstone, in which the stibnite occurs as irregular stringers, veins and pockets.

Uses   Stibnite is the main source of the antimony of commerce.

BISMUTHINITE, bismuth glance
C Bi_2S_3

Physical properties

CS Orthorhombic	**L** Metallic
F&H Small acicular crystals, but	**TR** Opaque
usually massive	**H** 2.0
COL Lead grey, usually tarnished	**SG** 6.4–6.5

Tests Heated before the blowpipe, bismuthinite fuses easily. Heated on charcoal with potassium iodide and sulphur, it gives a yellow and bright red encrustation. Heated in the open tube, it gives off sulphurous fumes. Bismuthinite is soluble in nitric acid, a white precipitate being formed on the addition of water.

Occurrence Bismuthinite occurs in veins associated with copper, lead, tin and other ores, as in Cornwall, Cumberland, Saxony, Bolivia, etc. Bismuthinite is an important ore of bismuth.

KERMESITE, red antimony
C Sb_2S_2O

Characters and occurrence
CS Orthorhombic or monoclinic
Occurs as red, acicular crystals, and results from the alteration of stibnite in the oxidized zones of antimony deposits, being a stage towards the formation of the oxides, senarmontite and valentinite, with which it is often associated.

Formula AX_2

PYRITE, PYRITES, IRON PYRITES, mundic
C FeS_2, with 46.6% iron

Physical properties
CS Cubic
F&H Cube and pentagonal dodecahedron forms occur in crystals. Faces are usually striated, the striae on one face being at right angles to those of the adjacent faces (this reduces the symmetry of the cube). Pyrite may occur massive, in nodules with a radiating structure, in finely disseminated crystals or particles, and as a replacement mineral for calcite in fossil shells.
COL Bronze yellow to pale brass yellow
S Greenish black or brownish black

CL None
F Conchoidal or uneven
L Metallic, splendent
TR Opaque
T Brittle. Sparks are produced when pyrite is hit with steel, for which reason it was formerly used instead of flint in the old wheel-lock guns, in which a steel wheel revolved rapidly, by clockwork, against a piece of pyrite and threw sparks into the pan of the weapon.
HD 6.0–6.5 (cf. chalcopyrite, 3.5–4.0)
SG 4.8–5.1

Tests Heated in the closed tube, pyrite gives a sublimate of sulphur, leaving a magnetic residue. In the borax bead it shows yellow in the oxidizing flame and bottle green in the reducing flame. Heated on charcoal with sodium carbonate, pyrite gives a magnetic residue. A black stain is produced in the silver coin test. Pyrite is soluble in nitric acid, but insoluble in hydrochloric acid.

Occurrence Pyrite is a common mineral in many parts of the world. It is rare as an accessory primary mineral in igneous rocks, although it often occurs as a secondary one. It is a frequent constituent of many ore veins, and large pyrite deposits are worked mainly for sulphur, but in some cases for the other metals contained in the pyrite, such as copper. The origins of many of the large pyrite masses are not known with certainty. Some are due to magmatic segregation, and in these the pyrite is accompanied by pyrrhotite. Others are the contact-metamorphic or pyrometasomatic origin, while a few are considered to be injected bodies. The great deposits of Rio Tinto in Spain, Mt Lyell in Tasmania, and Rammelsberg in the Harz, are low-temperature deposits associated with calcite, barite, quartz, etc. Pyrite deposits considered to be of sedimentary origin may be represented by the oolitic pyrite of the Cleveland Hills, England. In some cases, the pyrite contains enough gold to make extraction economic. Pyrite is not worked directly as an ore of iron, the sulphur which it contains making it comparatively worthless for that purpose; but a good deal of the sulphuric acid and sulphate of iron, copperas, is derived from its decomposition. When present in shales or clays, the decomposition and oxidation of pyrite give rise, upon the roasting of the pyritous clay or shale, to sulphuric acid. This combines with the alumina present to form, on the addition of potassium compounds, alum – a hydrous sulphate of aluminium and potassium. In this way some of the alum of commerce is made. Sulphur is also produced from pyrite, but the market for both sulphur and sulphuric acid is now largely supplied from native sulphur, or from sulphur impurities extracted during the processing of crude oil.

SPERRYLITE
C $PtAs_2$

Physical properties

CS Cubic		**S** Black
F&H Cubes or combinations of cube and octahedron; crystals usually small but may occur in aggregates		**L** Metallic
		HD 6.0–7.0
		SG 10.6
COL Tin white		

Occurrence Sperrylite occurs in the pyrrhotite deposits of Sudbury, Ontario, and as large crystals in the Bushveld norite and in many detrital platinum deposits.

COBALTITE
C CoAsS; iron is often present

Physical properties

CS Orthorhombic (pseudocubic)	**S** Greyish black
F&H Cubes or pentagonal dodecahe-	**L** Metallic
dra rare, usually occurring massive,	**HD** 5.5
granular and compact	**SG** 6.0–6.3
COL Silver white with a reddish tinge	

Tests Heated in the closed tube, cobaltite remains unaltered. Heated in the open tube, it yields a sublimate of arsenious oxide and gives off sulphurous fumes. It gives the blue cobalt bead with borax. Cobaltite is decomposed by nitric acid.

Occurrence Cobaltite is associated with copper–iron sulphides and copper–nickel arsenides in high- to medium-temperature deposits in veins, and as disseminations.

MARCASITE, white iron pyrites
C FeS_2 (same as pyrite)

Physical properties

CS Orthorhombic	**COL** Bronze yellow, paler than pyrite
F&H Tabular crystals common, often	**S** Greyish
showing repeated twinning leading to	**L** Metallic
pseudohexagonal crystals, called	**T** Brittle
cockscomb pyrites or spear pyrites	**HD** 6.0–6.5
when occurring as aggregates in the	**SG** 4.9
Chalk. Marcasite also occurs in radi-	
ating habits, externally nodular.	

Tests As for pyrite; it is paler than pyrite and decomposes more readily.

Occurrence Marcasite is found in low-temperature sulphide deposits, with pyrite, and as concretions in sedimentary rocks, such as the English Chalk.

Uses Marcasite is used for the same purposes as pyrite, and was formerly cut and polished for ornaments.

ARSENOPYRITE, MISPICKEL, arsenical pyrites
C FeAsS, with 46.0% arsenic

Physical properties

CS Monoclinic (pseudo-orthorhombic)

F&H Prism with {011} terminations, the (011) faces horizontally striated; also massive

TW Twinning is common on {100} and {001} giving pseudo-orthorhombic crystals; penetration twins on {101} and cruciform twins on {012} also occur

COL Silvery white, inclined to steel grey on fresh surfaces. It tarnishes to a pale copper colour on exposure

S Dark greyish black

CL {101} is distinct

F Uneven

L Metallic

T Brittle; it gives off sparks when struck with steel, and then emits a garlic odour

HD 5.5–6.0

SG 5.9–6.2

Tests Heated before the blowpipe, arsenopyrite gives rise to arsenical fumes, and fuses to a globule which is magnetic. Heated in the closed tube, it first gives the red sublimate of arsenic sulphide, and then the black sublimate of arsenic. Heated in the open tube, it gives sulphurous fumes and a white sublimate of arsenic oxide. When heated with hydrochloric acid, it gives sulphur.

Occurrence Arsenopyrite is considered to be typical of high-temperature hydrothermal veins with cassiterite, wolframite, chalcopyrite, pyrrhotite and gold being common associated minerals. It also occurs in most types of sulphide deposits.

MOLYBDENITE
C MoS_2

Physical properties

CS Hexagonal

F&H Usually as scales, but also found massive, foliaceous and sometimes granular

COL Lead grey

S Greenish lead grey; the greenish tint helps to distinguish it from graphite

CL Basal {0001} perfect

L Metallic

TR Opaque

T Sectile, almost malleable

HD 1.0–1.5 (easily scratched by the fingernail)

SG 4.7–4.8

Tests Molybdenite shows yellow green in the flame test; heated in the open tube, molybdenite gives sulphurous fumes, and a yellow sublimate of molybdenum oxide. Heated on charcoal, it yields a strong sulphurous odour, and coats the charcoal with an encrustation of molybdenum oxide, which is yellow when hot, and white when cold. This encrustation is copper red near the assay, and, if touched with the reducing flame, becomes blue. The microcosmic salt bead is green.

Occurrence Molybdenite is found in high-temperature hydrothermal veins and quartz pegmatites with bismuth, tellurium, gold, tin and tungsten minerals. It occurs in porphyry copper style deposits. It is an accessory late-stage mineral in acid igneous plutonic rocks; and it is occasionally found as a mineral of contact metamorphic or pyrometasomatic origin.

Uses It is the major ore of molybdenum.

CALAVERITE
C $(Au,Ag)Te_2$, with more than 25% gold

Physical properties

CS Monoclinic	COL Pale yellow
F&H Crystals small and elongated; also granular and massive	HD 2.5
	SG 9.0

Tests As for sylvanite (see below) calaverite has a higher percentage of gold than sylvanite.

Occurrence The gold tellurides occur mainly in veins and replacement deposits, in which they are associated with pyrite and other sulphides and often free gold. The main localities at which gold telluride ores are worked are Kalgoorlie (Western Australia), Cripple Creek and Boulder County (Colorado) and Nagyag (Transylvania). Telluride ores occur in the Porcupine (Ontario) gold veins, and in the gold–alunite deposits of Goldfield, Nevada.

SYLVANITE, graphic tellurium
C $(Ag,Au)Te_2$, with 24.5% gold and 13.4% silver by weight, if gold and silver are equal in amount in the formula; Sb and Pb may also be present in small quantities

Physical properties

CS Monoclinic	S Same as colour
F&H Crystals are arranged in regular lines rather like writing, and hence the name graphic tellurium; it also occurs granular and massive	CL Perfect {100}
	F Uneven
	L Metallic
	T Brittle
COL Steel grey to silver white, sometimes yellow	HD 1.5–2.0
	SG 8.0–8.2

Tests In the open tube, sylvanite behaves like native tellurium. On charcoal, before the reducing flame, it gives a yellow, malleable, metallic globule after long heating, with an encrustation of telluric oxide being formed on the charcoal. If the powdered mineral is heated with strong sulphuric acid and acid turns a reddish-violet colour. Sylvanite is decomposed by nitric acid, leaving gold powder; and when hydrochloric acid is added to the solution, a dense white precipitate of silver chloride forms.

Occurrence See previous page, under Calaverite.

Formula AX_3

SMALTITE, tin white cobalt
C $(Co,Ni)As_{3-n}$ (with Co > Ni); smaltite grades into chloanthite, $(Ni,Co)As_{3-n}$ (with Ni > Co); Fe is usually present

Physical properties

CS Cubic	**S** Greyish black
F&H {100}, {111} and {110} forms, often modified, seen in crystals; it also occurs massive or reticulated	**CL** Distinct {111}, less distinct {100}
	F Uneven, rather granular
	L Metallic
COL Tin white to steel grey, when massive; it tarnishes on exposure, sometimes with iridescence	**T** Brittle
	HD 5.5–6.0
	SG 6.4

Tests Heated in the closed tube, smaltite gives a sublimate of metallic arsenic. Heated in the open tube, it gives a sublimate of arsenious oxide. Heated on charcoal, smaltite gives off an arsenical odour, and fuses to a globule, which yields reactions for cobalt with borax. The presence of cobalt is often indicated by the occurrence of a pinkish coating (cobalt bloom) on the surface of the mineral.

Occurrence Smaltite occurs in hydrothermal veins, associated with calcite, barite, quartz and silver, nickel and copper minerals. The main source of supply is Cobalt, Ontario; other localities are Schneeberg, Freiberg and Annaberg (Saxony), and less important are several mines in Cornwall, Jachymov (Czechoslovakia), etc.

CHLOANTHITE, white nickel
C $(Ni,Co)As_{3-n}$; chloanthite grades into smaltite with cobalt replacing nickle; iron is usually present.

Physical properties

CS	Cubic	**S**	Greyish black
F&H	Crystals form cubes but usually massive	**HD**	5.5–6.0
COL	Tin white	**SG**	6.4–6.7

Tests As for niccolite.

Occurrence Usually occurs with smaltite at localities cited for that mineral (see above), especially at Cobalt, Ontario, and is a valuable nickel ore.

8.3.2 Multiple sulphides (or sulphosalts)

Formula $A_mB_nX_p$, with $(m + n) : p$ greater than $1 : 1$

POLYBASITE

C $(Ag,Cu)_{16}(Sb,As)_2S_{11}$, with about 70% silver

Physical properties

CS	Monoclinic	**S**	Black
F&H	Tabular prismatic crystals form, but usually massive	**L**	Metallic
COL	Iron black	**HD**	2.0–3.0
		SG	6.0–6.2

Variety **Pearcite**, an arsenical variety.

Tests Heated in open tube, polybasite gives sulphurous fumes and a sublimate of antimony and arsenic oxides. There is a copper residue and silver bead, given by lengthy heating with fluxes on charcoal.

Occurrence In silver veins, polybasite is associated with other primary silver ores. It also occurs in the argentite veins of Mexico, the argentite-gold-quartz veins of Tonopah (Nevada) and the Comstock Lode, in the replacement veins of Portland Canal (British Columbia) and in the silver deposits of San Juan (Colorado), Pribram (Czechoslovakia), Freiberg (Saxony), Andreasberg (Harz), Chile, etc.

STEPHANITE, brittle silver ore

C Ag_5SbS_4, with 68.5% silver

Physical properties

CS Orthorhombic

F&H Crystals are commonly flat, tabular prisms, often twinned, but also massive and disseminated

COL Iron black

S Iron black

F Uneven

L Submetallic

T Brittle

HD 2.0–2.5

SG 6.26

Tests Heated in the closed tube, stephanite fuses with decrepitation and gives a slight sublimate of antimony sulphide after long heating. On charcoal, it fuses to a dark metallic globule and encrusts the support with antimony oxide; when heated in the reducing flame with sodium carbonate, the globule yields metallic silver. Stephanite is decomposed by dilute nitric acid, leaving a residue of sulphur and antimony oxide. A clean strip of copper placed in the solution becomes coated with silver, and if hydrochloric acid is added to the solution a white precipitate of silver chloride is formed.

Occurrence Stephanite occurs with other primary silver ores in veins at many of the localities given for argentite: Mexico, Harz, Comstock Lode (Nevada), Freiberg (Saxony), Pribram (Czechoslovakia), Cornwall, etc.

RUBY SILVER GROUP

C **pyrargyrite** (dark red silver ore) Ag_3SbS_3

    proustite (light red silver ore) Ag_3AsS_3

Ruby silvers usually represent end-member compositions, but a complete series exists from the antimony to the arsenic compound. Pyrargyrite contains 59.9% silver, whereas proustite contains 65.4% silver.

Physical properties

CS Trigonal

F&H Pyrargyrite occurs as hexagonal prisms terminated by rhombohedra, often modified and twinned; it also occurs massive. Proustite forms pointed crystals but is commonly granular and massive

COL Pyrargyrite: black to cochineal red; proustite: cochineal red

S Cochineal red (scarlet)

CL Both minerals possess a $\{10\bar{1}1\}$ rhombohedral cleavage

F Conchoidal to uneven

L Metallic to adamantine

TR Translucent to opaque

HD 2.5 (pyrargyrite); 2.0–2.5 (proustite)

SG 5.82–5.85 (pyrargyrite); 5.55–5.64 (proustite)

Optical properties In thin section, the ruby silvers are deep red in colour with extreme relief (RIs = 2.79–3.09)

Tests Pyrargyrite. Heated in the open tube, it gives sulphurous fumes and a white sublimate of antimony oxide. Heated on charcoal, it spurts and fuses easily to a globule of silver sulphide, and coats the support white, and the globule heated with sodium carbonate and charcoal on charcoal yields metallic silver. Pyrargyrite is decomposed by nitric acid, leaving a residue of sulphur and antimony oxide.

Proustite. Heated in open tube, it gives off sulphurous fumes and yields a white sublimate of arsenic oxide. Heated on charcoal with sodium carbonate, it gives metallic silver. Proustite is decomposed by nitric acid.

Occurrence   Pyrargyrite is more common than proustite. The ruby silvers are associated with other sulphosalts, especially the tetrahedrite series in low-temperature lead–zinc mineralization and silver–nickel–cobalt veins; in Andreasberg (Harz), Freiberg (Saxony), Pribram (Czechoslovakia), Chanarcillo (Chile), Austin (Nevada) and Cobalt (Ontario). The ruby silvers often occur just below the enriched zone, as at Potosi (Bolivia).

TETRAHEDRITE SERIES

C tetrahedrite $(CuFe)_{12}Sb_4S_{13}$
 tennantite $(CuFe)_{12}As_4S_{13}$

There is a sulphide series from the antimony to the arsenic compound. Cu may be partly replaced by Fe, Zn, Ag or Hg. Sb or As may rarely be partly replaced by Bi. Tetrahedrite can contain up to 30% silver, replacing the copper, and then it is called argentiferous grey copper ore, fahlerz or **freibergite**.

Physical properties

CS Cubic
F&H Tetrahedral crystals, usually modified and frequently twinned; they also occur massive with a compact, granular, or cryptocrystalline structure
COL Steel grey to iron black
S Nearly the same as the colour

F Subconcoidal or uneven
L Metallic
TR Opaque, but very thin splinters are subtranslucent and appear cherry red by transmitted light
T Brittle
HD 3.0–4.5
SG 4.5–5.1

Tests   The tetrahedrite minerals vary in their chemical behaviour according to the different substances which the varieties contain. In the closed tube they all fuse, giving a deep red sublimate of antimony sulphide. In the open tube, tetrahedrite fuses and gives off sulphurous fumes, forming a white sublimate inside the tube. The mercurial varieties give minute globules of quicksilver. On charcoal, tetrahedrite fuses and, according to the constituents present, yields white encrustations of antimony oxide, arsenic oxide or zinc oxide, or a yellow encrustation of lead oxide – the zinc encrustation becoming green when moistened with cobalt nitrate and reheated. Heated before the blowpipe with sodium carbonate, tetrahedrite yields scales of metallic copper. Tetrahedrite is decomposed by nitric acid with arsenic oxide, antimony oxide and sulphur remaining.

Occurrence Tetrahedrite occurs in association with other ores of copper, and with galena in lead and zinc deposits, but is inexplicably abundant in some and absent in others. Tennantite is common in porphyry copper mineralization. Localities for tetrahedrite are Levant (Cornwall), Andreasberg (Harz), Freiberg (Saxony), Pribram (Czechoslovakia), Chile, Bolivia, and Montana and Colorado (USA).

ENARGITE GROUP

C	famatinite	Cu_3SbS_4
	enargite	Cu_3AsS_4

Sulphide isomorphous series from the antimony to the arsenic compound

Physical properties

CS	Orthorhombic	**CL**	Good {110} prismatic
F&H	Rare small crystals are often twinned, but they usually occur massive and granular	**L**	Metallic
		HD	3.0–3.5
		SG	4.5
COL	Famatinite: greyish to copper red; enargite: greyish black to iron black		

Tests Heated on charcoal, enargite fuses and yields encrustations of oxides of arsenic and antimony, and occasionally zinc. In the open tube it gives sulphur and arsenic fumes, and forms a white sublimate of arsenic oxide. It yields a copper residue with fluxes.

Occurrence An important ore of copper in copper veins of the Sierra Famatina in Argentina, and in Chile and Peru. It occurs in abundance with other ores of copper in veins in monzonite at Butte, Montana.

Formula $A_mB_nX_p$, with $(m + n) : p$ equal, or approximately equal, to 1 : 1

BOURNONITE, wheel ore, endellionite

C $CuPbSbS_3$

Physical properties

CS Orthorhombic

F&H Modified prisms, often twinned, producing a cruciform or cogwheel-like arrangement, thus giving bournonite its name 'wheel ore' or 'radelerz' (its German name); it also occurs massive

COL Steel grey or lead grey, and sometimes blackish

S Same as colour

F Conchoidal or uneven

L Metallic

TR Opaque

T Brittle

HD 2.5–3.0

SG 5.7–5.9

Tests On charcoal, bournonite fuses easily, at first giving a white encrustation of antimony oxide, and afterwards a yellow one of lead oxide. The residue, heated with sodium carbonate on charcoal, yields reddish flakes of metallic copper.

Occurrence Bournonite occurs with the other ores of copper. It was first found at St Endellion in Cornwall, and also occurs at Kapnic (Rumania), Clausthal, Andreasberg and Neudorf (in the Harz), Chile, Bolivia, etc.

FREIESLEBENITE

C $Pb_3Ag_5Sb_5S_{12}$, with about 22–23% silver

Physical properties

CS Monoclinic

F&H Prismatic crystals and also massive

COL Light steel grey to dark lead grey

F Subconchoidal or uneven

L Metallic

T Brittle, somewhat sectile

HD 2.0–2.5

SG 6.0–6.4

Tests Heated on charcoal, freieslebenite gives a white sublimate of antimony oxide near assay. It gives a yellow encrustation when roasted with potassium iodide and sulphur, indicating lead. When heated on charcoal in the oxidizing flame it gives lead.

Occurrence Found associated with other silver ores and galena in Spain, Saxony, Rumania, etc.

JAMESONITE

C $Pb_4FeSb_6S_{14}$, with 29.5% antimony

Physical properties

CS	Monoclinic
F&H	Acicular crystals, often in feather-like habits giving the variety plumosite or feather ore; it also occurs fibrous or massive
COL	Dark lead grey
S	Greyish black
CL	Basal {001} perfect
L	Metallic
HD	2.0–3.0
SG	5.5–6.0

Tests Jamesonite gives reactions for antimony in the open tube. When heated on charcoal with sodium carbonate and charcoal, it gives the metallic bead of lead. Lead is also given by the potassium iodide and sulphur test (see Appendix A). Jamesonite is soluble in hydrochloric acid, a precipitate of lead chloride being formed on cooling.

Occurrence Jamesonite occurs in veins, associated with stibnite, tetrahedrite and other lead–silver sulphosalts. Some localities are St Endellion (Cornwall), Foxdale (Isle of Man), Dumfriesshire (Scotland), etc.

8.4 Oxides

8.4.1 Simple oxides

Formula A_2X

CUPRITE, red oxide of copper
C Cu_2O, with 88.8% copper

Physical properties

CS	Cubic
F&H	Crystals show {111} or {110} forms, but sometimes massive or earthy and occasionally capillary
COL	Shades of red, especially cochineal red
S	Brownish red and shining
CL	Poor {111} octahedral
F	Conchoidal or uneven
L	Adamantine to submetallic to earthy
TR	Subtransparent to nearly opaque
T	Brittle
HD	3.5–4.0
SG	6.14–6.15

Optical properties Transparent varieties are red to orange–yellow in thin section, with extreme RI ($n = 2.849$), and isotropic.

Varieties **Ruby copper** is a crystallized cuprite. **Tile ore** is a red or reddish-brown earthy variety, generally containing iron oxide. **Chalcotrichite** (from the Greek *chalkos*, meaning copper, and *thrix*, hair) is a beautiful cochineal red variety consisting of delicate, straight, interlacing fibrous crystals.

Tests Before the blowpipe, cuprite colours the flame emerald green. On charcoal, it fuses to a globule of metallic copper. With the fluxes, it gives the usual copper reactions. It is soluble in acids.

Occurrence Cuprite is an important ore of copper, occurring in the upper oxidized zone of copper lodes as in Cornwall, Chessy (France), Linares (Spain), Bisbee (Arizona), Chile, Peru and Burra Burra (Australia).

Formula *AX*

PERICLASE, native magnesia
C   MgO; iron may replace magnesium in the structure

C   Cubic

Characters and occurrence Periclase occurs either as regular cubes showing a perfect {100} cleavage, or granular. It is white or pale coloured, colourless in thin section, with a high RI ($n = 1.736$–1.745), and isotropic. It may be found disseminated in masses of limestone caught up and contact metamorphosed by the lavas of Monte Somma, Vesuvius and elsewhere. The original limestone contained dolomite, and the magnesium carbonate of the dolomite dissociated upon heating into periclase and carbon dioxide, with the resulting rock being a periclase marble. However, periclase is easily converted by hydration into brucite, $Mg(OH)_2$, so that brucite marbles result, with fibrous brucite crystals often mantling cores of periclase. Periclase occurs with calcite, forsterite and serpentine.

ZINCITE, red oxide of zinc, spartalite
C   ZnO; up to 10% manganese may be present, which gives zincite its red colour, since chemically pure zinc oxide is white; traces of FeO are usually present

Physical properties
CS   Hexagonal

F&H   Crystals uncommon; usually found massive, foliaceous or granular, or as disseminated grains

COL   Deep red but deep yellow in thin scales; zincite weathers to a white crust of $ZnCO_3$

S   Orange–yellow

CL   Prismatic {10$\bar{1}$1} perfect; a basal parting may also be present

F   Subconchoidal

L   Almost adamantine

TR   Translucent to subtranslucent

T   Brittle

HD   4.0

SG   5.68

Optical properties   In thin section, zincite is deep red or yellow and non-pleochroic. It has a very high RI (n_o = 2.013, n_e = 2.020), and is uniaxial +ve.

Tests   Heated in the closed tube, zincite blackens, but on cooling it reverts to its original colour. It dissolves in acids without effervescence. Heated alone before the blowpipe, zincite is infusible. Heated in the reducing flame on charcoal, it yields a white encrustation of zinc oxide, which turns green when moistened with cobalt nitrate and reheated in the oxidizing flame. The presence of manganese is usually indicated by a reddish violet colour of the borax bead in the oxidizing flame.

Occurrence   Zincite occurs in zinc deposits with franklinite, willemite and calcite, as in the Franklin Furnace ore deposit in New Jersey, and is of pyrometasomatic or contact metamorphic origin.

TENORITE, melaconite
C   CuO, with 79.85% copper

Physical properties
F&H   Tenorite occurs mostly as a black powder, or in dull black masses, or in botryoidal concretions; it is sometimes seen as shining and flexible scales
HD   3.0–4.0
SG   6.25 when massive

Tests   Tenorite gives the usual copper reactions with the fluxes, but is infusible alone in the oxidizing flame. It is soluble in acids.

Occurrence   Tenorite occurs in the oxidized zone of weathered copper lodes. It is abundant in the copper mines of the Mississippi Valley and in Tennessee, USA.

Formula A_3X_4

MINIUM, red oxide of lead
C Pb_3O_4

Physical properties
F&H   Usually occurs as a fine powder   **TR**   Opaque
COL   Bright red, scarlet or orange–red   **HD**   2.0–3.0
S   Orange–yellow   **SG**   4.6
L   Faint, greasy or dull

Tests   Before the blowpipe in the reducing flame, minium yields globules of metallic lead. Oxygen, tested by glowing splinter, is given off on heating in a closed tube.

Occurrence   Minium occurs associated with galena and sometimes with cerussite, being derived by the alteration of these minerals.

Uses   The red lead of commerce, which has the same composition as minium, is artificially prepared by heating lead to form the yellow monoxide, and then again subjecting the cooled monoxide to heating, at a lower temperature. Red lead is used in the manufacture of glass and as a pigment.

Formula A_2X_3

CORUNDUM
C Al_2O_3

Physical properties
CS   Trigonal

F&H   Corundum occurs mostly as barrel-shaped or pyramidal crystals, due to the presence of various bipyramids and the basal plane. It may also form steep hexagonal bipyramids. Crystals from alluvial deposits are usually much water-worn and rounded. Corundum also occurs massive and granular.

COL   Commonly grey, greenish or reddish but occasionally colourless. The precious varieties ruby and sapphire are, of course, red and blue respectively. Some of the less well

known precious varieties of corundum, particularly oriental amethyst, oriental emerald and oriental topaz, were given these names because of their colours; namely, purple, green and yellow respectively.

CL   None, but an {001} basal parting is present

F   Conchoidal or uneven

L   Vitreous, but crystal faces are often dull

TR   Transparent to subtransparent

HD   9.0 (second highest on Moh's scale, below diamond)

SG   3.98–4.02

Optical properties
n_0 = 1.768–1.772
n_e = 1.760–1.763
δ = 0.008–0.009
Uniaxial −ve

COL Colourless, but some precious varieties may be pale coloured in thin section

P Non-pleochroic but sapphire may be weakly pleochroic, with e blue and o pale blue

H Well shaped crystals rare, usually occurring as small rounded crystals

CL None, but basal parting present

R High

A Corundum can alter to Al_2SiO_5 polymorphs during metamorphism if silica is present, or to muscovite if water and potassium are available as well as silica

B Low, usually first-order greys or whites

TW Lamellar twinning is common on $\{10\bar{1}1\}$, and simple twins can occur with $\{0001\}$ as the twin plane

DF Corundum can be distinguished from apatite by its higher relief and slightly higher birefringence

Varieties

(1) Corundum gemstones include **ruby**, **sapphire**, **oriental amethyst**, **oriental emerald** and **oriental topaz**, which are varieties of corundum coloured red, blue, purple, green and yellow respectively.

(2) The name corundum includes all ordinary types not of gem quality.

(3) **Emery** is a greyish-black variety of corundum, containing much admixed magnetite and hematite. It is crushed, powdered and sifted, and the powder used for polishing hard surfaces, either as a loose powder or as a surface layer attached to paper (emery paper).

Tests Corundum is not acted on by acids. Its hardness and physical properties are usually distinctive. Finely powdered corundum heated with cobalt nitrate on charcoal assumes a fine blue colour.

Occurrence Corundum occurs in silica-poor rocks such as nepheline syenites, and other undersaturated alkali igneous rocks, as in Canada. It occurs in anorthosites in India. In the Appalachian belt in the eastern USA, corundum occurs as veins and segregations associated with peridotites. It may occur in contact aureoles in thermally altered alumina-rich shales or impure limestones, and in aluminous zenoliths found within basic igneous rocks, where it is found in association with spinel, cordierite and orthopyroxene. Corundum occurs in metamorphosed bauxite deposits and in emery deposits. It was formerly mined at several localities, but now its production is from the USSR (8700 t in 1985), Zambia, India and South Africa, although information is very difficult to obtain. The corundum gemstones occur either as isolated crystals in crystalline limestone, or as rounded pebbles in alluvial deposits derived from such rocks. Important gemstone producers are Burma, Sri Lanka and Thailand. In the past,

Naxos (Greece) and Kayabachi (Turkey) produced corundum. The Peekshill emery of New York is principally a spinel deposit.

Uses Corundum is, with the exception of diamond, the hardest mineral known, and is used as an abrasive. Grinding wheels are made by the incorporation of a binding material with crushed corundum. Artificial corundum is made by fusing bauxite in an electric furnace. Emery is similarly used as an abrasive, and as a refractory material. The coloured varieties of corundum are important gemstones.

HEMATITE, specular iron ore, kidney ore

C Fe_2O_3, containing 70% iron; clay and sand impurities are sometimes present

Physical properties

CS Trigonal

F&H Crystals occur as rhombohedra, often modified and in thin, tabular habits. It occurs in foliaceous aggregates and also may be reniform, granular or amorphous. The internal structure is usually fibrous.

COL Steel grey to iron black. In transmitted light thin platy crystals appear blood red. Earthy aggregates are reddish brown.

S Cherry red to reddish brown

CL poor parallel to $\{10\bar{1}1\}$ and $\{0001\}$

F Subconchoidal or uneven

L Crystals, such as specular iron ore, are highly splendent; fibrous varieties are silky and amorphous varieties dull and earthy

TR Opaque, except in very thin plates

HD 5.5–6.5

SG 4.9–5.3

Optical properties In thin section, hematite is deep red or yellow–brown in colour and uniaxial −ve, with extreme relief ($n_o \approx 3.0$).

Varieties **Specular iron ore** is a variety occurring in rhombohedral crystals, black in colour, and with a metallic splendent lustre. **Micaceous hematite** includes the foliaceous and micaceous forms. **Kidney ore** is a reniform variety with a metallic lustre, especially on the mammillated surfaces, but beneath this surface, kidney ore usually displays a radiating or diverging columnar structure. **Reddle** is the most earthy variety of hematite, red in colour, and used in the manufacture of crayons, for polishing glass, and as a red paint. **Martite** is probably a pseudomorph of hematite after magnetite, and occurs in small black octahedra which give a reddish-brown streak. **Turgite** is a hydrous iron oxide of probable composition $2Fe_2O_3.H_2O$. It resembles limonite but has a red streak, and is distinguished from hematite by containing water. It is often found in association with limonite.

Tests Blowpipe reactions for hematite are the same as those for magnetite. Hematite becomes magnetic on heating. It is soluble in acids.

Occurrence Hematite is found with other iron – titanium oxide minerals in igneous and metamorphic rocks, as well as in sedimentary rocks such as banded ironstones. Hematite occurs in pockets and hollows, replacing limestone, such as in Ulverston, North Lancashire, where hematite occurs in irregular masses in and on the surface of the Carboniferous limestone, probably resulting from the replacement of the limestone by hematite from the overlying ferruginous Triassic sandstones. Metasomatic deposits of a similar nature are found in the Forest of Dean, and Cumberland, and much more important deposits of the same origin occur at Bilbao (Spain), Utah and elsewhere. The greatest hematite deposits in the world in the Lake Superior region are early Precambrian in age and include the Mesabi, Marquette and Menominee iron ranges, which have formed from the alteration and concentration of iron silicates and iron carbonates of sedimentary origin. Another notable iron ore deposit is the Clinton iron bed of Alabama and the neighbouring eastern states. There the hematitic ore is often oolitic in structure, and represents a thin sedimentary bed interstratified with Silurian rocks. The great Brazilian deposits are metamorphosed sedimentary ores. The hematite deposit of Elba is probably of contact metamorphic origin. Residual deposits rich in hematite are known, examples occurring in the Appalachian belt of the eastern USA, and including the far more important deposits in Cuba. The production of iron ores and steel is discussed in Chapter 7.

ILMENITE, menaccanite

C $FeTiO_3$; the Ti : Fe ratio varies due to the intergrowth of magnetite or hematite with the ilmenite. Some MgO is often present.

Physical properties

CS Trigonal	**F** Conchoidal
F&H Often as thin plates or scales, or massive; it may occur as heavy residue grains in sands	**L** Submetallic
	TR Opaque
	HD 5.0–6.0
COL Iron black	**SG** 4.5–5.0
S Black to brownish black	

Optical properties In thin section, ilmenite is opaque, but in reflected light it is often seen to be altered along three directions into a white substance called **leucoxene**, which is related to sphene.

Varieties **Menaccanite** is a variety of ilmenite occurring as a sand at Menaccan, Cornwall. **Iserine**, a variety occurring mostly in the form of loose granules or sand, and frequently in octahedral crystals (which are probably pseudomorphs), is most likely an iron-rich variety of rutile, and **kibdelphane** is a variety rich in titanium.

Tests Heated before the blowpipe, ilmenite is infusible, or nearly so, and gives iron reactions with the fluxes. Heated on charcoal with sodium carbonate, and dissolved in hydrocholic acid, a violet colour is produced on the addition of a small particle of tin. When ilmenite is heated on charcoal with sodium carbonate and the resulting mass dissolved in sulphuric acid and cooled, the solution, when diluted with an equal bulk of water, becomes amber coloured on the addition of a drop of hydrogen peroxide.

Occurrence Ilmenite occurs as an accessory constituent in gabbros and norites, where it may form very large magmatic segregations, as at Taberg and Ekersund (Norway), in Quebec and Ontario, and in the Adirondacks (USA). It also occurs in dyke-like bodies derived from such magmatic segregations, and as a minor constituent of certain copper veins. Many important deposits of ilmenite are of detrital character, particularly beach sands, and such deposits are worked at Travancore (India), in Australia, Florida, Tasmania and elsewhere. Detrital ilmenite is usually altered to leucoxene which is enriched in TiO_2. Magnesium-rich ilmenites occur in kimberlites and also in contact metamorphosed rocks.

ARSENOLITE, white arsenic, arsenious acid
C As_2O_3
CS Cubic

Characters and occurrence Arsenolite is not a mineral of common occurrence in nature, but is sometimes present as fine capillary, white crystals or crusts, resulting from the decomposition of arsenical ores. It is extensively manufactured and, on account of its very poisonous properties, is of great economic importance in the manufacture of insecticides, as explained under arsenic in Chapter 7.

SENARMONTITE
C Sb_2O_3
CS Cubic

Characters and occurrence Senarmontite is a white or greyish mineral occurring in octahedral crystals or crusts, and arising by the oxidation of primary antimony ores. It is a common mineral at Djebel-Haminate Mine, Algeria.

VALENTINITE
C Sb_2O_3
CS Orthorhombic

Characters and occurrence Valentinite is a white, greyish or reddish mineral, possessing a perfect {010} cleavage, and occurring in prismatic crystals or crusts and resulting from the oxidation of various antimony ores, as at the Djebel-Haminate Mine, Algeria.

BRAUNITE
C Mn_2O_3, with 64.3% manganese. Braunite usually contains about 10% silica, so that the mineral is sometimes considered to have the composition $[3MnMnC_3].MnSiO_3$ or Mn_7SiO_{12}.

Physical properties

CS Tetragonal	L Submetallic
F&H Octahedral crystals found, but also massive	TR Opaque
	T Brittle
COL Brownish black	HD 6.0–6.5
S Brownish black	SG 4.75–4.82
F Uneven	

Tests Braunite gives the usual manganese reactions with the fluxes. It gelatinizes when boiled with hydrochloric acid, and does not yield oxygen when heated in the closed tube.

Occurrence Braunite is usually of residual or secondary origin, but may occur as a primary mineral in veins.

Formula AX_2

RUTILE
C TiO_2; Fe^{2+}, Fe^{3+}, Nb and Ta may be present in rutile as major constituents, with lesser amounts of Cr, V and Sn

Physical properties

CS Tetragonal

F&H Crystals as elongate or acicular tetragonal prisms terminated by pyramids. Groups of crystals are often found enclosed within other minerals, expecially quartz. The crystals often show a longitudinal striation.

TW Crystals are twinned on {011}, giving geniculate or knee-shaped twins. Repeated twinning on {011} leads to the production of wheel-shaped multiple twins. Glide twins on {011} and {092} also occur.

COL Reddish brown, red, yellow or black

S Pale brown

CL Good prismatic on {110} poor on {100}; a parting on {011} may be present

F Subconchoidal or uneven

L Metallic to adamantine

TR Subtransparent to opaque

T Brittle

HD 6.0–6.5

SG 4.2–5.6

Optical properties

n_o = 2.605–2.616
n_e = 2.890–2.903
δ = 0.285–0.296
Uniaxial +ve

COL Deep reddish brown to yellowish brown

P Weak in brownish reds, yellows and occasionally greens

H Euhedral crystals common, often quite small, and, when enclosed in other minerals, acicular

CL Prismatic cleavages {110} distinct with {100} poor

R Extreme

A Rutile rarely alters

B Extreme with very high orders of colours shown

DF Extreme relief, birefringence, sign and reddish brown colour are diagnostic properties, and help to distinguish rutile from anatase, cassiterite and brookite

Tests Heated alone, rutile is infusible. The microcosmic salt bead in the oxidizing flame is yellow when hot and colourless when cold. In the reducing flame, the same bead is yellow when hot and violet when cold. Rutile reacts to the special tests for titanium described in Appendix A.

Occurrence Rutile occurs as an accessory constituent of igneous rocks of many kinds, such as granites and diorites and their metamorphic derivatives, such as gneisses and amphibolites. It forms fine needles, **clayslate needles**, in some slates and phyllites, where it results from the decay of titanium-bearing micas. It can be produced from ilmenite on wall-rock alteration by hydrothermal solutions. Decayed biotite often shows a lattice-like collection of rutile needles, called **saggenite**, the lattice arrangement arising by twinning. Rutile forms acicular needles within quartz and feldspar. The economically important rutile deposits are: first, segregations in igneous rocks such as syenite, gabbro and anorthosite, as in Virginia, Canada and Norway; secondly, a peculiar type of dyke, probably of pegmatitic character, composed of rutile, apatite and ilmenite, known as nelsonite and occurring associated with the rutile segregations of Virginia; and thirdly, as an important constituent of beach sands resulting from the denudation of rutile-bearing rocks, as in Australia, Florida and India.

Uses   Rutile is used as a source of titanium.

PYROLUSITE
C MnO_2, with 63% manganese

Physical properties

CS   Orthorhombic; usually pseudo-morphous

F&H   Occurs as pseudomorphs after manganite, polianite, etc. It usually occurs massive or reniform and often with a fibrous or radiate structure.

COL   Iron grey or dark steel grey

S   Black or bluish black

F   Hackly

L   Submetallic to dull

TR   Opaque

T   Brittle

HD   2.0–2.5 (it may soil the fingers)

SG   4.8

Tests   The borax bead is amethyst in the oxidizing flame and colourless in reducing flame. The microcosmic salt bead is red–violet in the oxidizing flame. The sodium carbonate bead is opaque blue–green in colour. Pyrolusite yields oxygen when heated in the closed tube, and it is soluble in hydrochloric acid, with the evolution of chlorine gas.

Occurrence   Pyrolusite occurs in secondary manganese deposits, where the manganese oxides aggregate together as nodules in the residual clay which forms on the outcrop of the weathered rock. Pyrolusite may form from alteration of rhodochrosite, $MnCO_3$, or rhodonite, $(Mn,Ca,Fe)_2$-Si_2O_6. Such deposits are known in India, Brazil and Arkansas (USA). Precipitation of manganese oxides in lakes forms bog-iron ores, as in Sweden, Spain and the USA. Pyrolusite is an abundant manganese ore.

POLIANITE
C MnO_2

Characters and occurrence

CS   Tetragonal

HD   6.0–6.5 (much harder than pyrolusite)

SG   5.0

Polianite resembles pyrolusite in appearance and colour, sometimes showing minute tetragonal crystals, isomorphous with cassiterite, SnO_2. The occurrence is as for pyrolusite.

WAD
C A complex aggregate of impure Fe- and Mn-oxides

Physical properties

F&H Amorphous, earthy, reniform, aborescent, encrusting, or as stains and dendrites; it is often loosely aggregated

COL Dull black, bluish, lead grey or brownish black

L Dull

TR Opaque

HD 5.0–6.0, but aggregates are usually soft

SG 3.0–4.28

Varieties **Earthy cobalt** or **asbolan** is a variety of wad containing up to 40% of cobalt oxide, and giving the blue borax bead due to cobalt. **Lampadite** is another variety which may contain as much as 18% of copper oxide. **Psilomelane** is similar to wad, but is a hydrated mineral.

Tests Wad gives the usual manganese reactions with the fluxes. It yields water in the closed tube.

Occurrence Wad results from the decomposition of other manganese minerals, and generally occurs in damp, low-lying places.

Uses Wad is not as valuable as pyrolusite and psilomelane, but is sometimes used in the manufacture of chlorine, and also serves for umber paint.

ASBOLITE, asbolan, earthy cobalt, black oxide of cobalt
C Asbolite is essentially wad with up to 40% cobalt oxide; cobalt sulphide and oxides of copper, iron and nickel may also be present

Physical properties

F&H Amorphous, earthy

COL Black or blue–black

S Black, shining and resinous

Tests Heated in the closed tube, asbolite yields water. The borax bead is blue, due to cobalt; and the sodium carbonate bead is opaque green, due to manganese. Asbolite is soluble in hydrochloric acid.

Occurrence Asbolite occurs with the major ores of cobalt in the oxidation zone, and with manganese ores, as at Mine La Motte, Missouri, USA. An important occurrence is in New Caledonia, where the asbolite deposits represent the superficial alteration of a cobaltiferous serpentine, analogous to the garnierite deposits formed from a nickeliferous serpentine.

CASSITERITE, tinstone
C SnO_2, with 78.6% tin

Physical properties

C   Tetragonal

F&H   Crystals occur as tetragonal prisms terminated by bipyramids. It also occurs massive, fibrous or as disseminated small grains. Water-worn grains occur in alluvial deposits.

TW   Twinning common on {011} gives knee-shaped twins

COL   Black or brown, rarely yellow or colourless

S   White or pale grey to brownish

CL   Poor prismatic cleavages {100} and {110}; there is also a {111} parting

F   Subconchoidal or uneven

L   Adamantine, brilliant

TR   Pale-coloured crystals are transparent, but dark-coloured crystals are opaque

T   Brittle

HD   6.0–7.0

SG   6.8–7.1

Optical properties

$n_0 = 1.990–2.010$
$n_e = 2.091–2.100$
$\delta = 0.100–0.090$
Uniaxial +ve

COL   Colourless, red brown

P   Variable, but present in dark-coloured varieties with o pale colours and e dark yellow, red or brown

H   Crystals are stubby prisms, rarely acicular, and usually euhedral to subhedral

CL   {100}, {110} prismatic cleavages are poor

R   Extremely high

B   Very high, but colour orders are often masked by body colour of mineral

TW   Simple and repeated twinning common on {011}

Z   Colour banding is very common

Varieties   **Wood tin** occurs in reniform masses and has a structure which is compact and fibrous internally, and exhibits concentric bands resembling wood. **Toad's eye tin** shows the characters of wood tin, but on a smaller scale. **Stream tin** is rolled and worn cassiterite, resulting from the erosion of tin veins or rocks containing the ore. It occurs in the beds of streams and in the alluvial deposits which border them. Much of the tin ore sent into the market is derived from this source.

Tests   Heated alone before the blowpipe, cassiterite is infusible. Heated with sodium carbonate and charcoal on charcoal, it gives a globule of metallic tin. The sublimate resulting from heating on charcoal with sodium carbonate, when moistened with cobalt nitrate anu strongly reheated, assumes a blue–green colour.

Occurrence Cassiterite is found mainly in acid igneous rocks with wolframite, tourmaline, topaz, arsenopyrite, molybdenite, pyrrhotite and bismuth-bearing minerals in high-temperature hydrothermal veins, pegmatites, greisens, stockworks and disseminations associated with acid igneous rocks, as in Cornwall, Saxony and Tasmania. Tin–silver veins of a rather different type are important in Bolivia, where they are associated with porphyries of hypabyssal or volcanic origin. Fully one-half of the world's supply of tin is obtained from placer deposits derived from the degradation of tin veins, this type of deposit occurring in Malaysia, Indonesia and Thailand. An interesting eluvial placer occurs immediately adjacent to the primary tin veins of Mount Bischoff, Tasmania. Cassiterite may occur in gossans over stanniferous sulphide deposits. Wood tin is found in the secondary oxidation zone.

ANATASE, octahedrite

C TiO_2; anatase is a low-temperature mineral polymorphous with brookite

Physical properties

CS Tetragonal

F&H Crystals are either slender, acute bipyramids (hence the name octahedrite) or tabular and flat

COL Brown, indigo blue or black

S Colourless

CL Perfect {001} basal and {111} pyramidal

L Adamantine

TR Transparent to opaque

T Brittle

HD 5.5–6.0

SG 3.82–3.95

Optical properties

$n_o = 2.561$
$n_e = 2.488$
$\delta = 0.073$
Uniaxial −ve

COL Colourless, pale yellow brown, deep red brown, green, dark blue

P Weak and variable with e > o, and sometimes o > e

H Euhedral to anhedral grains

CL Perfect pyramidal and basal cleavages

R Extremely high

A Anatase is an alteration product from ilmenite and other Ti minerals; it alters or inverts to rutile

B High in prismatic sections

Tests As for rutile. Form and blue colour, when present, are distinctive.

Occurrence Anatase is the low-temperature TiO_2 polymorph, occurring in veins and cavities of hydrothermal origin in granite, pegmatite, schist and gneiss. Anatase occurs associated with clay minerals in sedimentary rocks. It is a common alteration product of sphene, ilmenite and other titanium minerals.

BROOKITE
C TiO_2; brookite is polymorphous with anatase

Physical properties
CS Orthorhombic
F&H Thin, bladed crystals, tabular on {010}
COL Brownish, reddish iron black
S Colourless
CL Poor on {120}

L Adamantine to submetallic
TR Transparent to opaque
T Brittle
HD 5.5–6.0
SG 4.12

Optical properties
$n_\alpha = 2.583$
$n_\beta = 2.584$
$n_\gamma = 3.780$
$\delta = 0.117$
$2V = 0–30°$ +ve
OAP is parallel to (001)

COL Pale colours, yellow brown, deep brown
P Weak with α pale-coloured, and β and γ darker yellows or browns
H Small tabular crystals common
CL Poor indistinct prismatic cleavage occurs
R Extremely high
A None, but brookite may invert to rutile

B High, but brookite has very strong dispersion
IF Figure may be difficult to obtain due to small crystal size and high dispersion (crossed axial plane dispersion; that is, orientation of optic axial plane changes from red light (001) to blue light (100) – in green light brookite is approximately uniaxial).

Tests As for rutile

Occurrence Brookite occurs as an alteration product of other titanium-bearing minerals, as in the rotten dolerites of Tremadoc, North Wales, and in other decomposed rocks. It may be of contact metamorphic origin, as in Arkansas, and it occurs in mineral druses in the Alps.

TELLURITE
C TeO_2

Characters and occurrence
CS Orthorhombic
HD 2.0
SG 5.83–5.90
Tellurite occurs in small whitish and yellowish orthorhombic prisms, spherical masses and encrustations, resulting from the oxidation of tellurium or tellurides in the upper parts of veins, and has been recorded from Rumania, Colorado, etc.

CERVANTITE
C Sb_2O_4

Characters and occurrence
CS Orthorhombic
HD 4.0–5.0
SG 6.5–6.6
Cervantite occurs as acicular crystals or powdery crusts of a white or yellow colour, resulting from the oxidation of primary antimony ores.

BISMITE, bismuth ochre
C Bi_2O_3

Characters and occurrence A yellow earthy, powdery mineral, bismite is usually impure and hydrated, occurring as an alteration product of bismuth and bismuthinite.

PITCHBLENDE
C UO_2; oxidized in part to U_3O_8 and containing small amounts of Th, Zr, Pb, etc., and also some gases such as He, Ar, N, etc.
The specific names, **cleveite** and **broggerite**, have been given to pitchblendes of various compositions.

Physical properties
CS Cubic
F&H Usually occurs massive, botryoidal or as grains; many examples of pitchblende are as coloform gels
COL Velvet black, greyish or brownish
S Black, often with a brownish or greenish tinge

L Submetallic, greasy, dull
TR Opaque
HD 5.5
SG 6.4 (when massive) to 9.7 (when crystallized)

Tests Heated before the blowpipe alone, pitchblende is infusible. When heated with sodium carbonate, it is not reduced, but usually gives an encrustation of lead oxide, and an arsenical odour may be produced. With borax or microcosmic salt, pitchblende gives a yellow bead of uranium in the oxidizing flame, and a green bead of uranium in the reducing flame. It dissolves slowly in nitric acid when powdered.

Occurrence Uranium oxides are found in high-temperature pegmatitic to low-temperature hydrothermal vein and replacement deposits. There is an association with Ni–Co–Ag–Bi mineralization, with acid igneous rocks and with organic material in sedimentary rocks. Detrital uraninite is found in placer deposits. It occurs at Joachimsthal (Jachymov) in Czechoslovakia, Johanngeogenstadt in Saxony, in Cornwall, at Great Bear Lake in Canada, and in Katanga in Zaire.

Uses Pitchblende, and hydrous materials derived from it, are the chief sources of radium.

Thorianite (ThO_2), which also contains uranium as UO_2, is a valuable source of thorium, and is found in beach deposits at Madagascar and Sri Lanka.

Formula A_mX_p

TUNGSTITE, tungstic ochre
C WO_3; but some material has the composition $WO_3.H_2O$

Characters and occurrence Tungstite is a powdery and earthy mineral, of a bright yellow or yellowish green colour, resulting from the alteration of ores of tungsten. Heated before the blowpipe alone, it is infusible, but becomes black in the inner flame. The microcosmic salt bead is colourless or yellow in the oxidizing flame, and becomes violet (when cold) in the reducing flame. Tungstite is soluble in alkaline solutions, but not in acids.

8.4.2 Hydroxides

Formula AX_2

BRUCITE
C $Mg(OH)_2$

Physical properties

CS Trigonal

F&H Crystals, when present, have a prismatic and broad, tabular habit. Usually, brucite is found massive and foliaceous, sometimes fibrous, the laminae and fibres being easily separable.

COL White, often bluish, greyish and greenish

CL Perfect basal parallel to {0001}

L Pearly on cleavage planes, elsewhere waxy or vitreous; fibrous types silky

TR Translucent to subtranslucent

T Laminae are flexible, whereas fibres are elastic

HD 2.5

SG 2.39

Optical properties

n_o = 1.559–1.590

n_e = 1.580–1.600

δ = 0.02–0.01

Uniaxial +ve

COL Colourless

H Usually as fine foliaceous aggregates, as whorls and as fibres

CL Perfect {0001} basal

R Low to moderate

A Commonly to hydromagnesite, $MgCO_3.Mg(OH)_2.3H_2O$; less commonly to serpentine or deweylite

B Low to anomalous, with blue, red or brown colours showing, especially on fibrous varieties

Varieties Fibrous brucite is called **nemalite**.

Tests Heated in the closed tube, brucite gives off water and becomes opaque and friable. Before the blowpipe, it becomes brilliantly incandescent, and yields a pink mass when moistened with cobalt nitrate and strongly reheated. Brucite is soluble in hydrochloric acid, which distinguishes it from talc and gypsum. It is distinguished from heulandite and stilbite by its infusibility.

Occurrence Brucite is found in contact metamorphosed impure limestones, called pencatites or predazzites, typical examples coming from the Tyrol, Skye, and Assynt in Sutherland. It may occur in low-temperature hydrothermal veins associated with serpentinite, as in Unst, Shetland, and in chlorite schists. Brucite is used for the production of magnesium and refractories.

PSILOMELANE

C Possibly $Ba_3(Mn^{2+}Mn^{4+}O_{16}(OH)_4$; varying amounts of potassium may be present

Physical properties

CS Monoclinic

F&H Massive, botryoidal, reniform and stalactitic

COL Iron black to dark steel grey

S Brownish black and shining

L Submetallic; the mineral looks as if an attempt has been made to polish it

TR Opaque

HD 5.0–6.0

SG 3.7–4.7

Tests Heated in the closed tube, psilomelane gives water. With borax and microcosmic salt it gives the usual amethyst-coloured bead. The sodium carbonate bead is opaque bluish green. Psilomelane is soluble in hydrochloric acid, with the evolution of chlorine, the solution often giving a precipitate of barium sulphate on the addition of barium chloride.

Occurrence Psilomelane is found in secondary manganese deposits, where it was probably a colloidal precipitate.

Use Along with pyrolusite, psilomelane is an important ore of manganese.

Formula AX_3

SASSOLITE, native boric acid

C H_3BO_3 (boric acid)

Physical properties

CS Triclinic

F&H Prismatic crystals are rare, usually occurring as small glistening scales associated with sulphur

COL White, greyish, sometimes yellow if sulphur is present

L Pearly

TR Transparent to translucent

T Sectile and flexible

FEEL Smooth and oily

TASTE Acidulous, slightly saline and bitter

HD 1.0

SG 1.48

Tests Sassolite fuses easily in the blowpipe flame, tingeing the flame green. It is soluble in water and in alcohol. When dissolved in alcohol, it colours flame green by the presence of boron.

Occurrence Sassolite occurs with sulphur in the crater of Vulcano, Lipari. It is also found around fumaroles (small vents or outlets of sulphurous emanations), and in the steam or vapours which rise from the bottom of the small hot lakes or lagoons of Tuscany. It is condensed in the water, and afterwards separates out in large flakes, which contain about 50% boric acid. Sassolite also occurs in the natural waters of Clear Lake, California.

GIBBSITE, hydrargillite
C Al(OH)$_3$

Physical properties
CS Monoclinic
F&H Crystals rare, usually occurring
as concretions

COL White
HD 3.0
SG 2.35

Tests Gibbsite gives water when heated in the closed tube. If it is heated with cobalt nitrate a blue residue results.

Occurrence Gibbsite occurs in deposits of bauxite, and as an alteration product from aluminosilicates.

BAUXITE
C A mixture of the aluminium hydroxides, diaspore, boehmite and gibbsite, in different amounts, together with impurities of iron oxide, phosphorus compounds and TiO$_2$, the latter sometimes amounting to 4%.

Physical properties
F&H Amorphous in earthy granular
or pisolitic masses

COL Dirty white, greyish, brown,
yellow, or reddish brown

Tests General characters of bauxite are distinctive. It gives the aluminium reaction when heated with cobalt nitrate. It does not give a silica skeleton in the microcosmic salt bead test.

Occurrence Bauxite results from the decay and weathering of aluminous rocks, often igneous but not necessarily so, under tropical conditions. It may form residual deposits replacing the original rock, or it may be transported from its place of origin and form deposits elsewhere. It occurs in pockets of Cretaceous limestone in France and must therefore be the result of pre-Tertiary tropical weathering.

Uses The principal use of bauxite is for the manufacture of aluminium, but considerable quantities are also used as abrasives and in the manufacture of aluminium compounds. Lower grades of bauxite are used as refractories, as refractory bricks and for furnace and converter linings. Commercially speaking, there are two kinds of bauxite, red and white. For chemical purposes the white bauxite containing only a trace of iron is used, but for the manufacture of metallic aluminium, iron is not harmful, although the presence of more than 3% of silica or of titania is objectionable. Calcined bauxite is used to produce non-skid aggregate, which is laid on top of normal road surfaces in areas of exceptionally high traffic densities, such as the Heathrow link from the M4 motorway

8.4.3 Multiple oxides

Formula ABX_2

The formulae of boehmite or goethite, etc. can be written as $HAlO_2$ and $HFeO_2$ and can be included here, although they are normally written $(Fe,Mn,Al)O(OH)$.

BOEHMITE

C γ–AlO(OH); it forms a series with diaspore but has a different structure

Characters and occurrence

CS Orthorhombic

HD 3.5–4.0

SG 3.01–3.06

Boehmite occurs in tiny flakes or aggregates which are white in colour. Boehmite yields water in the closed tube test and, when heated on charcoal moistened with cobalt nitrate and strongly reheated, gives a blue infused residue. Boehmite is an important constituent of most bauxites with diaspore.

DIASPORE

C α–AlO(OH); it forms a series with boehmite

Characters and occurrence

CS Orthorhombic

HD 6.5–7.0

SG 3.3–3.5

Diaspore occurs as prismatic crystals, foliaceous and scaly in habit and white in colour. It is found with corundum and emery, and probably results from the alteration of these. It is a basic constituent of bauxite deposits.

MARGANITE

C MnO(OH), with 62.5% manganese

Physical properties

CS Monoclinic, pseudo-orthorhombic	**S** Reddish brown to nearly black
F&H Prismatic crystals, often grouped in bundles and striated longitudinally; it is also found in columnar habits	**F** Uneven
	L Submetallic
	TR Opaque
	HD 4.0
COL Iron black or dark steel grey	**SG** 4.2–4.4

Tests Heated in the closed tube, manganite gives water. It gives the usual manganese reactions with the fluxes (see Appendix A).

Occurrence In veins, and associated with other manganese oxides.

GOETHITE

C α–FeO(OH), with 62.9% iron. Goethite forms a series with lepidocrocite, or γ–FeO(OH), which has very similar properties to goethite. Their structures are analogous to those of diaspore and boehmite.

Physical properties

CS Orthorhombic	**S** Brownish yellow or ochre yellow
F&H Crystals occur as flattened prisms which are longitudinally striated, and assume a tabular habit; it is also found massive, stalactitic and fibrous	**CL** Good {010} prismatic
	L Adamantine in crystals
	TR Subtranslucent to opaque
	HD 5.0–5.5
COL Brownish black, yellowish or reddish	**SG** 4.1–4.3

Optical properties

In rare basal sections goethite is blood red, with extreme relief ($n_\beta = 2.4$) extreme dispersion. It is biaxial negative, with a low $2V$.

Tests As for limonite, but distinguished from limonite by being crystalline.

Occurrence Goethite forms from the surface weathering of iron-rich materials, especially sulphides and oxides. It is stable in a humid environment at normal temperatures and pressures. In laterites, goethite is associated with hematite, limonite, and aluminium and manganese hydroxides. It is precipitated from ground waters in springs and bogs and is an important constituent of bog iron ores. Low-temperature hydrothermal deposits can yield goethite as a primary mineral. In these deposits it is associated with quartz (amethyst variety), siderite and calcite. Some important localities are Lostwithiel and Botallack in Cornwall, Altenberg in Saxony, Lake Onega in Russia, and Jackson Iron mine, Michigan, USA.

LIMONITE, brown hematite

C $FeO(OH).nH_2O$, with clay minerals and other impurities

Physical properties

CS   None (amorphous)

F&H   Limonite occurs in mammillated or stalactitic habits, with a radiating fibrous structure resembling that of hematite; it also occurs as a dull, earthy mineral and in concretions

COL   Shades of brown on fracture surfaces but yellow or brownish yellow when earthy. The external surfaces of mammillated and stalactitic habits often exhibit a blackish, glazed or shiny coating

S   Yellowish, yellow brown

L   Submetallic to silky in some varieties, but usually dull

HD   4.0–5.5

SG   2.7–4.3

Optical properties   In some varieties limonite in thin section is yellowish, non-pleochroic and with a very high relief ($n \approx 2.0$).

Varieties   **Bog iron ore** is a loose, porous earthy form of limonite, found in swampy and low-lying ground, often impregnating and enveloping fragments of wood, leaves and mosses. **Pea iron ore** is a variety of limonite having a pisolitic structure. The **ochres** are brown or yellow earthy forms of limonite used for paint.

Tests   Heated in the closed tube, limonite gives water. It gives the iron reactions with the fluxes and gives a magnetic residue on heating. Limonite is soluble in hydrochloric acid. It is distinguished mainly by its streak and habit.

Occurrence   Limonite is always secondary resulting from the alteration of other iron minerals, and is always intimately associated with goethite, and frequently with hematite and manganese hydroxides. Weathered residual deposits consisting largely of ferric hydroxide, plus clay and other impurities, may be formed by degradation of a highly ferruginous rock. These iron caps are common over the outcrops of pyrites and iron oxide deposits, and the lateritic iron ores are formed in an analogous manner. Examples of important deposits of this character are the Bilbao (Spain) and the Cuba deposits. The bog iron ores are formed on the floors of some lakes, as in Sweden, where a layer 150 mm thick accumulated in 26 years. The deposition of limonite in lakes or spring deposits is the result of organic or inorganic colloidal deposition. Some limonite beds are true chemical precipitates, while others result from the alteration of chalybite. Limonite enters into the ore of some of the Jurassic iron-ore fields of England, and is the dominant iron mineral in the great 'minette' iron ores of Alsace-Lorraine.

Formula AB_2X_4

Spinel group

The spinel group is represented by at least 12 minerals, all conforming to the general formula $A^{2+}B_2^{3+}O_4$. Three series occur, each being defined by the dominant trivalent cation. These are:

spinel series, in which Al^{3+} is the dominant trivalent ion;
magnetite series, in which Fe^{3+} is the dominant trivalent ion;
chromite series, in which Cr^{3+} is the dominant trivalent ion.

SPINEL SERIES

C	spinel ·	$MgAl_2O_4$	**RI**	$n = 1.719$
	hercynite	$FeAl_2O_4$		$n = 1.835$
	gahnite	$ZnAl_2O_4$		$n = 1.805$
	galaxite	$MnAl_2O_4$		$n = 1.920$

Complete solid solution series exist between spinel and hercynite, gahnite, galaxite and **magnesiochromite** ($MgCr_2O_4$). Spinel containing appreciable Fe^{2+} (Mg : Fe = 1–3) is called **pleonast** or **ceylonite**, and if it contains Fe^{2+} and Cr^{3+} it is called **mitchellite**.

Hercynite forms complete solid solution with chromite and spinel but only limited solid solution with magnetite. **Picotite** is Cr-rich hercynite. Ti^{4+} is a common element in many spinels.

Physical properties

CS Cubic
F&H Crystals as octahedra, rarely rhombic dodecahedra
TW Common on {111}, the spinel law
COL Variable, with colourless, red and green varieties. Hercynite is dark green to black, gahnite is dark blue green, brown or yellow, and galaxite is deep red brown to black

CL None, but a parting exists on {111}
F Conchoidal
L Vitreous to submetallic (for dark-coloured spinels)
HD 7.5–8.0
SG 3.5–4.1

Optical properties

$n = 1.7–1.9$ (see beginning)
COL Spinel is colourless to deep colours (red, blue, green, etc). Hercynite is deep green, gahnite is pale to deep blue green or brownish, and galaxite is deep red brown. Pleonast is deep green.

Isotropic
H Spinels usually occur as independent euhedral or subhedral grains, with square or triangular cross sections
CL None
R Very high
A Spinels are stable but can alter to serpentine, talc, etc.

Varieties **Ruby-spinel** or magnesia spinel is the clear red variety. **Spinel-ruby**, **balas-ruby** and **rubicelle**, are deep red, rose red and yellow varieties, used as gemstones.

Tests Spinels are infusible. If black in colour, they resemble magnetite, but they are not magnetic. If red, they resemble garnet, but do not fuse. If brown, they resemble zircon, but are harder. The form, hardness and infusibility are characteristic of the spinel series.

Occurrence Common spinel occurs in schists and gneisses with sillimanite, talc, garnet, chondrodite, phlogopite, corundum and cordierite. In contact or regionally metamorphosed limestones spinel occurs with chondrodite, forsterite and orthopyroxene. Pleonast is found in emery deposits with magnetite, corundum and hematite. Spinel is incompatible with quartz and occurs with it only when surrounded by a reaction rim. Spinel is developed in aluminous xenoliths in mafic and silica-poor igneous rocks. Spinel occurs also in alluvial deposits resulting from the degradation of the parent rocks. The gem varieties come from Sri Lanka, Burma, Thailand and Afghanistan.

Hercynite forms in metamorphosed iron-rich argillaceous sediments (such as laterites), in granitic granulites where it can exist in the presence of quartz. Gahnite is found in granite pegmatites, and metasomatic zinc ores. Galaxite occurs in manganese vein deposits with rhodenite, etc.

MAGNETITE SERIES

C		
	magnetite	$Fe^{2+}Fe_2^{3+}O_4$
	magnesioferrite	$MgFe_2^{3+}O_4$
	franklinite	$ZnFe_2^{3+}O_4$
	jacobsite	$MnFe_2^{3+}O_4$
	trevorite	$NiFe_2^{3+}O_4$

Small amounts of other elements may be present. Magnetite usually contains Ti and Mg, with minor amounts of Mn, Ca and Ni. Franklinite usually contains appreciable amounts of Mn and Fe^{2+}

Physical properties

CS Cubic

F&H Octahedra common, rounded at the edges in franklinite. Combinations of {111} and {110} may occur. It is also found granular and massive

COL Black

S Black

CL Parting parallel to {111} common, especially in magnetite

F Subconchoidal

L Metallic to submetallic

TR Opaque

T Brittle

Magnetism Magnetite, magnesioferrite and trevorite are strongly magnetic; franklinite and jacobsite less so

HD 5.5–6.5

SG 5.2 (magnetite) to 4.6 (magnesioferrite)

Optical properties Most magnetites appear grey in reflected light in thir. section, and may be euhedral (square grains) or anhedral.

Tests Heated before the blowpipe, magnetite is very difficult to fuse. With borax in the oxidizing flame, it gives a bead which is yellow when hot and colourless when cold. If much material is added to the bead, the bead is red when hot and yellow when cold. In the reducing flame, the borax bead is bottle green. Magnetite is soluble in hydrochloric acid. It is strongly magnetic and often exhibits polarity.

Franklinite has a borax bead which is amethyst coloured in the oxidizing flame, due to manganese, and bottle green in the reducing flame, due to iron. The sodium carbonate bead is bluish green. When heated with cobalt nitrate on charcoal, a greenish mass forms due to the presence of zinc.

Occurrence Magnetite occurs as a primary constituent of most igneous rocks, being titaniferous in mafic rocks. Large deposits are considered to be the result of magmatic segregation, as in the Urals, and Kiruna and Gällivaare in northern Sweden. In this last case, however, it is suggested that the magnetite has moved after its segregation. Workable magnetite deposits also occur as lenses in crystalline schists, as in the Adirondack belt in the eastern USA. Magnetite occurs in metamorphic rocks derived from ferruginous sediments, and is associated with quartz and chlorite. Magnetite may form by metasomatic replacement of limestone and in contact metamorphic deposits, which are common but rarely of commercial value. Examples include the 'skarn' ores of Scandinavia and the contact skarns of the Beinn an' Dubhaich granite in Isle of Skye (Scotland), and certain deposits in the western USA and the Urals. Magnetite may be deposited by fumerolic gases and may appear in high-temperature hydrothermal sulphide veins. Magnetite is also a constituent of many veins, and is found in residual clays, and in placer deposits, the 'black sands', formed by the degradation of earlier deposits.

Franklinite occurs at Franklin Furnace, New Jersey, in association with willemite, zincite and calcite, an average ore being 50% franklinite, 25% willemite, 5% zincite and 20% calcite. The zinc minerals occur as rounded grains and lenses in a crystalline limestone and are considered to be the result of pyrometasomatism, but may possibly result from the contact metamorphism of previously existing hydrothermal zinc deposits.

Uses Magnetite is one of the most valuable ores of iron.

CHROMITE SERIES

C	chromite	$Fe^{2+}Cr_2O_4$
	magnesiochromite	$MgCr_2O_4$

Physical properties

CS Cubic

F&H Occurs as octahedra, but is commonly found massive with a granular or compact structure

COL Iron black or brownish black

S Brown

CL None

F Uneven, but sometimes flat

L Metallic to submetallic

TR Opaque; may be translucent on thin edges

T Brittle

Magnetism Weak

HD 5.5 (chromite) to 7.0 (magnesiochromite)

SG 5.1 (chromite) to 4.2 (magnesiochromite)

Tests Chromite is infusible in the oxidizing flame, but, in the reducing flame, it becomes slightly rounded on the edges of splinters, which are magnetic on cooling. Heated with sodium carbonate on charcoal, chromite is reduced to magnetic oxide. The borax and microcosmic salt beads are a beautiful chrome green, this colour being rendered more intense if the mineral is first fused on charcoal with metallic tin.

Occurrence Chromite occurs as a primary mineral of ultramafic plutonic igneous rocks and serpentinites derived from them. Most peridotites and dunites contain chromite (usually iron-rich magnesiochromite), often in layers or bands as the result of double diffusive convection. The chromite usually occurs as small grains but, by the segregation of these grains, ore bodies consisting of a peridotite extremely rich in chromite may be formed, such as those of Norway, Rhodesia, Smyrna, and New Caledonia. Being very obdurate, chromite occurs in detrital deposits, and is a well known mineral in meteorites.

Uses Chromite is the essential ore of chromium.

HAUSMANNITE

C $Mn^{2+}Mn^{4+}O_4$, with 72% manganese

Physical properties

CS Tetragonal

F&H Pyramidal crystals are common, often twinned; also granular and massive

COL Brownish black

S Chestnut brown

F Uneven

L Submetallic

TR Opaque; thin splinters are deep brown

HD 5.0–5.5

SG 4.86

Tests Hausmannite gives usual manganese reactions with the fluxes. Chlorine is released when it is dissolved in hydrochloric acid. Oxygen is not given off when hausmannite is heated.

Occurrence Hausmannite is a primary manganese mineral occurring in high-temperature hydrothermal veins connected with acid igneous rocks, and in contact zones with other manganese minerals.

CHRYSOBERYL, alexandrite
C $BeAl_2O_4$

Physical properties
CS Orthorhombic
F&H Prismatic crystals common, often with vertical striations
TW Repeated twinning on {031} produces stellate and six-sided trillings; simple contact twins on {031} yield kite-shaped or heart-shaped crystals
COL Shades of green

CL Distinct prismatic on {011}; indistinct on {010} and poor on {100}
F Conchoidal and uneven
L Vitreous
TR Transparent to translucent
HD 8.5
SG 3.68–3.75

Optical properties
$n_\alpha = 1.732–1.747$
$n_\beta = 1.734–1.745$
$n_\gamma = 1.741–1.758$
$\delta = 0.008–0.011$
$2V = 70°–10° +ve$
COL Colourless, but thick fragments are greenish and pleochroic
P Thick fragments show α purple, β orange, and γ green
H Short prismatic crystals common
CL Distinct on {011}; also {010} and {100}
R High

A Resistant mineral but may eventually break down to clays
B Low, with first order whites or yellows
IF A single isogyre is required to obtain $2V$ sign (+ve) and size (very variable)

Varieties **Alexandrite** is a greenish variety, reddish in artificial light, and is used as a gemstone.

Tests Recognized by its physical properties, especially hardness. When chrysoberyl is heated with cobalt nitrate on charcoal, a blue mass is formed, indicating aluminium.

Occurrence Chrysoberyl is a fairly common mineral in complex granite pegmatites in association with tourmaline, topaz, columbite, apatite and other beryllium-bearing minerals such as beryl. It may occur in contact metamorphosed dolomitic limestones, and in some mica schists. Chrysoberyl occurs in alluvial placer deposits with many other gemstones. Gem varieties come from the Urals, Sri Lanka, Madagascar, etc.

Formula $A_mB_nX_p$ with $(m + n) : p \approx 2 : 3$

PYROCHLORE, ellsworthite, hatchettolite
C $(Na,Ca,U)_2(Nb,Ta,Ti)_2O_6(OH,F)$; REEs may also be present

Physical properties

CS Cubic	**CL** Distinct octahedral {111}
F&H Octahedral crystals common, also massive and granular	**TR** Translucent to opaque
	HD 5.0-5.5
COL Brown to black; Nb-rich types are lighter coloured	**SG** 4.31–4.48

Optical properties   In thin section, grains are brownish in colour where transparent.

Occurrence   Pyrochlore is found in alkaline igneous rocks and related pegmatites, and also in carbonatites. It is worked in deposits in Canada, Brazil and elsewhere.

PEROVSKITE
C $CaTiO_3$; REEs may replace Ca, and Nb may replace Ti

Physical properties

CS Monoclinic or orthorhombic; pseudocubic	**CL** Cubic {100} poor
F&H Tiny cubic or octahedral crystals	**L** Adamantine to metallic
COL Yellow, reddish brown, black	**HD** 5.5
	SG 3.98–4.26

Optical properties

$n = 2.30$–2.38	Isotropic
COL Colourless, pale yellow, pale reddish brown to dark brown	**A** Perovskite may be an alteration product from sphene or ilmenite; it may alter to leucoxene
H Small cubic or octahedral crystals	**TW** Lamellar twinning is common on {111}
CL See above	**Z** Colour zoning occasionally found
R Extremely high	

Occurrence   Perovskite is a common accessory mineral in silica-deficient igneous rocks. In highly alkaline rocks it forms in association with nepheline or melilite. Perovskite is common in some ultramafic plutonic igneous rocks, and may occur in contact metamorphosed impure carbonate sediments.

PSEUDOBROOKITE
C Fe_2TiO_5

Physical properties

CS Orthorhombic	**CL** {010} cleavage distinct
F&H Prismatic or acicular crystals found, but usually tabular	**HD** 6.0
	SG 4.39
COL Reddish brown to black	

Optical properties

$n_\alpha = 2.38$
$n_\beta = 2.39$
$n_\gamma = 2.42$
$\delta = 0.04$
$2V = 50° $ +ve

Pseudobrookite is brown in thin section, with moderate to high birefringence and extreme relief

Occurrence Psendobrookite is found in cavities in basalt or rhyolite often with topaz.

COLUMBITE, TANTALITE, tantalite–niobite

C $(Fe,Mn)(Ta,Nb)_2O_6$; when Ta > Nb the mineral is called tantalite, and when Nb > Ta the mineral is called columbite

Physical properties

CS Orthorhombic	**F** Subconchoidal to uneven
F&H Prismatic or tabular crystals are common; it often occurs massive	**L** Submetallic to resinous, sometimes iridescent
TW Contact twins common on {201}, with trillings on {010}	**TR** Opaque
	HD 6.0
COL Grey, black or brown	**SG** 5.15–6.50 (increasing with increasing Ta content)
S Dark red to black	
CL {010} distinct, {100} poor	

Tests Columbite is distinguished from black tourmaline by its higher specific gravity and by the shape of the crystals, and from wolframite by its poorer cleavage.

Occurrence Columbite occurs as a constituent of certain granitic pegmatites, as in the Black Hills of South Dakota, where very large crystals of columbite have been mined, and in similar rocks in Western Australia and elsewhere. Columbite also occurs associated with cassiterite and wolframite in certain alluvial deposits, and such deposits have been worked on a small scale. The main producers are Western Australia, Rhodesia, Nigeria, South Dakota and Brazil.

Uses   Columbite is an important source of tantalum and niobium.

8.5 Carbonates

This section includes all natural carbonates. They are dealt with in order of increasing A to B ratio.

8.5.1   Anhydrous normal carbonates

Formula AX ($X = CO_3$)

CALCITE, calc spar, carbonate of lime
C $CaCO_3$

Physical properties

CS   Trigonal

F&H   Well formed crystals are common and exhibit three main habits: (1) nail-head spar, a combination of flat rhombohedron $\{10\bar{1}2\}$ and prism $\{10\bar{1}0\}$; (2) dog-tooth spar, a combination of scalenohedron $\{21\bar{3}1\}$ and prism $\{10\bar{1}0\}$; and (3) rhombohedron $\{10\bar{1}1\}$. Calcite also occurs as shapeless grains, fibrous, lamellar, massive and in stalactitic and stalagmitic habits. It may be found nodular, compact or earthy.

TW   Lamellar twins are common $\{01\bar{0}2\}$ and also contact twinning on $\{0001\}$; rare twins on $\{10\bar{1}1\}$ and $\{02\bar{2}1\}$

COL   Colourless or white, sometimes with grey, yellow, blue, red, brown or black tints

S   White

CL   Perfect rhombohedral $\{10\bar{1}1\}$; powdered calcite consists of minute cleavage rhombohedra

F   Conchoidal, but rarely observed because of perfect cleavage

L   Vitreous to dull

TR   Transparent to opaque

HD   3.0 (a mineral on Moh's scale)

SG   2.715 (if pure) to 2.94

Optical properties

$n_o = 1.658$ }   increasing with increasing Fe or
$n_e = 1.486$ }   Mg content
$\delta = 0.172$
Uniaxial −ve

COL Colourless

H Usually as anhedral grains, but occasional rhombohedron seen

CL Perfect {10$\bar{1}$1} rhombohedral; three cleavages seen in some sections

R Moderate but large variation (see RI values); crystals 'twinkle' during rotation

A Calcite is easily dissolved by weakly acidic waters

B Maximum birefringence is extreme with very high fourth order and higher colours

IF A normal negative uniaxial interference figure is seen on basal sections, but in some metamorphic marbles, calcite is strained and shows biaxial interference figures with a small ($\sim$15°) negative 2V

Varieties As already discussed under form and habit, **nail-head spar** and **dog-tooth spar**, are common habits of calcite. Others include **Iceland spar**, a very pure transparent form of calcite first brought from Iceland which cleaves into perfect rhombohedra, and **satin spar**, a compact finely fibrous variety with a satin-like lustre, which is displayed to great advantage when polished. It is formed in veins or crevices in rocks, the fibres stretching across the crevices. The term 'satin-spar' is more commonly applied to the fibrous form of gypsum. **Aphrite** and **argentine** are unimportant lamellar varieties of calcite.

Stalactites are pendant columns formed by the dripping of water enriched in calcium carbonate from the roofs of caverns in limestone rocks and other favourable situations. Successive layers of calcite are deposited one over another, so that a cross section of the stalactite displays concentric rings of growth. The surplus dripping of the water gives rise to a similar deposit which forms in crusts one above the other on the floors of the caverns, this deposit being called a **stalagmite.** Beneath these stalagmitic crusts in some caves, the remains of prehistoric humans and animals have been found. **Oriental alabaster** and **Algerian onyx** are stalagmitic varieties of calcite characterized by well marked banding, and were used by the ancients for making ointment jars. Both names, however, are misleading, since true alabaster is calcium sulphate, and onyx is a cryptocrystalline banded variety of silica.

Calcareous tufa, **travertine** and **calc tufa** are more or less cellular deposits of calcium carbonate derived from waters charged with calcareous matter in solution. At Matlock, Knaresborough and many other places where natural springs are thus highly charged, twigs, birds' nests and other objects become encrusted with a hard coating of tufa when immersed in the spring. Calcareous tufa sometimes forms thick beds, as in Italy, and is then used as a building stone. **Thinolite** is a rock consisting of interlacing crystals of yellow or brown calcite, occurring as tufa deposits in Nevada, Australia, etc. The crystals are often skeletal.

Agaric mineral, **rock milk** and **rock meal** are white earthy varieties of calcite, softer than chalk, and deposited from solution in caverns.

Chalk is soft, white earthy carbonate of lime, forming thick and extensive beds in various parts of the world. It has been formed from the remains of marine microfossils, of Cretaceous age in Europe. Chalk sometimes consists to some extent of the remains of microscopic organisms, such as foraminifera. It has been suggested that the sea in which the English Chalk was laid down was surrounded by a desert area, so that no clayey or sandy material was contributed to the deposit forming in that sea.

Limestone, including chalk, is a general term for carbonate of lime. It may be crystalline, oolitic, or earthy, and, when impure, argillaceous, siliceous, bituminous, ferruginous or dolomitic. **Marbles** are limestones which have been crystallized by heat or pressure during metamorphic processes, but the name **marble** is often applied to some special type of non-metamorphic limestone. The different names of limestones and marbles are derived from the locality where they are found, the formation in which they occur, the fossils which make up their substance, or from some peculiarity of structure, colour, etc. Examples are shell marble, ruin marble, crinoidal limestone, Carboniferous limestone, etc. **Lithographic stone** is a very fine grained variety of limestone used in printing. **Pisolite** and **oolite** are granular varieties of limestones produced by the deposition of calcium carbonate in successive layers around small nuclei. Pisolite differs from oolite in the larger size of the granules. The oolitic structure is considered to be the result of a purely inorganic process in which the granules are washed backwards and forwards on beaches in sea water saturated with calcium carbonate, especially in lagoons near the reef. **Anthraconite**, or stinkstone, is a dark-coloured limestone containing bituminous matter, and emitting a fetid odour when struck.

Fontainbleau sandstone is a name given to calcite which contains a large admixture of sand, up to 80% when concretionary, and as much as 65% even when crystallized. It was originally found at Fontainbleau, France.

Tests Calcite is infusible, but becomes highly luminous when heated. It effervesces, with the evolution of carbon dioxide, in cold dilute hydrochloric acid. The brick red calcium flame is seen. Physical properties, especially cleavage and habit, are very distinctive and help to distinguish between calcite and aragonite.

Occurrence Calcite is one of the most common and most widespread minerals on or near the Earth's surface, where it is the only stable form of $CaCO_3$. It is a principal constituent of sedimentary limestones, and occurs in carbonate shells, as fine precipitates and as clastic material.

Aragonite is usually the initial carbonate material of which shells are composed, but it eventually changes to calcite through geological time. Under metamorphism pure limestones change to pure calcite marbles, whereas if the limestone is impure the calcite combines with impurities to give new minerals such as wollastonite. In metamorphosed impure limestones calcite may be found in association with diopside, grossular (Ca-garnet), idocrase, Mg-rich olivine, etc.

Calcite may be deposited in vugs or cavities in igneous rock by the action of late-stage hydrothermal solutions. In hydrothermal veins calcite is a common gangue mineral, together with fluorite, quartz or barite found in association with sulphide ore minerals such as sphalerite and galena. Calcite can occur as a primary mineral in some alkali igneous rocks and carbonatites, and is a common secondary mineral in basic igneous rocks after the alteration of ferromagnesium minerals by late-stage hydrothermal solutions carrying CO_2. The ferromagnesian minerals change to serpentine, and serpentine changes to talc and magnesite ($MgCO_3$), or to calcite if enough calcium is present.

Uses Calcite finds many different uses according to its purity and character. The varieties containing some clayey matter are burnt for cement, while the purer varieties provide lime that is used in many industrial processes, such as the manufacture of bleaching powder, calcium carbide, glass, soap, paper and paints. Enormous quantities of limestone of various kinds are used with clay in cement manufacture, and limestone is an important road metal. Over 90 million tonnes of limestone are quarried in the UK each year. Marbles and crystalline limestones, and the more resistant calcareous rocks generally, are important building and ornamental stones. Calcium carbonate is used as a flux in smelting, and certain varieties of limestone are used in printing processes. Crushed limestones have a major use as agricultural lime, provided their 'neutralizing value' is suitable. The lime is used to neutralize the natural acids in the soils, and 14% of all limestone produced in the UK is used for this purpose.

MAGNESITE

C $MgCO_3$; a complete solid solution series can be formed with siderite ($FeCO_3$), and some limited solid solution is common

Uses Magnesite is used in the production of carbon dioxide, magnesium and magnesium salts. Its most important use is for refractory bricks, furnace linings and crucibles. It is employed in the manufacture of special cements, and in the paper and sugar industries.

Physical properties

CS Trigonal

F&H Crystals rare but occur as rhombohedra with curved faces. Magnesite is usually massive and fibrous, sometimes compact and sometimes granular.

COL White, greyish white, yellow or brown; often chalk-like

CL Perfect rhombohedral $\{10\bar{1}1\}$

F Flat or conchoidal

L Vitreous, or pearly, but dull in massive or earthy varieties

TR Transparent to opaque

HD 3.5–4.5

SG 3.0–4.8 (depending on the amount of Fe^{2+} present)

Optical properties

n_o = 1.700
n_e = 1.509
δ = 0.191
Uniaxial −ve

COL Colourless, sometimes cloudy

H Aggregates of anhedral grains, sometimes massive

CL Perfect $\{10\bar{1}1\}$ rhombohedral

R Low to high, giving characteristic carbonate 'twinkle' on rotation

A None, but may be dissolved

B Extreme, increasing with increasing Fe content

Tests The compact chalk-like variety of magnesite is quite distinctive. It effervesces with hot acids. If heated on charcoal, magnesite gives an incandescent mass which, when moistened with cobalt nitrate and strongly reheated, turns pink.

Occurrence Economically important deposits of magnesite occur in two forms. The first is as irregular veins and fracture zones in serpentine masses, from which it has presumably been derived by low-grade metamorphism in the presence of CO_2, and such deposits are worked in Greece and India and elsewhere. The second type of deposit is found where magnesite replaces dolomite and calcite, and is most probably due to the alteration of these rocks by solutions coming from an igneous magma. The Austrian deposits are of this type and supply an important part of the world production, and other similar deposits are worked in Manchuria, Washington State and Quebec. Certain bedded deposits of magnesite have been interpreted as saline residues. Magnesite may also occur as veins or disseminations in talc, schists, chlorite schists or mica schists.

Uses Magnesite is used in the production of carbon dioxide, magnesium and magnesium salts. Its most important use is for refractory bricks, furnace linings and crucibles. It is employed in the manufacture of special cements, and in the paper and sugar industries.

SIDERITE, chalybite, spathose iron

C $FeCO_3$; with 48.3% iron. Often a little manganese and calcium are present. Limited solid solution with magnesite may occur.

Physical properties

CS Trigonal

F&H Rhombohedra $\{10\bar{1}1\}$ common, often with curved faces; also massive and granular

COL Pale yellowish, brownish, brownish black or brownish red

S White

CL Perfect $\{10\bar{1}1\}$ rhombohedral

F Uneven

L Pearly or vitreous

TR Translucent (rare) to opaque

T Brittle

HD 4.0–4.5

SG 3.7–3.96 (pure Fe)

Optical properties

$n_o = 1.875$

$n_e = 1.633$

$\delta = 0.242$

Uniaxial −ve

COL Colourless to pale yellow

H Euhedral rhombohedra occur, but usually found as granular aggregates

CL Typical rhombohedral cleavage is perfect

R Moderate to very high, with twinkling noted on rotation

A Siderite alters to goethite or limonite; pseudomorphs after siderite are common

B Extreme, with some masking of colours by siderite body colour

Varieties **Clay ironstone** is an impure iron carbonate occurring as beds and nodules, especially in the Carboniferous Coal Measures of many countries. It is common in most of the British coalfields and in those of the USA. It occasionally exhibits a curious radiately disposed, subcolumnar structure, causing it, when struck, to fall to pieces in conical masses which envelop or cap one another, and to which the name of 'cone-in-cone' structure has been given. These clay ironstones formerly constituted valuable ores of iron. As well as occurring in the Coal Measures, clay ironstones are also found in layers and nodules in other formations. **Blackband** is a dark, often carbonaceous, type of clay ironstone.

Oolitic ironstone is an iron carbonate which has replaced the calcium carbonate of an oolitic limestone, retaining the structure of the original rock, as in the celebrated Cleveland iron ore. In many examples of oolitic ironstone, however, it has been shown that the iron carbonate did not replace calcium carbonate, but was formed at the same time as the oolitic structures, these resulting from colloidal processes.

Tests Heated before the blowpipe, siderite blackens and becomes magnetic. It gives iron reactions with the fluxes. It is very slowly affected by cold acids, but in hot hydrochloric acid it effervesces very briskly.

Occurrence Siderite is found in bedded sedimentary deposits, in association with clays, cherts and chamosite, as in the ironstones of the island of Raasay in Scotland, and is a common constituent in ironstone nodules and other concretions. Sedimentary siderite represents the precipitation of soluble ferrous bicarbonates in oxygen-poor environments, sometimes by biogenic processes, and may be interbedded with coal, fireclays or seat earths (underclays), and is common in bog iron ore deposits. Siderite has contributed to the great Lake Superior hematite deposits. The sedimentary siderites include those of the Coal Measures already mentioned, typical localities being South Wales, South Staffordshire and other British coalfields, and the coalfields of the eastern USA. Mg-rich siderite is a fairly common gangue mineral in high-temperature hydrothermal veins, along with calcite, ankerite (iron-rich dolomite), fluorite and barite, and is associated with ores such as sphalerite, galena, cassiterite, silver, etc. It is a very rare mineral in late-stage pegmatites and may be present in carbonatites.

Metasomatic siderites are extremely important deposits. In these, iron carbonate has replaced the calcium carbonate of limestones, retaining many of the original features of the rock (oolitic structure, fossils, etc.), such as occurs in the Cleveland ores. Vein deposits of siderite have been worked in Germany, and the minette ores of Alsace-Lorraine contain siderite, as do the Mesozoic ores of central England.

RHODOCHROSITE, dialogite
C $MnCO_3$; with 47.8% manganese, often with varying quantities of iron, calcium and magnesium

Physical properties
CS   Trigonal
F&H   Rhombohedral crystals rare, usually massive, globular, botryoidal or encrusting
COL   Shades of rose red, yellowish grey, brownish
S   White
CL   The usual rhombohedral {1011} perfect

F   Uneven
L   Vitreous to pearly
TR   Transparent to subtranslucent
T   Brittle
HD   3.5–4.0
SG   3.45–3.6

Optical properties

$n_o = 1.816$
$n_e = 1.597$
$\delta = 0.219$

Uniaxial −ve

COL Pale pink, weakly pleochroic

H Coarse aggregates common, as are banded botryoidal masses

CL Three rhombohedral cleavages seen

R Moderate to high

A Complete alteration yields pyrolusite, MnO_2 or manganite, $MnO(OH)$

B Extreme, but may be masked by body colour

Varieties **Ponite** is Fe-rich rhodochrosite, but when the Fe^{2+} : Mn ratio becomes 1 : 1 it is called **manganosiderite**. **Kutnohorite** $CaMn(CO_3)_2$ is an intermediate type between rhodochrosite and calcite (similar to dolomite).

Tests Heated before the blowpipe, rhodochrosite is infusible, but the mineral changes to grey brown and black, and decrepitates strongly. It gives the usual manganese purple colour to the borax and microcosmic salt beads. It dissolves with effervescence in warm hydrochloric acid. On exposure to air, the red varieties lose colour.

Occurrence Rhodochrosite occurs as a gangue mineral in low- to medium-temperature hydrothermal veins and in limestone replacement bodies with calcite, siderite, dolomite, fluorite, barite and quartz, and with the ore minerals sphalerite, galena, tetrahedrite, and silver minerals. It may occur in metamorphic zones in association with other manganese minerals such as hausmannite, spessartine, etc. Sedimentary rhodochrosite may be syngenetic or diagenetic, and is associated with iron silicates and iron carbonates.

SMITHSONITE (formerly CALAMINE in the UK)

C $ZnCO_3$; Zn is often partly replaced by Fe^{2+} and Mn, and some Ca, Mn, Cd, Co or Cu is often present

Physical properties

CS Trigonal

F&H Rare rhombohedra; usually massive, reniform, botryoidal, stalactitic, encrusting, granular or earthy

COL White, greyish, greenish, brownish white

CL Perfect rhombohedral {10Ī1}

F Uneven

S White

L Vitreous to pearly

TR Subtransparent to opaque

T Brittle

HD 4.0–4.5

SG 4.43

Optical properties
n_o = 1.850
n_3 = 1.625
δ = 0.225
Uniaxial −ve
COL   Colourless
H   Crystals rare; usually massive, botryoidal, etc.
CL   Perfect rhombohedral
R   Moderate to very high

A Smithsonite may alter to hemimorphite and limonite. It is often seen pseudomorphed by minerals such as quartz, etc., and may itself be seen as a pseudomorph after calcite or dolomite.
B Extreme, with whites of a very high order

Tests   Heated in the closed tube, the smithsonite gives off CO_2 and turns yellow when hot, and white when cold. Heated alone before the blowpipe it is infusible. Heated on charcoal, moistened with cobalt nitrate and strongly reheated, it assumes a green colour on cooling. Heated with sodium carbonate on charcoal, smithsonite gives zinc vapours and forms the usual encrustation of zinc oxide. It is soluble in hydrochloric acid, with effervescence.

Occurrence   Smithsonite is a secondary mineral found in ore deposits containing primary zinc minerals, especially sphalerite. It is usually associated with sphalerite, hemimorphite, galena, and iron and copper ores. Localities include the Mendip Hills, Matlock and Alston Moor in England and Leadhills in Scotland, and it is also found in most lead- and zinc-mining centres.

Uses   Smithsonite is an important ore of zinc. In commerce, the term calamine includes the zinc silicates as well as the carbonate.

ARAGONITE

C $CaCO_3$, identical to calcite, with up to 2% strontium carbonate, or other impurities, present.
The atomic structure of aragonite differs from that of calcite in that the $[CO_3]^{2-}$ ions are not in the same positions, and the symmetry is lowered from trigonal to orthorhombic. Other carbonates with a similar atomic structure are strontianite, witherite and cerussite.

Physical properties

CS Orthorhombic

F&H Prismatic or acicular crystals, often in radiating clusters, terminated by dome forms: {011}, {021}, {201}, etc. Aragonite also occurs globular, stalactitic, coralloidal or encrusting, and brachiopod or bivalve shells normally consist of aragonite which inverts to calcite after a time.

COL White, grey, yellowish; rarely green or violet

CL Poor, parallel to {010}

F Subconchoidal

L Vitreous

TR Transparent to translucent

T Brittle

HD 3.5–4.0

SG 2.95

Optical properties

$n_\alpha = 1.530$
$n_\beta = 1.680$
$n_\gamma = 1.685$
$\delta = 0.155$
$2V = 18°$ −ve

OAP is parallel to (100)

COL Colourless

H Thin prismatic, acicular or fibrous crystals occur, for example in shell structures

CL {010} cleavage imperfect

R Low to moderate

A Aragonite inverts to calcite, but high Sr content inhibits inversion

B Extreme with pale colours of a high order

IF A basal section gives a good figure, but it is difficult to obtain because of the small size of crystals

E Straight, crystals are length fast

TW Lamellar twins common on {110}

Varieties Aragonite occurs in crystallized, crystalline, massive or stalactitic varieties. **Flos-ferri** is a stalactitic coralloidal variety, which consists of beautiful snow white divergent and ramifying branches, in many cases encrusting hematite. **Pisolites** deposited at some hot springs are of aragonite.

Tests Heated before the blowpipe, aragonite whitens and crumbles, changing to calcite. Aragonite gives off CO_2 when attacked by cold dilute hydrochloric acid. It is brick red in the flame test. Aragonite is distinguished from calcite by the following: (1) its crystal shape; (2) it possesses different cleavage and differently shaped cleavage fragments; (3) its hardness and SG (4 and 2.95, against 3 and 2.71 for calcite); (4) Neigen's test, in which aragonite is stained with a solution of cobalt nitrate, whereas calcite is not (in Neigen's test the mineral under observation is boiled with cobalt nitrate solution for a quarter of an hour, and then washed, and a pink staining indicates aragonite); (5) Leitmeier and Feigl's test, in which a solution of manganese sulphate of 11.8g $MnSO_4.7H_2O$ in 100 ml water is prepared, some solid silver sulphate introduced, the whole heated, cooled and filtered, one or two drops of dilute caustic soda solution added, and the precipitate filtered off after 1–2 hours. This solution is usually kept in an opaque bottle. To distinguish between aragonite and calcite, the powder or slice of the mineral is covered by the solution. Aragonite at once turns grey and finally black, while calcite only becomes greyish after more than an hour. This is a good test for fine intergrowths of the two minerals.

Aragonite can easily be distinguished from calcite by X-ray diffraction techniques, although about 0.5 g of each mineral, or both mixed, must be available.

Occurrence Aragonite occurs with beds of gypsum, or associated with iron ore in the form of flos-ferri, or as a deposit from the waters of hot springs in oolitic or pisolitic forms. It is formed in low temperature near-surface deposits of rather recent geological origin. Aragonite is originally the material for many invertebrate shells, and gradually inverts to calcite on diagenesis: thus pre-Mesozoic fossil shells will inevitably consist of calcite. Aragonite occurs in amygdales in basic extrusive igneous rocks along with zeolites, and also in veins in serpentine. It is a widespread metamorphic mineral in glaucophane schist rocks (blue schist facies), at low temperatures ($<300°C$) and high pressures (6–10 kbar), but may invert in these rocks to calcite as the rock recovers to normal temperatures and pressures.

WITHERITE

C $BaCO_3$

Physical properties

CS Orthorhombic

F&H Crystals are stubby pseudo-hexagonal dipyramids with horizontally striated faces, but sometimes massive, often with a columnar structure, or globular or botryoidal

TW Repeated twinning is always present on {110}, giving pseudo-hexagonal forms

COL White, yellowish, greyish

S White

CL Distinct parallel to {010}; poor on {110} and {012}

F Uneven

L Vitreous, but resinous on fracture surfaces

TR Subtransparent to translucent

T Brittle

HD 3.5

SG 4.29

Optical properties

$n_\alpha = 1.529$
$b_\beta = 1.676$
$n_\gamma = 1.677$
$\delta = 0.148$
$2V = 16°$ −ve
OAP is parallel to (010)

COL Colourless

H Stubby pseudo-hexagonal dipyramids, but sometimes massive

CL {010} good, {110} and {012} poor

R Low to moderate

A Witherite may alter to barite, or from barite

B Extreme with high order whites and pale colours

IF Best seen in a basal section, where the whole figure is seen with a very small $2V$

Tests Witherite colours the flame yellowish green and effervesces with hydrocholoric acid, the solution giving a dense white precipitate on the addition of sulphuric acid. The density of a specimen of this mineral is noticeable.

Occurrence Witherite occurs in low-temperature hydrothermal veins as a gangue mineral, associated with galena and barite in many of the veins of the North of England. Important localities are Settlingstones and Fallowfield in Northumberland, New Brancepeth colliery in Durham, many of the Alston veins of Cumberland, and near St Asaph in North Wales.

Uses Witherite is the source of barium salts, and the finely divided sulphate required in industry is produced from it. Small quantities are employed in the pottery industry.

STRONTIANITE

C $SrCO_3$; a small proportion of Ca may be present up to a Ca : Sr ratio equal to 1 : 3

Physical properties

CS Orthorhombic

F&H Prismatic or acicular crystals, often divergent; also found fibrous or granular

TW Twinning on {110} produces crude, pseudo-hexagonal forms similar to those of witherite

COL Pale green, yellow, grey or white

CL Perfect parallel to {110}

F Uneven

L Vitreous, but resinous on fracture surfaces

TR Transparent to translucent

T Brittle

HD 3.5–4.0

SG 3.6–3.75 (if pure)

Optical properties
$n_\alpha = 1.516$
$n_\beta = 1.664$
$n_\gamma = 1.666$
$\delta = 0.150$
$2V = 7° -ve$
OAP is parallel to (010)

COL Colourless
H Prismatic crystals common
CL Good {110} prismatic cleavages, but poor on {021} and {010}
R Low to moderate

A Strontianite may alter to celestite and is easily dissolved in acid solutions
B Extreme, with whites and pale colours of a high order
IF Basal section gives a good interference figure, almost uniaxial

Tests Strontianite gives the crimson flame of strontium. It effervesces with hydrochloric acid and dissolves, the dilute solution giving a precipitate of strontium sulphate on the addition of sulphuric acid.

Occurrence The major commercial source of strontianite is from low-temperature hydrothermal veins traversing Cretaceous marls and limestones in Westphalia (Germany). A 1 m wide vein of strontianite cuts the Carboniferous limestone at Green Laws mine, Weardale, Durham. It occurs in veins with galena and barite, as in the original locality at Strontian in Argyllshire, and elsewhere. It also occurs as nodules, nests and geodes in limestones, where it may be an original deposit, a replacement deposit, or formed by alteration of celestite.

Uses As a source of strontium salts.

CERUSSITE, ceruse, white lead ore
C $PbCO_3$; minor amounts of Zn and Sr may be present

Physical properties
CS Orthorhombic
F&H Prismatic crystals variously modified, flattened parallel to the basal plane; crystal aggregates are common, but it also occurs granular, massive, and columnar
TW Repeated twinning common on {110}, giving pseudo-hexagonal cyclic forms or polysynthetic lamellae
COL White or greyish, but sometimes tinged blue or green by copper salts

S Colourless
CL Distinct on {110} and {021}; poor on {010} and {012}
F Conchoidal
L Adamantine, occasionally vitreous or resinous
TR Transparent to translucent
T Very brittle
HD 3.0–3.5
SG 6.56 (cerussite is commonly found pure)

Optical properties

$n_\alpha = 1.803$
$n_\beta = 2.074$
$n_\gamma = 2.076$
$\delta = 0.273$
$2V = 9°$ −ve
OAP is parallel to (010)

COL Colourless
H Crystals common, either as prisms or grains, or aggregates
CL Good on {110} and {021}
R Very high to extremely high

A Cerussite is an alteration product of galena, with anglesite representing the intermediate stage. Cerussite may be replaced by malachite, limonite and many other secondary minerals.
B Extreme, with white of a high order shown
IF A basal section shows a good and very small (9°) interference figure

Tests Cerussite is soluble in hydrochloric acid, with effervescence. Heated before the blowpipe, it decrepitates and fuses. Heated on charcoal with sodium carbonate and charcoal it yields the lead bead, which is malleable and marks paper. Heated with potassium iodide and sulphur on charcoal, cerussite gives a brilliant yellow encrustation.

Occurrence Cerussite is a secondary mineral occurring in the oxidation zone of lead veins, associated with anglesite, smithsonite, limonite, malachite and other secondary ores of lead, zinc and copper. It may result from the decomposition of anglesite by water charged with bicarbonates. It occurs at most localities where lead ores occur, such as Cornwall, Derbyshire, Durham, Cardigan and Leadhills in the UK

Uses When found in quantity, cerussite is a valuable ore, ranking next to galena. The white lead of commerce is hydrated lead carbonate, $2Pb\text{-}CO_3.Pb(OH)_2$, and is artificially prepared by various processes, the most general in the UK being the Old Dutch Process. In this, white lead is produced by the action of acetic acid on metallic lead cast in the form of gratings. Lead acetate is produced and converted into the carbonate by CO_2 liberated by fermenting tan-bark, etc. White lead may also be prepared by passing CO_2 through a solution of basic lead acetate, but this product is usually considered inferior. White lead is used as a pigment, and is sometimes adulterated with barite.

Formula ABX_2

DOLOMITE–ANKERITE SERIES

C   **dolomite** $CaMg(CO_3)_2$

       ankerite $Ca(MgFe)(CO_3)_2$

Fe^{2+} may replace Mg in dolomite, but when Mg < Fe the mineral is called ankerite. Co, Zn and Mn may replace Mg, and Ba and Pb may replace some Ca.

Physical properties

CS Trigonal; the presence of Mg cations in the structure reduces the symmetry from trigonal holosymmetric in calcite, to a lower crystal class in the trigonal system for dolomite

F&H Rhombohedra common, usually with curved faces. Dolomite also occurs massive and granular, as in many geological formations, in which it has a saccharoidal texture

TW Glide twin lamellae occur on $\{02\bar{2}1\}$. Simple twinning is common on $\{0001\}$, $\{10\bar{1}0\}$ and $\{11\bar{2}0\}$; rarely on $\{10\bar{1}1\}$

COL White, yellowish, brown and sometimes red, green or black

CL Perfect $\{10\bar{1}1\}$ rhombohedral

F Conchoidal or uneven

L Vitreous to pearly of crystals; dull of massive varieties

TR Translucent to opaque

T Brittle

HD 3.5–4.0

SG 2.86 (dolomite to 3.10 (ankerite)

Optical properties

	Dolomite	Ankerite
n_o	1.679	1.690–1.750
n_e	1.500	1.510–1.548
δ	0.179	0.180–0.202

Uniaxial −ve

COL Colourless

H Rhombohedra or grains common and also massive

CL Rhombohedral $\{10\bar{1}1\}$ perfect

R Low to moderate or high

A Dolomite is itself a common secondary mineral, replacing calcite, and it may be replaced by several other minerals including quartz, calcite, etc.

B Extreme

TW Glide twin lamellae seen on many sections

Varieties **Pearl spar** is a white, grey, pale yellowish or brownish variety, with a pearly lustre, occurring in small rhombohedra with curved faces, and frequently found associated with sphalerite and galena. **Brown spar**, **rhomb spar** and **bitter spar** comprise the iron-bearing varieties which turn brown on exposure. **Miemite** is a yellowish brown fibrous variety found at Miemo in Tuscany. **Magnesian limestone** is a crystalline granular dolomite, occurring in massive beds of considerable extent in, for example, the Permian rocks of England.

Tests Before the blowpipe, dolomite behaves like calcite. Cold acid acts very slightly on fragments, but in warm acid the mineral is readily dissolved with effervescence. In Lemberg's test, if calcite is boiled for 15 minutes with a solution of aluminium chloride and logwood, it is stained pink, but dolomite undergoes no such staining (ferric chloride may be used instead of aluminium chloride). When ankerite is heated before the blowpipe on charcoal, it becomes black and magnetic.

Occurrence Dolomite occurs in extensive beds at many geological horizons. Dolomite may be deposited directly from sea water, but most dolomite beds have been formed by the alteration of limestones, the calcite of which is replaced by dolomite. Dolomitization is often related to joints and fissures through which the solutions penetrated, and thick beds, as in the Dolomite Alps of Tyrol, may be completely changed to dolomite. As a result of this change, a shrinkage takes place and useful minerals may afterwards be deposited in the cracks so caused. The solutions giving rise to dolomitization are mainly derived from the sea, and an example of the change is seen in the conversion of the aragonite and calcite of coral reefs into dolomite by reaction with the magnesium salts contained in the sea water. Dolomite may be precipitated as a primary mineral from abnormally saline isolated waters in association with anhydrite, gypsum, halite, sylvite and other evaporite minerals. Under metamorphism, dolomite may recrystallize to give a dolomite marble, but at high temperatures it may decompose to calcite and periclase or brucite, or if silica is present it may combine with the magnesia to give magnesian silicates such as forsterite, diopside, tremolite, etc. Dolomite occurs in hydrothermal veins with calcite, siderite, fluorite, etc., and with the ore minerals sphalerite, chalcopyrite and galena. Dolomite may occur as a secondary mineral in hydrothermally altered ultramafic igneous rocks.

Ankerite may occur as veins or concretions in iron-rich sediments with siderite and iron oxides. It may occur in high-grade regionally metamorphosed schists with garnet and cummingtonite. Ankerite is a characteristic mineral in ore zones with fluorite, galena, sulphosalts and siderite. Both dolomite and ankerite may appear in some ultra-alkaline dyke rocks, and also in carbonatites.

Uses Dolomite is an extremely important building material. It is also used for making refractory furnace linings, and as a source of carbon dioxide.

ALSTONITE, bromlite

C (Ba,Ca)CO$_3$, with Ca replacing Ba in varying proportions

HD 4.0–4.5   **SG** 3.67–3.71

Alstonite is similar to witherite in properties and mode of occurrence. It is found in a few of the barite–witherite veins of the North of England.

BARYTOCALCITE

C BaCa(CO$_3$)

Physical properties

CS Monoclinic	**F** Uneven
F&H Prismatic crystals and also massive	**L** Vitreous or slightly resinous
	TR Transparent to translucent
COL White, greyish or yellowish	**HD** 4.0
CL {210} perfect, {001} imperfect	**SG** 3.65–3.71

Optical properties

In thin section, barytocalcite is colourless with low to high relief ($n_\beta = 1.684$) and extreme birefringence ($\delta = 0.161$). It is biaxial negative with a small $2V$ (15°).

Tests Barytocalcite gives the yellowish-green flame of barium, and rarely the brick red flame of calcium. It effervesces with hydrocholoric acid. When heated on charcoal, the barium carbonate fuses and sinks into the block, leaving the calcium carbonate as an infusible mass.

Occurrence Barytocalcite occurs in barite and lead veins in limestones.

8.5.2 Hydrated carbonates

THERMONATRITE

C Na$_2$CO$_3$.H$_2$O

Physical properties

CS Orthorhombic	**HD** 1.0–1.5
F&H Platy, acicular crystals and also efflorescent crusts	**SG** 2.25–2.26

Optical properties

It is colourless in thin section, with low to moderate negative relief ($n_\beta = 1.506$–1.509) and high birefringence ($\delta = 0.10$). It is biaxial negative, with a moderate $2V$ (48°), but interference figures are difficult to obtain.

Occurrence Thermonatrite is found as a saline residue.

Uses All sodium carbonate minerals are extensively employed in the manufacture of chemicals, glass, soap, detergents and paper, and in the bleaching, dyeing and printing of various fabrics.

NATRON
C $Na_2CO_3.10H_2O$

Physical properties
CS Monoclinic
F&H Usually in solution but where solid it occurs as efflorescent crusts or rarely tabular crystals
COL White, grey or yellowish
CL Basal cleavage present in crystals

L Vitreous or dull
TASTE Alkaline
SOLUBILITY Very soluble in water
HD 1.0–1.5
SG 1.44–1.46

Optical properties
In thin section, natron is a colourless mineral with a moderate negative relief ($n_\beta = 1.425$), moderate birefringence ($\delta = 0.035$), and a large negative 2V (80°).

Tests Natron effervesces with acid. It gives water on heating in the closed tube and shows the yellow sodium flame.

Occurrence Natron is found in solution in the soda lakes of Egypt, the East African Rift Valley, the western USA, and elsewhere. It occurs in saline residues, as in British Columbia.

TRONA, urao
C $Na_3H(CO_3)_2.2H_2O$

Physical properties
CS Monoclinic
F&H Fibrous or columnar layers or masses
COL Grey or yellowish white
CL Perfect parallel to {100}
L Vitreous

TR Translucent
TASTE Alkaline
SOLUBILITY Soluble in water
HD 2.5–3.0
SG 2.11–2.13

Optical properties
In thin section, trona is a colourless mineral with low negative relief ($n_\beta = 1.492$), extreme birefringence ($\delta = 0.128$), and a large negative 2V (76°).

Tests Trona effervesces with acids. It gives water on heating in the closed tube, and shows the yellow flame of sodium.

Occurrence Trona occurs in saline residues, with other minerals formed in this way, as in California, Mexico, Fezzan and Egypt.

GAYLUSSITE
C $Na_2Ca(CO_3)_2.5H_2O$

Physical properties

CS Monoclinic
F&H Flattened wedge-shaped crystals
COL White, colourless

CL Prismatic {110} perfect, {001} imperfect
HD 2.5–3.0
SG 1.99–2.00

Optical properties
In thin section, gaylussite is a colourless mineral with low negative relief ($n_\beta = 1.516$), very high birefringence ($\delta = 0.077$) and a small negative $2V$ (34°).

Occurrence Gaylussite occurs with other evaporites in muds and other sediments.

8.5.3 Hydroxyl-bearing carbonates

ZARATITE, emerald nickel
C $Ni_3(CO_3)(OH)_2.4H_2O$; in the paler varieties, some Ni may be replaced by Mg

Physical properties

CS Cubic
F&H Occurs as an encrustation sometimes minutely mammillated and stalactitic; it also occurs massive, compact and fibrous
COL Emerald green

S Pale green
L Vitreous
TR Transparent to translucent
HD 3.5
SG 2.57–2.69

Optical properties In thin section, zaratite is green and isotropic, with low relief ($n = 1.56–1.60$).

Tests Heated in the closed tube, zaratite gives off water and CO_2, leaving a dark, magnetic residue. It gives the usual nickel reactions in the borax bead test. It dissolves, with effervescence, when heated in dilute hydrochloric acid.

Occurrence Zaratite occurs as a coating to other nickel minerals, and is associated with chromite-bearing serpentinites, as in Unst, in the Shetland Islands.

HYDROZINCITE
C $Zn_5(CO_3)_2(OH)_6$

Characters and occurrence
HD   2.0–2.5   SG   3.6–3.8

Hydrozincite is monoclinic, white in colour and usually occurs massive, fibrous or encrusting. It results from the alteration of sphalerite and is found with smithsonite in the oxidation zones of zinc deposits, as near Santander (Spain), and elsewhere.

MALACHITE
C $Cu_2CO_3(OH)_2$, with 57.3% copper

Physical properties
CS   Monoclinic

F&H   Prismatic crystals are common. It also occurs massive, encrusting, stalactitic or stalagmitic and often with a smooth mammillated or botryoidal surface. The internal structure is often divergently fibrous and compact Malachite also occurs granular and earthy.

TW   Simple on {100} common, sometimes polysynthetic

COL   Bright green, shades of green, often concentrically banded

S   Green of massive varieties, paler than colour

CL   Perfect on {201}; good on {010}

L   Silky on fibrous surfaces. Massive varieties are dull. Crystals have an adamantine to vitreous lustre.

TR   Translucent to subtranslucent (crystals); opaque in massive varieties

HD   3.5–4.0

SG   3.9–4.1

Optical properties
n_α = 1.655
n_β = 1.875
n_γ = 1.909
δ = 0.254
$2V$ = 43° −ve
OAP is parallel to (010)

COL   Green

P   Distinct, with α colourless to pale green, β yellowish green and γ dark green

H   Single crystals rare, usually massive or fibrous

CL   {201} perfect, {010} good

R   High to very high

A   Rare

B   Extreme, but usually masked by body colours

IF   Difficult to obtain size because of body colour and dispersion

E   Maximum extinction angle is $\alpha\hat{\,}c$ = 24° in an (010) section

Tests   Heated in the closed tube, malachite gives off water and blackens. It dissolves, with effervescence, in acids. Before the blowpipe, alone, it fuses and colours the flame emerald green. On charcoal malachite is reduced to metallic copper. It colours the borax bead green.

Occurrence Malachite is a secondary mineral found in the zone of weathering or oxidation of copper deposits, lodes or other types, where it is closely associated with azurite. Some localities are Redruth (Cornwall), Chessy (France), Nishni Tagilsk (Siberia), Burra Burra mine (South Australia), Chile, Pennsylvania and Tennessee. The colour banding marking successive deposits of malachite results from the percolation of water through copper-bearing rocks and the subsequent deposition of the dissolved carbonate in fissures or cavities. The solution has apparently dripped in slowly, the water evaporated, and the series of layers formed in the same way that stalactites and stalagmites form by the percolation of water through limestone. Very large masses of malachite have been procured from Siberia and Australia. The most remarkable deposit of malachite is probably that of the Katanga region of Zaire and the adjacent part of Zambia. The ores are malachite, with other ores of the oxidized zone, such as azurite, chrysocolla, melaconite, and chalcocite, along with the gangue minerals calcite and chalcedony. At some mines, chalcopyrite and bornite also occur. The ores are disseminated through sedimentary rocks such as dolomites, dolomitic sandstones and feldspathic sandstones, and probably represent the weathered upper part of an enormous disseminated deposit, since chalcopyrite and other sulphides have been encountered at depth. The origin of the ores may be either due to emanations from a granite mass, or sedimentary.

Uses Malachite is a valuable ore of copper. It is also cut and polished and used for ornamental purposes.

PHOSGENITE, cromfordite, horn lead
C $Pb_2(CO_3)Cl_2$

Physical properties
CS Tetragonal
F&H Prismatic crystals; also granular and massive
COL White, grey or yellow
S White

L Adamantine
TR Transparent to translucent
HD 2.0–3.0
SG 6.13

Optical properties
Phosgenite is colourless in thin section, with extreme relief ($n_o = 2.118$) and moderate birefringence ($\delta = 0.026$), and is uniaxial positive.

Tests Phosgenite dissolves, with effervescence, in hydrochloric acid. It is soluble in nitric acid, giving a solution which reacts for chloride with silver nitrate. When heated in the closed tube, colourless globules of lead chloride are obtained.

Occurrence Prosgenite is a rare mineral formed in the oxidation zone of lead deposits, where it is associated with cerussite.

BISMUTITE, basobisutite
C $Bi_2O_2(CO_3)$; water may be present in the formula

Characters and occurrence Bismutite is a white, grey or yellowish mineral, occuring as fibrous or earthy crusts, and resulting from the alteration of native bismuth and bismuthinite.

AZURITE, chessylite, blue carbonate of copper
C $Cu_3(CO_3)_2(OH)_2$; with 55.1% copper

Physical properties
CS Monoclinic

F&H Crystals as modified prisms or tabular parallel to basal plane, but usually massive or earthy. Radiated aggregates of columnar crystals may assume globular or stalactitic habits

COL Deep azure blue, hence the name

S Blue, lighter than the colour

CL Perfect on {001}, distinct on {100} and poor on {110}

F Conchoidal

L Adamantine to vitreous of crystals; silky or resinous for radiating aggregates; dull if massive

TR Transparent (crystals) to opaque (massive)

T Brittle

HD 3.5–4.0

SG 3.77–3.89

Optical properties
$n_\alpha = 1.730$
$n_\beta = 1.754$
$n_\gamma = 1.835$
$\delta = 0.105$
$2V = 68°$ +ve
OAP is perpendicular to (010)

COL Blue

P Strongly pleochroic, with α clear blue, β azure blue and γ deep blue

H Tabular or prismatic crystals common; also radiating crystals and massive

CL {011} perfect, also {100} and {110}

R High to very high

B Very high, but masked by body colour

IF Interference figure is difficult to obtain because of high dispersion

E An (010) section gives an extinction angle of α(fast)^cl = 15°

Tests As for malachite. It is distinguished from malachite by its azure blue colour.

Occurrence Azurite is common in the oxide or weathered zone of copper deposits, where it forms by the interaction either of carbonated solutions with copper minerals, or of soluble copper salts with carbonate rocks. It is always found in association with malachite but it is less common, and it is also found with cuprite, tenorite, limonite, chrysocolla, native copper, etc. Some localities are Redruth (Cornwall), Chessy (France) and Katanga (Zaire).

8.5.4 Compound carbonates

LEADHILLITE
C $PbSO_4.2PbCO_3.Pb(OH)_2$ or $Pb_4SO_4(CO_3)_2(OH)_2$

Characters and occurrence
HD 2.5–3.0 **SG** 6.55–6.57
Leadhillite is a greyish white monoclinic mineral, found as tabular crystals with a pearly or resinous lustre. It has a good basal cleavage, and splits into flexible laminae.

Leadhillite occurs in lead deposits as an alteration product of galena or cerussite in the zone of oxidation, as at Leadhills (Scotland) and Matlock (Derbyshire).

8.6 Nitrates

Crystallographically, the nitrates are similar to the carbonates. Only two nitrates are important, and both are simple nitrates with the formula *AX*.

SODA-NITRE (soda niter), nitratine, nitrate of soda, Chile saltpetre
C $NaNO_3$

Physical properties
CS Trigonal
F&H Rhombohedral crystals common, but usually as efflorescent crusts, or granular and massive
TW Common on $\{01\bar{1}2\}$, $\{02\bar{2}1\}$ and $\{0001\}$
COL White, grey, yellow, greenish, purple and reddish brown

CL Rhombohedral $\{10\bar{1}1\}$ perfect; imperfect on $\{01\bar{1}2\}$ and $\{0001\}$
L Vitreous
TR Transparent
TASTE Cooling
SOLUBILITY Soluble in water
HD 1.5–2.0
SG 2.25–2.27

Optical properties
$n_o = 1.587$
$n_e = 1.336$
$\delta = 0.251$
Uniaxial −ve
Soda-nitre is colourless in thin section and occurs as rhombohedra or massive. Relief is extremely variable and birefringence is extreme.

Tests Soda-nitre deflagrates less violently than nitre when heated, and colours the flame yellow, by which colour and its deliquescence it may be distinguished from that mineral.

Occurrence A mineral as soluble as soda-nitre can occur in workable quantities only in regions of very low rainfall. Economically important deposits are found in the Atacama Desert of northern Chile, and Chile produced 327 000 t of soda-nitre in 1985. The soda-nitre occurs, mixed with sodium chloride, sulphate and borate, and with clayey and sandy material, in beds up to 2 m thick. The sodium nitrate forms 14–25% of this **caliche,** as the material is termed, and is accompanied by 2–3% of potassium nitrate, and up to 1% of sodium iodate, the last being an important source of iodine. These remarkable deposits have most probably been leached from surrounding volcanic rocks, and owe their preservation to the very dry climate of the area.

Uses Soda-nitre is a source of nitrates used in explosives and fertilizers.

NITRE, SALTPETRE, nitrate of potash (niter)
C KNO_3

Physical properties

CS Orthorhombic	**L** Vitreous
F&H Found as acicular crystals; also in silky tufts and thin crusts	**TR** Subtransparent
	T Brittle
TW Cyclic twins occur on {110}	**TASTE** Saline and cooling
COL White, colourless, blue, yellow	**HD** 2.0
CL {011} perfect; imperfect on {110} and {010}	**SG** 2.08–2.11

Optical properties
$n_\alpha = 1.332$
$n_\beta = n_\gamma = 1.504$
$\delta = 0.172$
$2V = 7°$ −ve
In thin section, nitre is colourless, with very variable relief and extreme birefringence.

Tests Nitre shows the lilac flame of potassium. It is soluble in water and deflagrates on heating on charcoal.

Occurrence Nitre occurs in considerable quantities in the soil of certain countries including India, Egypt, Algeria, Iran and Spain. It also occurs in the loose earth forming the floors of natural caves, as in Kentucky, Tennessee and the Mississippi valley. The sodium nitrate deposits of Chile contain some 2–3% of potassium nitrate, and Chile supplies 113 500 t of nitre each year. Nitre is artificially manufactured from refuse animal and vegetable matter, which is mixed with calcareous soil. The calcium nitrate thus formed is treated with potassium carbonate and yields nitre.

Uses Nitre is used in the manufacture of explosives, in metallurgical and chemical processes and as a fertilizer.

8.7 Borates

KERNITE, rasorite
C $Na_2B_4O_7.4H_2O$

Physical properties

CS Monoclinic	**L** Pearly to vitreous
F&H Massive	**TR** Transparent to translucent
COL White	**HD** 2.5–3.0
CL Perfect {100} and {001}	**SG** 1.91–1.93

Optical properties
In thin section, kernite is colourless, with moderate negative relief ($n_\beta = 1.472$) and moderate birefringence ($\delta = 0.034$). It is biaxial negative, with $2V \approx 80°$.

Tests As for borax (see below).

Occurrence Kernite is a most important source of industrial borates. The important deposits are playa-lake evaporites which are worked in California.

BORAX, tincal
C $Na_2B_4O_7.10H_2O$

Physical properties

CS Monoclinic

R&H Squat prisms occur and also granular aggregates

COL White, sometimes tinged with blue, green or grey

S White

CL Perfect on {100}, good on {110}

F Conchoidal

L Vitreous, sometimes dull

TR Transparent to translucent

T Soft and brittle

TASTE Sweetish alkaline

SOLUBILITY Borax is soluble in water

HD 2.0–2.5

SG 1.71

Optical properties

$n_\alpha = 1.447$

$n_\beta = 1.469$

$n_\gamma = 1.472$

$\delta = 0.025$

$2V = 49° \, -ve$

COL Colourless

H Stubby prisms with octagonal or rectangular sections, resembling pyroxene

CL Perfect on {100}, good on {110}

R Moderate negative; RIs are less than 1.54

A Borax dehydrates to tincalconite ($Na_2B_4O_7.5H_2O$)

B Moderate with low second order colours, but sometimes anomalous colours seen

IF Bx_a figures seen on (010) sections, but dispersion is strong

DF Borax is distinguished from pyroxenes by its negative relief, negative $2V$ and by its solubility in water

Tests Heated before the blowpipe, borax bubbles up and fuses to a clear glassy bead. Borax is soluble in water, producing an alkaline solution. It colours the flame yellow, due to sodium, and when moistened with sulphuric acid gives a green flame due to boron. All boron minerals can be dissolved in dilute HCl, perhaps after fusion with sodium carbonate, and the solution tested on turmeric paper, by moistening the paper with the solution, drying it at 100°C by placing it on a flask filled with boiling water. The turmeric paper assumes a red–brown colour which changes to inky black on being moistened with ammonia.

Occurrence Borax, together with other borates, ulexite and colemanite, is deposited in playa deposits, alkaline flats and borax marshes formed by the drying up of saline lakes. Deposits of this type are well developed in California, where the borates have been leached out from bedded colemanite deposits of Tertiary age. Borax accompanies other borates in the lake deposits of Tertiary age in the same area. Borax also occurs in Tibet, on the shores and in the waters of lakes, and is there called 'tincal'. It may also be formed as a deposit by thermal springs.

ULEXITE, boronatrocalcite, natroborocalcite
C $NaCaB_5O_9.8H_2O$

Physical properties

CS Triclinic
F&H Globular or reniform masses with a fibrous structure; rarely acicular
COL White
CL {010} perfect, {110} good

L Silky fibres
TR Translucent to opaque
TASTE None
HD 2.5
SG 1.96

Optical properties

In thin section, ulexite is colourless with low relief (n_β = 1.504–1.506), moderate birefringence (δ = 0.023) and is biaxial positive with a large $2V$ (78°).

Tests Ulexite gives water on heating. It fuses to a clear glass, colouring the flame yellow. When moistened with sulphuric acid, it colours the flame green for an instant.

Occurrence Ulexite occurs associated with borax in lake deposits of California. It is also found in Chile and Argentina, as white reniform masses and in lagoon deposits associated with gypsum, glauberite and halite.

COLEMANITE
C $Ca_2B_6O_{11}.5H_2O$

Physical properties

CS Monoclinic
F&H Stubby prismatic crystals, growing within cavities; it is also found massive and granular
COL Colourless, white, greyish
S White

CL {010} perfect, {001} good
F Hackly
L Vitreous to adamantine
TR Transparent to translucent
HD 4.0–4.5
SG 2.42

Optical properties

n_α = 1.586
n_β = 1.592
n_γ = 1.614
δ = 0.028
$2V$ = 55° +ve
OAP is approximately parallel to (001)

COL Colourless
H Squat prisms or massive
CL {010} perfect, {001} good, meeting at ~90° on (100) section
R Moderate

B Moderate with middle second order colours
IF Single isogyres needed for sign and size; these are found on prismatic sections
E On an (010) section, γ (slow)^cl = 26° (max)

Tests Colemanite resembles feldspar and calcite in appearance, but may be readily distinguished from these by blowpipe tests. It yields water on heating. When heated before the blowpipe it decrepitates and colours the flame yellowish green. Heated on charcoal it becomes white and, moistened with cobalt nitrate and reheated, it turns blue. Colemanite is soluble in hydrochloric acid, with separation of boric acid on cooling. Test with turmeric paper as described in the tests for borax.

Occurrence Colemanite is found in the evaporites of borate playas, probably from the alteration of borax or ulexite. It is found in association with the borates as well as gypsum, calcite and celestite. Colemanite occurs in deposits of Tertiary age in San Bernadino, Los Angeles, Kern and Inyo counties, California. The colemanite is present as nodules in clays, and also in beds 3–15 m thick, resting on rhyolitic tuffs.

Priceite and **pandermite** are hydrated calcium borates related to colemanite; priceite is a soft white earthy mineral and pandermite is a somewhat harder mineral.

β-BORACITE, boracite, stassfurtite
C $Mg_3B_7O_{13}Cl$

Physical properties

CS Orthorhombic, pseudocubic	**F** Uneven or conchoidal
F&H Pseudo-isometric crystals; also massive, columnar or granular	**L** Vitreous
	TR Transparent to translucent
COL Colourless, white, yellow, greenish, greyish	**T** Brittle
	HD 7.0–7.5
S White	**SG** 2.95–2.97
CL None	

Optical properties
In thin section, β-boracite is colourless with moderate relief ($n_\beta = 1.662$–1.667) and low birefringence ($\delta = 0.010$). It is biaxial positive with a large $2V$ (82°).

Tests Heated on charcoal, boracite fuses and forms a bead which solidifies on cooling to a crystalline mass. It gives the green flame of boron. Heated on charcoal, moistened with cobalt nitrate and reheated, it yields a pink mass, due to magnesium. The chloride is given by the copper oxide microcosmic salt bead test. Boracite is insoluble in water, but soluble in hot hydrochloric acid.

Occurrence Borocite is an evaporite deposit occurring in the Stassfurt saline deposit in Germany, associated with halite, gypsum and anhydrite.

There the boracite occurs as small crystals or concretions. It is also found at Panderma (Turkey), at Kalkberg and Schildstein (Hanover, Germany), and Luneville, La Meurthe (France), where it is associated with similar minerals as at Stassfurt.

8.8 Sulphates

8.8.1 Anhydrous sulphates

Formula A_2X ($X = SO_4$)

MASCAGNITE
C $(NH_4)_2SO_4$

CS Orthorhombic

Characters and occurrence Mascagnite occurs as yellowish grey, pulverulent, mealy crusts in the neighbourhood of volcanoes. It also occurs in guano deposits accompanied by other ammonium sulphates such as taylorite $K_{2-x}(NH_4)_xSO_4$ with $x \approx 0.35$. Mascagnite has a vitreous lustre, and is easily soluble in water.

THENARDITE
C Na_2SO_4

Physical properties

CS Orthorhombic	**L** Vitreous
F&H Prismatic or tabular crystals common, and also as crusts	**TR** Transparent, but becomes opaque when exposed to moisture in the air
TW Common on {110}	**TASTE** Slightly saline
COL Whitish	**SOLUBILITY** Soluble in water
CL {010} perfect, {101} and {100} imperfect	**HD** 2.5–3.0
	SG 2.67

Optical properties
In thin section thenardite is colourless with low RIs ($n_\beta = 1.473$–1.477) and low birefringence ($\delta = 0.014$–0.017). It is biaxial positive with a large $2V$ (83°).

Occurrence Thenardite occurs in playa-lake evaporites, as in the alkali lakes of the western USA and Canada. It may occur as a deposit in fumeroles.

Formula AX

BARITE, BARYTES, heavy spar
C $BaSO_4$; Sr and Ca are often present as impurities

Physical properties
CS Orthorhombic

F&H Crystals prismatic elongated parallel to the a axis, but usually tabular in thick or thin plates. Barite also occurs massive, coarsely lamellar, granular and compact, often with a radiating fibrous structure.

COL Colourless or white, often tinged with red, yellow or brown; rarely blue

S White

CL Perfect on {001} and {210}; imperfect on {010}

F Uneven

L Vitreous to resinous; sometimes pearly

TR Transparent to opaque

T Brittle

HD 3.0–3.5

SG 4.50 (when pure)

Optical properties
$n_\alpha = 1.636$
$n_\beta = 1.637$
$n_\gamma = 1.647$
$\delta = 0.012$
$2V = 37° +ve$
OAP is parallel to (010)

COL Colourless

H Clusters of tabular crystals common; also massive

CL Basal cleavage {001} is perfect, as {210}; imperfect on {010}

R Moderate

A Barite may alter to witherite or be replaced by a large number of minerals

B Low, first order yellows or mottled colours seen

IF Bx_a figure seen on (100) section, with two good cleavages present

Varieties   **Cockscomb barytes** has tabular crystals arranged nearly parallel to one another. **Caulk** and **boulder** are terms used in the Derbyshire mines, caulk being the white massive variety and boulder the crystallized type. **Bologna stone** is a nodular and concretionary form of barite.

Tests   Heated before the blowpipe, barite decrepitates and fuses with difficulty, colouring the flame yellowish green. It is absorbed by the charcoal when fused with sodium carbonate, and the saturated charcoal when placed on a silver coin and moistened leaves a black stain. The high specific gravity of barite is distinctive.

Occurrence   Barite is a very common gangue mineral in metalliferous hydrothermal veins in association with fluorite, sphalerite, galena, calcite, etc., deposits of this nature being worked in the North of England, the USA, etc. It also occurs as residual nodules, resulting from the decay of limestones containing barite veins, as in Virginia and Derbyshire. Some barite veins appear to form in limestones rising by leaching of barium compounds from rocks containing these and being deposited by either magmatic or meteoric solutions. The cement of some sandstones, as for example the Triassic sandstone of Elgin and the Hemlock Stone of Nottingham, is barite which has been deposited by sedimentary processes.

Uses Barytes are used in the manufacture of white paint, especially to give weight to paper, for dressing poor quality calico, etc., in the production of wallpaper and asbestos goods, and as a weighting agent in drilling mud, this last use accounting for 85–90% of the annual world output of 5.88 Mt in 1985.

CELESTITE, celestine
C $SrSO_4$

Physical properties
CS Orthorhombic

F&H Tabular crystals parallel to basal plane, and also prismatic crystals parallel to a axis; it may be fibrous, granular or massive

COL White, sometimes with a pale blue tint

CL As barite; {001} perfect, {210} and {010} good

F Imperfect conchoidal

L Vitreous to pearly

TR Transparent to substranslucent

T Brittle

HD 3.0–3.5

SG 3.96

Optical properties
$n_\alpha = 1.621$
$n_\beta = 1.623$
$n_\gamma = 1.630$
$\delta = 0.009$
$2V = 50° +ve$
OAP is parallel to (010)

COL Colourless

H Tabular or prismatic crystals

CL Prismatic cleavage {210} good and basal cleavage {001} perfect

R Moderate

A Celestite alters to strontianite

B Low, with first order whites or greys

IF Figure too large to fit into field of view; single isogyre needed, found on a section approx parallel to (101)

DF Celestite has larger $2V$ and lower SG than barite

Tests Celestite gives the crimson flame of strontium. It is insoluble in acids. It fuses to a milk white globule which gives an alkaline reaction.

When fused with sodium carbonate, it gives a mass which blackens silver when moistened.

Occurrence Celestite is most common as a deposit in carbonate sedimentary rocks and evaporite deposits. It occurs in fissures with strontianite, dolomite, gypsum and fluorite. Deposits of this type are worked in Mexico and Morocco. Celestite may occur as disseminated grains or cement in various fine grained terrigenous rocks. Celestine occurs also in the sulphur deposits of Sicily and in the 'cap rock' of salt domes of the Gulf states of the USA.

Uses Celestite is used as a source of strontium salts.

ANGLESITE, lead vitriol
C $PbSO_4$; some Sr or Ba may replace Pb

Physical properties

CS Orthorhombic

F&H Prismatic, bipyramidal or tabular crystals found, variously modified. It is also found granular, massive or nodular – often around a galena core.

COL White, sometimes with a grey, green, blue or yellow tint

CL Fair on {001} and {210}; poor on {010}

F Conchoidal

L Adamantine to vitreous or resinous

TR Transparent to opaque

T Brittle

HD 2.5–3.0

SG 6.38

Optical properties

$n_\alpha = 1.878$
$n_\beta = 1.883$
$n_\gamma = 1.895$
$\delta = 0.017$
$2V = 60°–75°$ +ve
OAP is parallel to (010)

COL Colourless

H Prismatic or tabular crystals common variously modified; it is also massive

CL {001} and {210} fair, {010} poor

R High

A Anglesite may alter to cerussite

B Low, with first order red seen

IF A single isogyre is seen in a section approximately parallel to (101)

Tests Heated before the blowpipe, in the oxidizing flame, anglesite decrepitates and fuses to a clear globule, which on cooling becomes milk white. Heated in the reducing flame, it effervesces and yields metallic lead. Anglesite dissolves in hydrochloric acid, the solution yielding a dense precipitate of barium sulphate on the addition of barium chloride solution, indicating the presence of a sulphate; sulphate is also detected by the silver coin test (p. 455).

Occurrence Anglesite, when found in sufficient quantity, is a valuable lead ore. It is usually associated with galena and results from the decomposition of that mineral in the upper portion of lead veins. Localities include Parys mine in Anglesey, Cornwall, Derbyshire, Cumberland, Leadhills, Broken Hill (NSW), etc.

ANHYDRITE
C   CaSO₄

Physical properties

CS Orthorhombic

F&H Prismatic crystals tabular parallel to basal plane and elongated along *a* axis. It may occur as cubes pseudomorphing halite. It is also commonly fibrous, lamellar, granular and as aggregate masses.

COL Colourless or white often with a grey, bluish or reddish tint

CL Three mutually perpendicular cleavages on {010} and {100} perfect, and {001} good

F Uneven, splintery in lamellar and fibrous varieties

L Vitreous although pearly in cleavage planes

TR Transparent to subtranslucent

HD 3.0–3.5

SG 2.93–3.0

Optical properties

$n_\alpha = 1.570$
$n_\beta = 1.576$
$n_\gamma = 1.614$
$\delta = 0.040$
$2V = 42°–44° +ve$
OAP is parallel to (100)

COL Colourless

F Prismatic crystals with aggregates common

CL {010} and {100} perfect; {001} good

R Low to moderate

A Anhydrite can hydrate to form gypsum; this involves an increase in volume, and thick beds can suffer distortion

B High, with third order colours

IF Bx₍ₐ₎ figure is seen as a basal section and is just larger than the field of view

TW Repeated twinning common on {011}

Varieties **Vulpinite** is a scaly granular variety, found at Vulpino in Lombardy. It is sometimes harder than common anhydrite, owing to the presence of silica, and is occasionally cut and polished for ornaments.

Tripestone is a contorted concretionary form of anhydrite, and **muriacite** is a name sometimes applied to some of the crystallized varieties.

Tests Before the blowpipe, anhydrite turns white, but does not exfoliate like gypsum, and after a time yields an enamel-like bead. Fused with sodium carbonate and charcoal, anhydrite gives a mass which blackens silver when moistened. Anhydrite is soluble in boiling hydrochloric acid, a white precipitate being given on the addition of barium chloride. Anhydrite is harder than gypsum, has three cleavages while gypsum has one, has a greater specific gravity, and does not yield water when heated in the closed tube.

Occurrence Anhydrite occurs as a saline residue associated with gypsum and halite, as in the Stassfurt (Germany) and in many similar deposits. It has been shown that anhydrite forms from gypsum in sea water at 25°C, and so the alternating bands of gypsum and anhydrite found, for example, in the German deposits may possibly be annual layers. Whether gypsum or anhydrite was the original mineral in many deposits has been much discussed. Primary precipitation of anhydrite is favoured by warm and highly saline waters, but many anhydrite beds form by dehydration of gypsum. Anhydrite is associated with gypsum and sulphur in the 'cap rock' overlying salt domes.

Uses Anhydrite is of importance as a fertilizer, in the manufacture of plasters and cements, and of sulphates and sulphuric acid. The UK produced 3.2 Mt of gypsum and anhydrite in 1985.

Formula $A_m B_n X_p$ with $(m + n) : p > 1$

GLAUBERITE
C $Na_2Ca(SO_4)_2$

Physical properties

CS Monoclinic	**F** Conchoidal
F&H Prismatic crystals, tabular parallel to basal plane	**L** Vitreous
	TR Transparent to translucent
COL White, yellow, red, grey	**HD** 2.5–3.0
S White	**SG** 2.79
CL {001} perfect; {110} poor	

Optical properties
In thin section, glauberite is colourless with low relief ($n_\beta = 1.536$) and moderate

birefringence ($\delta = 0.021$) and with low second order colours. It is biaxial negative with a small $2V$ (7°).

Occurrence Glauberite occurs as a saline residue associated with the other sodium minerals of this type at the localities already cited. Deposits of glauberite that were formerly of great economic importance occur in the valley of the Ebro (Spain), and bear the characters of a saline residue.

8.8.2 Hydrated sulphates

Formula $A_2X.xH_2O$

MIRABILITE, glauber salt
C $Na_2SO_4.10H_2O$

Physical properties

CS Monoclinic	**CL** {100} perfect, others very poor
F&H Squat prisms or elongate acicular crystals found; it also occurs in efflorescent crusts and in solution in mineral waters	**L** Vitreous
	TR Translucent to opaque
	TASTE Cooling, saline and bitter
	SOLUBILITY Soluble in water
COL White, yellow, colourless	**HD** 1.5–2.0
	SG 1.465

Optical properties In thin section, mirabilite is colourless with moderate negative relief ($n_\beta = 1.410$) and moderate birefringence ($\delta = 0.023$). It is biaxial negative with a high $2V$ (76°) and extinction angle $\gamma\hat{}cl = \sim30°$ on an (010) section.

Tests Mirabilite gives water on heating in the closed tube. It shows the yellow flame of sodium, and produces a black stain when the residue obtained by heating on charcoal is moistened with water on a silver coin. When exposed to dry air, mirabilite loses water and goes to powder.

Occurrence Mirabilite occurs in the residues of alkali lakes, as at the Great Salt Lake of Utah, the alkaline lakes of Wyoming and other western states, and in Saskatchewan, Canada, and elsewhere. Sulphur is produced from mirabilite obtained from the Gulf of Karabugas.

Formula $A_mB_nX_p.xH_2O$ with $(m + n) : p > 1$

POLYHALITE
C $K_2(Ca_2Mg)(SO_4)_4.2H_2O$

Physical properties

CS Triclinic   **CL** {101} perfect
F&H As compact lamellar masses   **HD** 3.5
COL Pinkish, reddish, sometimes   **SG** 2.68–2.78
    colourless or white

Optical properties

In thin section, polyhalite is colourless with low relief ($n_\beta = 1.560$) and moderate birefringence ($\delta = 0.020$). It is biaxial negative with a high $2V$ ($\approx 65°$).

Occurrence   Polyhalite occurs in marine salt beds or saline residues, as at Stassfurt, where it forms a layer about 50 m thick, above the halite layer and below the carnallite and kieserite ($MgSO_4.H_2O$) layer. Other associated minerals include sylvite and anhydrite.

Formula $ABX_2.xH_2O$

POTASH ALUM, ALUM, kalinite
C $KAl(SO_4)_2.12H_2O$

Characters and occurrence

HD 2.0–2.5   **SG** 1.750–1.760

Alum crystallizes as cubic octahedra but usually occurs massive and is found in shales (alum shales) which contain pyrites and are undergoing decomposition. Such shales occur at Whitby in Yorkshire, and elsewhere potash alum also occurs in the neighbourhood of volcanoes. In neither mode of occurrence is the mineral sufficiently plentiful to be of much economic value nowadays. Alum is readily soluble in water, and has a characteristic astringent taste.

Formula $AX.xH_2$

KIESERITE, wathlingenite
C $MsSO_4.H_2O$

Characters and occurrence

HD 3.5   **SG** 2.57

A white monoclinic mineral, massive, granular or compact, occurring as a saline residue in Stassfurt (Germany), and also in other salt deposits elsewhere.

GYPSUM
C   CaSO$_4$.2H$_2$O

Physical properties
CS Monoclinic

F&H Prisms common, flattened parallel to (010). Crystals are combinations of {010}, {110} and {011}. It also occurs in laminated, granular and compact masses, and on fibrous habits.

COL Crystals colourless; massive varieties colourless or white – occasionally grey, yellowish, pink or buff coloured

TW Twins on {100} common, giving 'swallowtail' or 'butterfly' contact twins. Twinning on {101} is less common, giving 'arrowhead' types. Crystals sometimes occur in stellate interpenetrant groups.

CL {010} perfect; {100} and {$\bar{1}$11} imperfect

L Some faces and all cleavage faces shining and pearly; other faces subvitreous. Massive varieties glistening, but mostly dull and fibrous types are silky.

TR Gypsum is transparent (glass-like) to translucent, and even opaque

T Sectile; cleavage plates are flexible and non-elastic

HD 1.5–2.0

SG 2.31

Optical properties
$n_\alpha = 1.519$
$n_\beta = 1.522$
$n_\gamma = 1.529$
$\delta = 0.010$
$2V = 58°$ +ve
OAP is parallel to (010)

COL Colourless

H Usually crystalline, but may occur in aggregate masses

CL {010} perfect, {100} and {$\bar{1}$11} good

R Low

B Low (similar to quartz)

IF Single optic axis is best for obtaining sign and size

E Straight in (010) cleavage

Varieties
Broad transparent plates of gypsum are called **selenite**. **Alabaster** is a very fine grained snow white or light-coloured massive variety. **Satin spar** is the fibrous variety and has a silky lustre; **gypsite** is gypsum mixed with sand and earth.

Tests
Heated in the closed tube, gypsum gives water. In the flame test, it gives the calcium flame, but not readily. When fused with sodium carbon-

ate it yields a mass which blackens silver when moistened. It is readily soluble in dilute hydrochloric acid.

Occurrence Gypsum occurs mainly as thick, stratified sedimentary beds associated with beds of dolomite, limestone, halite and other evaporite deposits. It is a chemical precipitate of isolated massive basins, both recent and ancient, where it is preceded by clays and limestones and followed by anhydrite, halite and magnesium and potassium salts. Some gypsum probably forms from the hydration of anhydrite. In the normal evaporite sequence gypsum precipitates first. Gypsum is associated with anhydrite and sulphur in the cap rocks of salt domes, the sulphur possibly coming from the reduction of sulphates by the action of bacteria. Gypsum occurs in salt pans and dry lake beds, and ascending ground waters may deposit gypsum as single crystals or aggregates ('desert-roses') or as efflorescences in desert soils. Gypsum may form near fumeroles and volcanic vents, and it may occur in the gossan or oxide zones of metalliferous sulphide deposits if carbonate rocks are present. It is found in commercial quantities at Stassfurt, and also in the USA. The formation of gypsum by the action of sulphuric acid, generated by the decomposition of pyrite, on the calcium carbonate of shells, etc. in shales or mudrocks (good crystals of selenite are found in many clay formations, such as the London Clay, Oxford Clay, etc.) should be noted, as this will constitute serious problems for the engineer, especially with regard to concrete infill (if needed) or the stability of these clay slopes.

Production and uses World production of gypsum in 1985 was 83 Mt, with the two largest producers being the USA (13.8 Mt) and Canada 8.4 Mt). Gypsum is produced by a large number of countries, including the USSR (4.9 Mt) and the UK (3.2 Mt, including anhydrite). It is used primarily in the manufacture of plasterboard and other plaster products for the building industry.

CHALCANTHITE, blue vitriol, cyanosite, copper vitriol
C $CuSO_4.5H_2O$; with 25.4% copper

Physical properties

CS Triclinic	**TR** Subtransparent to translucent
F&H Tabular prismatic crystals found, and it also occurs massive, stalactitic, fibrous and encrusting	**T** Brittle
	TASTE Nauseous and metallic
COL Sky-blue, greenish	**SOLUBILITY** Soluble in water
CL {110} poor	**HD** 2.5
L Vitreous	**SG** 2.12–2.30

Optical properties
In thin section, chalcanthite is pale blue in colour with low relief ($n_\beta = 1.539$), and moderate birefringence ($\delta = 0.030$). It is biaxial negative with a moderate $2V$ ($56°$).

Tests Heated in the closed tube, chalcanthite gives water. It is soluble in water, the solution coating a clean strip of iron with metallic copper. Copper may be procured by this means from water pumped from copper mines. Heated on charcoal with sodium carbonate and carbon, chalcanthite yields metallic copper

Occurrence Chalcanthite results from the alteration of chalcopyrite and other copper sulphides, and is found in the zone of weathering of copper lodes, as in Cornwall, the Rammelsberg mine in the Harz, and in Chile.

Brochanthite, or waringtonite, and **langite** are other hydrated sulphates of copper, of an emerald green colour, and occurring in copper gossans.

MELANTERITE, green vitriol, copperas
C $FeSO_4.7H_2O$

Physical properties

CS Monoclinic

F&H Crystals rare, occurring as acute prisms but commonly massive or pulverulent, botryoidal reniform or stalactitic

COL Shades of green or white on fresh surfaces, but exposed surfaces are yellowish brown with a vitrified or glazed appearance, resembling a furnace slag

S Colourless

CL Perfect {001} basal

F Conchoidal

L Vitreous

TR Subtransparent or translucent

T Brittle

TASTE Sweetish, astringent, metallic and nauseous

HD 2.0

SG 1.90

Optical properties
In thin section, melanterite is colourless or pale green with a low negative relief ($n_\beta = 1.478$) and a low birefringence ($\delta = 0.015$). It is biaxial positive with a very large $2V$ ($85°$).

Tests Melanterite is soluble in water and, on the addition of barium chloride to the solution, a white precipitate is thrown down. Heated before the blowpipe, melanterite becomes magnetic. It gives a green glass with borax.

Occurrence Melanterite results from the decomposition of pyrites in the zone of oxidation, and is found in small quantities wherever pyrite occurs, notable localities being Copperas Mount, Ohio, USA, and Goslar in the Harz.

Uses Melanterite is used by tanners, dyers and ink manufacturers, as it yields a black colour with tannic acid. When treated with potassium ferrocyanide (yellow prussiate or potash) it forms the pigment known as Prussian blue.

EPSOMITE, EPSOM SALTS, seelandite
C $MgSO_4.7H_2O$

Physical properties
CS Orthorhombic

F&H Crystals rare occurring as prisms, but usually in fibrous crusts or botryoidal masses

COL White

L Vitreous

TR Transparent to translucent

TASTE Bitter and saline

HD 2.0–2.5

SG 1.68

Optical properties
In thin section, epsomite is colourless with a moderate negative relief (n_β = 1.452–1.462) and moderate birefringence (δ = 0.028). It is biaxial negative with a moderate $2V$ (50°).

Tests Epsomite is soluble in water. Heated in the closed tube, it gives water. Heated with cobalt nitrate on charcoal, it gives a pink residue. If epsomite is heated with sodium carbonate on charcoal, and the mass obtained is moistened and placed on a silver coin, a black stain is produced.

Occurrence Epsomite occurs in solution in sea water and in mineral waters. It is deposited from the waters of saline lakes, as in British Columbia and Saskatchewan. It occurs as efflorescent crusts and masses, as in the limestone caves of Kentucky, and encrusting serpentinite and other rocks rich in magnesium.

Uses In medicine and in tanning.

GOSLARITE, white vitriol
C $ZnSO_4.7H_2O$

Physical properties

CS Orthorhombic

F&H Prismatic crystals found, but usually stalactitic or encrusting

COL White

S White

CL {010} perfect

L Vitreous

TR Transparent to translucent

TASTE Astringent, metallic and nauseous

HD 2.0–2.5

SG 1.94–1.98

Optical properties

In thin section, goslarite is colourless with low negative relief (n_β = 1.465–1.480) and moderate birefringence (δ = 0.023). It is biaxial negative with a moderate $2V$ (46°–65°).

Tests Heated in the closed tube, goslarite boils and gives off water, fusing to an opaque white mass. Heated on charcoal, it fuses with ebullition, and gives an encrustation of zinc oxide, yellow when hot and white when cold; and when this encrustation is moistened with cobalt nitrate and reheated a green mass is produced. Goslarite is readily soluble in water, the solution yielding a white precipitate of barium sulphate on the addition of barium chloride solution, indicating the presence of a sulphate.

Occurrence Goslarite results from the decomposition of sphalerite and is found sparingly in some of the Cornish mines and at Holywell, Flintshire. The type locality is the Rammelsberg mine, Goslar (Germany).

MORENOSITE, nickel vitriol
C $NiSO_4.7H_2O$

Physical properties

CS Orthorhombic

F&H Accicular crystals occur, and also fibrous and efflorescent crusts

COL Apple green to pale green

HD 2.0–2.5

SG 1.91–1.95

Optical properties

In thin section, morenosite is pale green with low negative relief (n_β = 1.49) and moderate birefringence (δ = 0.02). It is biaxial negative with a moderate $2V$ (42°).

Tests Morenosite gives nickle reactions in the borax bead. It is soluble in hydrochloric acid, the solution giving a dense white precipitate on the addition of barium chloride solution.

Occurrence Morenosite is associated with other nickle blooms, as a weathering product of primary nickel minerals.

Formula $A_2X_3.xH_2O$

ALUNOGEN
C $Al_2(SO_4)_3.18H_2O$

Physical properties

CS Triclinic

F&H Crystals rare, usually occurring in fibrous masses or massive and in crusts

COL White, occasionally yellowish or reddish

L Vitreous, but silky on delicate fibrous masses

TR Subtransparent to subtranslucent

TASTE Astringent, similar to potash alum

SOLUBILITY It is soluble in water

HD 1.5–2.0

SG 1.64–1.78

Optical properties

In thin section, alunogen is colourless with low negative relief ($n_\beta = 1.463$–1.478) and low birefringence ($\delta = 0.014$). It is biaxial positive with a variable $2V$ ($31°$–$69°$).

Tests Heated in the closed tube, alunogen gives off water. It gives a blue mass when heated with cobalt nitrate on charcoal.

Occurrence Alunogen occurs in the neighbourhood of volcanoes. It is found in shales, especially alum shales, where iron pyrite is decomposing.

8.8.3 Anhydrous or hydrated sulphates with hydroxyl (or halogen)

Formula $A_2X(OH)_q$ or $A_4X_2(OH)_q$

LINARITE
C $PbCuSO_4(OH)_2$

Characters and occurrence

HD 2.5   SG 5.30–5.32

A rare monoclinic mineral. It is deep azure blue in colour with a pale blue streak. It occurs in the zone of oxidation of lead–copper veins, and has been found in Cumberland, Leadhills, etc.

ALUNITE, alumstone
C $KAl_3(SO_4)_2(OH)_6$; extensive solid solution exists between alunite and natroalunite ($NaAl_3$, etc.) and therefore most alunite contains appreciable Na

Physical properties

CS Trigonal

F&H Crystals rare, but found as rhombohedra, often flattened parallel to basal plane, but usually occurring fibrous, granular, massive and columnar

COL White, greyish, reddish

CL Good {0001} basal

F Flat conchoidal or uneven of crystals; splintery of massive varieties

L Vitreous of crystals; dull of massive varieties

TR Transparent to translucent

HD 3.5–4.0

SG 2.6–2.9

Optical properties

n_o = 1.568–1.585

n_e = 1.590–1.601

δ = 0.010–0.023

Uniaxial +ve

COL Colourless

H Found as aggregates of euhedral and anhedral crystals; also massive

CL Basal cleavage present

R Low to moderate

B Low to moderate

Tests Heated in the closed tube, alunite gives water, and on intense heating sulphurous fumes are given off. Heated with cobalt nitrate it gives a blue colour. Sulphur is given by the silver coin test. Alunite is insoluble and therefore has no taste.

Occurrence Alunite occurs as an alteration or replacement of alkali feldspars by sulphur gases and sulphate hydrothermal solutions in trachytes and rhyolites, in which it forms seams and pockets, as in Italy, Spain, Korea, New South Wales, the USSR, and in Nevada, Utah and Colorado.

Uses As a source of potassium and aluminium salts

PLUMBOJAROSITE

C $PbFe_6(SO_4)_4$

Characters and occurrence

HD Soft **SG** 3.60–3.64

A dark brown trigonal mineral occurring as tiny tabular crystals in certain mines in Utah. It is rare.

Formula $A_nB_nX_p(OH,Cl)_q.xH_2O$

KAINITE

C $KMgSO_4(Cl).3H_2O$

Occurrence   Kainite occurs in the upper parts of saline residues, such as at Stassfurt, where it is in part due to the leaching out of magnesium chloride from the carnallite zone.

ALUMINITE, websterite
C $Al_2(SO_4)(OH)_4.7H_2O$

Physical properties

CS Monoclinic or orthorhombic

F&H It occurs as friable, nodular masses of fibres in veins

COL White, yellowish

L Dull

TR Opaque

OTHERS It adheres to the tongue

HD 1.0–2.0

SG 1.66–1.82

Optical properties

In thin section, it is colourless, with moderate negative relief (n_β = 1.464) and low birefringence (δ = 0.011). It is considered to be biaxial positive with a very large $2V$.

Tests   Heated in closed tube, aluminite gives water. Heated on charcoal with cobalt nitrate it gives a blue mass. Heated on charcoal with sodium carbonate, and the resulting mass transferred to a silver coin and moistened, a black stain is produced. The physical properties are distinctive.

Occurrence   Aluminite is usually found in clays of Tertiary age, and sometimes in clay-filled pipes or pot holes in the surface of the Chalk, as at Newhaven in Sussex, and elsewhere.

ZIPPEITE, uraconite
C $(UO_2)_2SO_4(OH)_2.xH_2O$; zippeite has a slightly different composition to that of uraconite

Characters and occurrence

HD Soft   SG 3.66–3.68

Zippeite occurs as earthy or powdery crusts to other uranium minerals, as at Joachimsthal, and in Cornwall. Zippeite may also contain some copper oxide. It occurs in delicate acicular orthorhombic crystals, rosettes and warty crusts, and accompanies uraconite. When zippeite is heated before the blowpipe with microcosmic salt, a yellowish green bead is given in the oxidizing flame, and an emerald green bead in the reducing flame.

8.9 Chromates

CROCOITE, crocoisite, crocoise
C PbCrO₄

Physical properties

CS Monoclinic
F&H Prismatic crystals occur with longitudinal striations, and also as granular aggregates
COL Hyacinth red, shades of red
S Orange yellow
CL {100} good, others poor

F Conchoidal or uneven
L Adamantine to vitreous
TR Translucent
T Sectile
HD 2.5–3.0
SG 5.99–6.11

Optical properties

In thin section, crocoite is red or yellow in colour with extreme relief ($n_\beta = 2.36$) and moderate birefringence masked by mineral colour. It is biaxial positive with a large $2V$ (57°).

Tests Heated in the closed tube, crocoite decrepitates and blackens, but reverts to its original colour on cooling. Heated on charcoal with sodium carbonate, it is reduced to metallic lead, and coats the charcoal with an encrustation of chromium and lead oxides. The microcosmic salt bead is emerald green in both oxidizing and reducing flames.

Occurrence Crocoite occurs where lead lodes traverse rocks containing chromium, as at Beresof in Siberia, in the Urals, Hungary and the Philippines.

8.10 Phosphates, arsenates and vanadates

There are partial or complete substitution series between P and As, As and Sb, As and V and, less commonly between P and V, which is why these oxysalts are taken together.

8.10.1 Anhydrous phosphates

XENOTIME
C YPO₄, wiith REEs replacing Y

Physical properties

CS Tetragonal
F&H Stubby bipyramidal or elongate prismatic crystals are most common; detrital grains common
COL Yellowish or brownish

CL {110} good
L Vitreous or adamantine to resinous
TR Transparent
HD 4.0–5.0
SG 4.3–5.1

Optical properties
$n_o = 1.719–1.724$
$n_e = 1.816–1.827$
$\delta = 0.095–0.107$
Uniaxial positive (length slow)

COL Pale yellow to colourless
P Pleochroic with o pale yellow, pale brown, and e yellow, brownish or brownish green
H Euhedral crystals common

CL Prismatic {110} cleavages intersect at 90° in basal section
R High to very high
A None
B Very high to extreme, with pale colours of high orders often masked by mineral colour

Tests Similar to monazite (see below).

Occurrence Xenotime is a widespread accessory mineral in granites, syenites and other late-stage magmatic rocks. Often misidentified as zircon, it is associated with zircon, monazite, allanite and other REE minerals. It is a common detrital heavy mineral in sands and placer deposits.

MONAZITE

C (Ce,La,Th)PO$_4$; with Th^{4+} replacing Ce^{3+} and [SiO$_4$]$^{4-}$ groups replacing [PO$_3$]$^{3-}$ groups. Solid solution may exist with huttonite (ThSiO$_4$).

Physical properties

CS Monoclinic
F&H Small, blocky, euhedral crystals common elongated along b or flattened parallel to (100)
TW Contact twins on {100} are common
COL Pale yellow to dark reddish brown

S White
CL {100} good, {010} poor; a basal parting may be present
F Conchoidal or uneven
L Resinous to vitreous or waxy
TR Transparent
HD 5.0–5.5
SG 4.6–5.4

Optical properties
$n_\alpha = 1.770–1.800$
$n_\beta = 1.777–1.801$
$n_\gamma = 1.828–1.851$
$\delta = 0.045–0.075$
$2V = 6°–19°$ +ve
OAP is perpendicular to a axis

COL Colourless to pale yellow
P Weak in yellows (β is stronger yellow than α and γ)
H Euhedral blocky crystals common; it is also found in detrital grains
CL {100} good, {001} parting, others very poor
R Very high
A None

B High to very high, with upper third and fourth order colours seen
IF Basal sections give excellent Bxa figures with small $2V$
E An (010) section gives $\gamma\hat{}cl = 2°–7°$
DF Sphene is similar but has much larger $2V$ and extreme dispersion Xenotime and zircon are uniaxial.

Tests Monazite is infusible before the blowpipe. If it is fused with sodium carbonate, and the resulting mass is dissolved in nitric acid, a few drops of ammonium molybdate added to the solution will give a white precipitate, indicating phosphate.

Occurrence Primary monazite occurs as an accessory component of acid igneous rocks such as granites, and large crystals and masses have been found in pegmatites. It is found as a heavy residue in sediments, and it is obtained on a commercial scale from sands where natural processes of concentration have occurred, the source of the monazite being a neighbouring monazite-bearing granitic rock. The mineral occurs as a constituent of the seashore sands, or monazite sands, near Prado, in the south of the State of Bahia, Brazil, and at various parts of the coast of the states of Espirito Santo and Rio de Janeiro, the concentrate from such sands containing 5–7% ThO_2. Less important inland deposits yield monazite with 4–5.7% ThO_2. Brazil produced 6000 t in 1985. Travancore, Madras (India), has become an important producer of monazite at 4000 t per annum, the mineral occurring with ilmenite, rutile and zircon, as a beach sand. The mineral from this locality is richer in ThO_2 than that of Brazil, the oxide rising to 14% in selected specimens. The mineral is also mined in Australia (14 000 t per annum), which is the biggest producer of REEs in the world, along with China which mines bastnaesite – a hydrated REE carbonate mineral. Monazite sands usually consist of monazite, naturally concentrated with other heavy obdurate minerals such as garnet, magnetite, rutile, ilmenite, zircon, etc. Separation is effected by electromagnetic separators, the magnets of which are adjusted to varying intensities; magnetite and ilmenite are removed first, and monazite, usually the most feebly magnetic, last. Rutile, zircon and siliceous matter are rejected and are further treated for the recovery of rutile and zircon. The manufacture of ThO_2 from the separated monazite is a complicated and purely chemical operation.

8.10.2 Hydrated phosphates, etc.

Vivianite Group

VIVIANITE, blue-iron earth

C $Fe_3(PO_4)_2.8H_2O$) iron peroxide is sometimes present

Physical properties

CS Monoclinic

F&H Very small crystals occur, either as modified prisms or bladed, often in divergent aggregates of crystals; it also occurs radiating, reniform and as encrustations

COL White or colourless if fresh, but usually deep blue or green

S Bluish white, sometimes colourless or changing to indigo blue. The dry powder is dark brown in colour.

CL {010} perfect

L Pearly to vitreous

TR Transparent to translucent, turning opaque on exposure to air

T Sectile; thin laminae are flexible

HD 1.5–2.0

SG 2.66–2.71

Optical properties

In thin section, vivianite is pleochroic in blues, with α blue, β pale bluish or yellowish green and γ pale yellow–green. It has a moderate relief ($n_\beta = 1.602$–1.656) and a high birefringence ($\delta = 0.051$). It is biaxial positive with a large $2V$ (63°–83°).

Tests Heated before the blowpipe, vivianite fuses, loses its colour, and becomes converted to a greyish black magnetic globule. With the fluxes it gives reactions for iron. Heated in the closed tube, it whitens, exfoliates and yields water. Vivianite is soluble in hydrocholic acid. If fused with sodium carbonate, ignited with magnesium and moistened, phosphoretted hydrogen is produced.

Occurrence Vivianite is found associated with iron, copper and tin ores. It also may occur in clay, mud and peat, and especially in bog iron ore. Sometimes it is found in or upon fossil bones or shells. Localities include several mines in Cornwall and Devon, peat swamps in Shetland, the Isle of Man (occurring with the horns of elk and deer), Bodenmais in Bavaria, Orodna in Transylvania, etc.

ERYTHRITE–ANNABERGITE, cobalt bloom – nickel bloom.

C   **erythrite** $Co_3(AsO_4)_2.8H_2O$

   annabergite $Ni_3(AsO_4)_2.8H_2O$

Ca and Fe^{2+} may be present in minor amounts.

Physical properties

CS Monoclinic

F&H Rare prismatic crystals, but usually earthy, pulverulent, encrusting, and sometimes globular or reniform

COL Peach red or crimson (erythrite); white, greyish or greenish (annabergite)

S Same as colour but rather paler. When the powder is dry it is lavender blue in colour.

CL {010} perfect

L Pearly of cleavage planes but adamantine or vitreous of crystal faces; dull and lustreless of massive varieties

TR Subtransparent to opaque

HD 1.5–2.5

SG 3.07–3.25

Optical properties In thin section, erythrite is pleochroic in pinks and reds whereas annabergite is colourless to pale green. Relief is moderate ($n_\beta = 1.658$) and birefringence is very high ($\delta = 0.059$–0.072). They are biaxial positive with $2V \approx 90°$.

Tests When heated slightly in the closed tube, erythrite yields water, and on additional heating it gives a sublimate of arsenious oxide. Heated alone before the blowpipe, it fuses and colours the flame light blue. With borax, it gives a deep blue bead. It is soluble in hydrochloric acid, and forms a rose red solution. Annabergite, the nickel-rich member, gives nickel reactions in the borax bead test and when heated in the closed tube with charcoal gives water and an arsenic mirror.

Occurrence Erythrite is the cobalt bloom formed in the upper parts of veins, etc., or by the weathering of cobalt ores, and is found associated with smaltite and cobaltite in the oxide zones of cobalt deposits. Annabergite occurs as a coating of apple green capillary crystals, and results from the decomposition of nickel minerals. It is found in the oxide zones of nickel deposits.

8.10.3 Anhydrous phosphates, etc. containing hydroxyl or halogen

Formula A_5X_3:(OH,Cl,F) with $X = PO_4$, AsO_4, VO_4

APATITE

C $Ca_5(PO_4)_3(F,OH,Cl)$; the F-bearing apatite is called **fluorapatite**, the Cl-bearing one is called **chlorapatite**, and the (OH)-bearing one is called **hydroxylapatite**. Complete solid solution probably exists between all three types.

Physical properties

CS Hexagonal

F&H Crystals occur as prisms with pyramids, and the basal plane is sometimes present. Apatite also occurs as fibrous to columnar aggregates and may be coarsely granular. When it occurs globular, botryoidal or in other colloform structures, it is called collophane – a secondary cryptocrystalline mineral of apatite composition.

COL Pale sea green, bluish green, yellowish green and greenish yellow are the usual colours, but it may appear in shades of blue, grey red and brown; occasionally it is colourless and transparent

S White

CL Basal {0001} is poor

F Conchoidal and uneven

L Subresinous to vitreous

TR Transparent to opaque

T Brittle

HD 5 (one of the minerals on Mohs' scale)

SG 2.9–3.5

Optical properties

n_o = 1.633–1.667 F (lowest) to Cl (highest)

n_e = 1.629–1.665

δ = 0.007–0.017

Uniaxial −ve (crystal is length fast)

COL Colourless

H Small prismatic crystals with hexagonal cross section, often found with amphiboles and micas

R Moderate

B Low, first order greys usual (reds in Cl-bearing types)

Varieties The two major varieties of natural phosphates are (a) **apatite**, which has a definite chemical composition, and (b) **rock phosphates**, such as phosphorite, phosphatic limestone, guano, bone beds, etc., which have no definite chemical composition. **Phosphorite** is a variety of natural phosphate resulting, in some important occurrences, from the accumulation of organic remains and droppings upon desert islands, the calcite of the island rock being replaced by phosphates, forming a mixture of calcium phosphate and unaltered calcite. Phosphorite may show traces of the original structure of the parent rock, or it may be concretionary or mamillated, in which case it is known as **staffelite**. **Coprolite** is a term applied more particularly to those masses of phosphate found in sedimentary rocks which exhibit a corrugated or convoluted form, corresponding with what is supposed to have been the form of the internal casts of the intestines of certain saurians, fishes, etc., coprolites being consequently regarded as the fossil excrement of those animals. The name coprolite has been loosely applied to phosphatic concretions which have formed round fossil shells or bones, and which have been worked at several geological horizons such as, for example, in the Greensand, Gault and Crags of

England. **Asparagus stone** is the translucent greenish-yellow crystallized variety of apatite. **Osteolite** is a massive impure altered phosphate, usually having the appearance of lithographic stone.

Tests   The resinous lustre is rather distinctive. Reactions to blowpipe and other tests depend on the composition of apatite. It is soluble in hydrochloric acid, the solution giving a precipitate of calcium sulphate on addition of sulphuric acid. The red flame of calcium is sometimes given and, when moistened with sulphuric acid, apatite may show the blue–green flame of phosphorus. When heated with magnesium and moistened, phosphoretted hydrogen is given off. When heated with sulphuric acid, greasy bubbles of hydrofluoric acid are sometimes produced. The presence of chloride is sometimes given by the copper oxide – microcosmic salt bead test.

Occurrence   Apatite is an important accessory mineral of igneous rocks, particularly the acid igneous rocks such as granites, pegmatites and veins, but it is also common in diorites and gabbros. Apatite is present in many chlorite schists and amphibole-bearing schists and gneisses. Apatite occurs as a detrital mineral in sedimentary rocks. Sedimentary phosphatic deposits commonly contain a cryptocrystalline phosphate mineral called '**collophane**'. Apatite is also present in crystalline limestones. Workable deposits of apatite occur in pegmatitic and pneumatolytic veins, as in Ontario and Norway. In the case of the Canadian deposits it is possible that a part of the apatite rock represents a thermally altered limestone. Large apatite deposits are found in the alkaline syenites of the Kola peninsula in northern Russia. Dyke-like masses of apatite, rutile and ilmenite, called nelsonite, have been worked in Virginia. Rock phosphates, phosphorite, etc. occur in marine-bedded deposits, where the phosphate forms thick and extensive beds, layers of nodules, or the material of bone beds, in residual deposits, and in replacement deposits in which the original phosphate has been leached from overlying guano beds and has replaced the underlying calcareous rocks. The production of phosphate is discussed in Chapter 7 under Phosphorus.

Uses   The most important use of apatite and phosphate rock is as fertilizers, only minor amounts being employed for the production of phosphorus chemicals.

Pyromorphite series
Three lead minerals with related properties were included here. Their compositions are:

pyromorphite	$Pb_5(PO_4)_3Cl$	(phosphate)
mimetite	$Pb_5(AsO_4)_3Cl$	(arsenate)
vanadinite	$Pb_5(VO_4)_3Cl$	(vanadate)

These minerals are related to the apatite family, and certain intermediate compounds are known.

The three minerals considered here are isomorphous; they belong to the hexagonal system and crystallize as long prismatic crystals, with vivid colours.

They occur in the zone of oxidation of lead deposits, as at numerous mines in Cornwall, Derbyshire, Cumberland, Flintshire, Leadhills, Saxony, Harz, Mexico and the USA, and sometimes serve as minor ores of lead.

PYROMORPHITE, green lead ore
C $Pb_5(PO_4)_3Cl$; As or Ca may also be present

Physical properties
CS Hexagonal

F&H Crystals form as prisms with pyramidal terminations and are usually aggregated; it also occurs as crusts, and reniform and botryoidal masses

COL Shades of green, yellow and brown, often vivid

S white or yellowish white

F Subconchoidal or uneven

L Resinous

TR Subtransparent to subtranslucent

T Brittle

HD 3.5–4.0

SG 7.04–7.10

Optical properties
In thin section, pyromorphite is colourless or shows pale colours. It has extreme relief ($n_o = 2.058$), low birefringence ($\delta = 0.010$), and is uniaxial negative.

Tests Heated in the closed tube, pyromorphite gives a white sublimate of lead chloride. Heated alone before the blowpipe it fuses easily, colouring the flame bluish-green. Heated on charcoal it fuses to a globule which, when cool, assumes a crystalline angular form, but without being reduced to metallic lead. At the same time, the charcoal becomes coated with white lead chloride and yellow lead oxide. Heated with sodium carbonate, it yields a metallic lead bead. When heated with magnesium in the closed tube, and then moistened, a smell of phosphoretted hydrogen is given. Pyromorphite is soluble in acids.

Occurrence Pyromorphite is found in company with other ores of lead in the oxidized zone of lead veins, as at the localities cited in the introduction to the pyromorphite series.

MIMETITE, green lead ore
C $Pb_5(AsO_4)_3Cl$

Physical properties

CS Hexagonal; but may be monoclinic, pseudohexagonal

F&H Similar crystals to those of pyromorphite; it also occurs in botryoidal and crusty varieties

COL Pale yellow, brown or white
S Whitish
L Resinous
HD 3.5–4.0
SG 7.28

Optical properties

In thin section, mimetite is colourless or shows pale yellow colours. It has extreme relief (n_o = 2.124), low to moderate birefringence (δ = 0.018–0.024), and is uniaxial negative.

Variety **Compylite** is a brown or yellowish variety occurring as barrel-shaped crystals.

Tests Heated in the closed tube, mimetite behaves like pyromorphite. Heated on charcoal in the reducing flame, it yields metallic lead, gives off an arsenical odour, and coats the charcoal with lead chloride, lead oxide and arsenious oxide. Mimetite colours the flame blue and green. It is soluble in hydrochloric acid.

Occurrence Mimetite occurs in the oxidation zone of lead deposits. It is a minor ore of lead.

VANADINITE
C $Pb_5(VO_4)_3Cl$; PO_4 may replace VO_4, and As may sometimes be present in small amounts

Physical properties

CS Hexagonal

F&H Prismatic or acicular crystals occur, similar to those of mimetite and pyromorphite; it also occurs in crusts

COL Ruby red, orange brown, yellowish
S White to yellowish
CL None
L Resinous
HD 2.5–3.0
SG 6.88–6.93

Optical properties

In thin section, vanadinite shows pale colours in yellow, orange or brown and pleochroic with o pale colours and e colourless. It has extreme relief (n_o = 2.416) and very high birefringence (δ = 0.066). It is uniaxial negative.

Tests   Vanadinite gives the reactions for lead. It is soluble in nitric acid, a precipitate of silver chloride being given with silver nitrate solution. The presence of vanadium is given by the microcosmic salt bead; in the oxidizing flame it is yellow, and in the reducing flame bright green.

Occurrence   Vanadinite occurs in the zone of oxidation of lead veins, associated with lead oxysalts and the other members of the pyromorphite series, as at many lead-mining localities. Vanadinite also accompanies other vanadium minerals in sediments, as in the Triassic sandstone of Alderley Edge in Cheshire.

Uses   As a source of vanadium, and as a minor ore of lead.

8.10.4   *Hydrated phosphates containing hydroxyl or halogen*

Formula $A_m B_n X p$ (OH,Cl)$_q$ xH$_2$O with $(m + n) : p \approx 3 : 2$

TURQUOISE, turquois, henwoodite
C   CuAl$_6$(PO$_4$)$_4$(OH)$_8$.+5H$_2$O

Physical properties

CS Triclinic		
F&H Fine granular aggregates, massive stalactitic or encrusting	**L** Waxy to dull
	TR Translucent to opaque
COL Turquoise blue or bluish green	**HD** 5.0–6.0
F Conchoidal	**SG** 2.60–2.91

Optical properties
In thin section, turquoise is blue and pleochroic, with α and β colourless and γ pale blue to pale green. It has moderate relief (n_β = 1.62), high birefringence (δ = 0.04), and is biaxial positive with moderate $2V$ (40°).

Tests   Turquoise gives water when heated in the closed tube. It gives reactions for copper, and it is soluble in hydrochloric acid after ignition.

Occurrence   Turquoise occurs in thin veins, patchy deposits and seams in rocks, such as trachytes and other igneous rocks and Al-rich rocks that have been profoundly altered. The phosphate in the mineral may be derived from apatite. Gem production comes from the USA, Egypt and the USSR.

Uses Turquoise is used in jewellery. Fossil bones and teeth, coloured by vivianite, and termed **odontolite**, or bone turquoise, are frequently cut and polished for the same purpose.

WAVELLITE
C $Al_3(PO_4)_2(OH)_3.5H_2O$; some F and Fe may also be present

Physical properties
CS Orthorhombic

F&H Crystals rare, usually occurring as fibrous spherules, up to 10 mm diameter, with a radiating structure. Single crystals are stout or thin prisms.

COL White, yellowish or brownish

CL {110} perfect

L Vitreous, pearly or resinous

TR Translucent

HD 3.5–4.0

SG 2.32–2.37

Optical properties
In thin section, wavellite is colourless or weakly pleochroic in pale blues and browns. It has low relief (n_β = 1.526–1.543), moderate birefringence (δ = 0.025), and is biaxial positive with a large 2V (72°).

Tests Wavellite yields water when heated in the closed tube. It gives a blue mass when heated with cobalt nitrate on charcoal, and also usually gives reactions for fluorine. The radiating fibrous spherulitic structures are characteristic.

Occurrence Wavellite occurs in residual deposits formed from igneous rocks, as at St Austell in Cornwall. It also occurs as nodular masses associated with manganese ores and limonite, as at Holly Spring, Pennsylvania.

TORBERNITE, copper uranite
C $Cu(UO_2)_2(PO_4)_2.8–12H_2O$; some PO_4 may be replaced by AsO_4

Physical properties
CS Tetragonal

F&H Square, tabular crystals, often modified on the edges; it also occurs scaly, foliaceous and earthy

COL Emerald to grass green

S Green, paler than mineral colour

CL {001} perfect, giving very thin laminae, similar to micas, whence its occasional name uran-mica

L Sub-adamantine of crystals, pearly of cleavage planes

TR Transparent to subtranslucent

T Cleavage laminae are brittle and not flexible

HD 2.0–2.5

SG 3.22–3.28

Optical properties
In thin section, torbernite is green and pleochroic with o green and e colourless to

pale greenish blue. It has low relief ($n_o = 1.592$, $n_e = 1.581$), low birefringence ($\delta = 0.01$), and is uniaxial negative.

Tests Heated in the closed tube, torbernite yields water. It is soluble in nitric acid. When heated alone, it fuses to a blackish mass, and colours the flame green. The microcosmic salt bead is green, due to the presence of copper.

Occurrence Torbernite occurs with other uranium minerals as a secondary product, as at Jachymov (Czechoslovakia), Schneeberg (Saxony), Cornwall and in South Australia, etc.

AUTUNITE, lime uranite

C $Ca(UO_2)_2(PO_4)_2.10$–$12H_2O$; autunite is related to torbernite

Physical properties

CS Tetragonal

F&H Autunite sometimes occurs as square, tabular crystals; it also occurs scaly foliaceous and earthy

COL Lemon yellow to sulphur yellow

S Yellowish

CL Basal {001} perfect, {100} poor

L Sub-adamantine of crystal faces, pearly of cleavage planes

TR Transparent to subtranslucent

T Cleavage laminae brittle, not flexible

HD 2.0–2.5

SG 3.05–3.20

Optical properties

In thin section, autunite is pale yellow in colour, with a low relief ($n_o = 1.577$, $n_e = 1.553$), moderate birefringence ($\delta = 0.024$) and is uniaxial negative.

Tests Autunite gives no reactions for copper; otherwise its behaviour with reagents resembles that of torbernite.

Occurrence Autunite, is together with torbernite, a secondary product from other uranium minerals.

8.10.5 Vanadium oxysalt

CARNOTITE

C $K_2(UO_2)_2(VO_4)_2.1$–$3H_2O$

Physical properties

CS Monoclinic

F&H Minute crystal plates; also foliaceous, earthy or powdery

COL Canary yellow

CL {001} perfect

L Resinous

HD Soft

SG 4.70–4.95

Optical properties
In thin section, carnotite is pleochroic, with α colourless and β, γ pale yellow. It has very high relief (n_β = 1.90–2.06), extreme birefringence (δ = 0.17–0.30), and is biaxial negative with a variable 2V (43°–60°).

Occurrence Carnotite occurs as an impregnation, or as lenses, in Jurassic sandstone in Colorado and Utah, where it was formerly mined for radium. It has been reported from pitchblende deposits in South Australia, Zaire and elsewhere.

8.11 Molybdates and tungstates

8.11.1 *Anhydrous molybdates, etc.*

WOLFRAM
C (Fe,Mn)WO$_4$; wolframite is a mineral within the solid solution series between ferberite (FeWO$_4$) and hubnerite (MnWO$_4$)

Physical properties

CS Monoclinic

F&H Prismatic, tabular or bladed crystals common; it is also found massive

COL Chocolate brown, dark greyish black, reddish brown

S Chocolate brown

CL {010} perfect

F Uneven

L Submetallic; brilliant and shining on cleavage surfaces, dull on other surfaces

TR Subtransparent to opaque

HD 5.0–5.5

SG 7.12–7.60

Optical properties
In thin section, wolframite is pleochroic, with α yellow, β greenish yellow or red brown and γ olive green or dark red brown. The relief is extreme (n_β = 2.2–2.40), extreme birefringence (δ = 0.130–0.150), and is biaxial positive with a large 2V (75°±).

Other tungstates include **hubnerite**, MnWO$_4$, usually contains some iron, and resembles wolframite in appearance. **Ferberite**, FeWO$_4$, may contain appreciable manganese for iron. **Reinite,** an iron tungstate, is probably a pseudomorph after scheelite.

Tests Wolframite is fusible. The microcosmic bead is reddish, the borax bead is green, and the sodium carbonate bead is green in the oxidizing flame. When heated with sodium carbonate and carbon, and the fused mass treated with hydrochloric acid and powdered tin, a blue solution is formed. When heated with sodium carbonate it yields a magnetic mass.

—

Wolframite is characterized by its high specific gravity, streak, cleavage and the two lustres.

Occurrence Wolframite occurs in pneumatolytic veins surrounding granite masses, associated with cassiterite and quartz, as in Cornwall, Zinnwald, Malaysia, Bolivia, etc. It also occurs in veins formed at lower temperatures, and in certain gold-bearing quartz veins. The disintegration of these tin–wolframite veins results in the formation of alluvial or placer tin and wolfram deposits, which are worked at many localities, as in Burma.

SCHEELITE
C $CaWO_4$

Physical properties

CS	Tetragonal	**CL**	{101} fair
F&H	Dipyramid or tabular crystals common; it also occurs in reniform masses with a columnar structure, and also massive and granular	**F**	Uneven
		L	Vitreous to adamantine
		TR	Transparent to translucent
		T	Brittle
COL	Yellowish white or brownish, sometimes orange–yellow	**HD**	4.5–5.0
		SG	6.10–6.12
S	White		

Optical properties
In thin section, scheelite is colourless or pale brown, with extreme relief (n_o = 1.921, n_e = 1.938), and low birefringence (δ = 0.017), and is uniaxial positive.

Tests The microcosmic salt bead is green when hot and blue when cold in the reducing flame. The fused mass obtained by heating scheelite on charcoal with sodium carbonate, when dissolved in hydrochloric acid and with tin added, gives a blue solution. The high specific gravity is distinctive.

Occurrence Scheelite occurs under the same conditions as wolfram; that is, in pneumatolytic and other veins associated with cassiterite, quartz, topaz, and other minerals of pneumatolytic origin, as at Caldbeck Fell (Cumberland), Cornwall, Zinnwald, the Harz, Dragon Mountains (Arizona) and Trumbull (Connecticut). It also occurs in pyrometasomatic deposits, at the contacts of acid igneous rocks with limestones, as in California and elsewhere.

WULFENITE
C $PbMnO_4$

Physical properties

CS Tetragonal

F&H Square, tabular crystals occur and it is also found massive and granular

COL Yellow, orange yellow, yellowish grey, greyish white, brown and shades of orange, red or green

S White

CL {011} good; {001} and {013} poor

F Subconchoidal

L Waxy or adamantine

TR Subtransparent to subtranslucent

T Brittle

HD 2.5–3.0

SG 6.5–7.0

Optical properties

In thin section, wulfenite is yellow to pale orange in colour, with extreme relief (n_o =2.405, n_e =2.283) and extreme birefringence (δ = 0.123). It is uniaxial negative.

Tests Heated before the blowpipe, wulfenite decrepitates and fuses. With borax in the oxidizing flame, it becomes opaque, black or dirty green with black specks. With microcosmic salt it gives a yellowish-green bead in the oxidizing flame, which becomes dark green in the reducing flame. When heated with sodium carbonate on charcoal, it yields metallic lead. Wulfenite is decomposed when heated in hydrochloric acid, and the addition of a piece of zinc to the solution causes it to assume a deep blue colour.

Occurrence Wulfenite is found in the oxides zones of lead deposits with pyromorphite, mimetite and cerussite.

8.11.2 Hydrated molybdates, etc.

FERRIMOLYBDITE, molybdite, molybdic ochre, molybdena

C $Fe_2(MoO_4)_3.8H_2O$

Physical properties

CS Orthorhombic

V&H Very fine silky, fibrous crystals occur in radiating aggregates; it also occurs earthy or as encrustations

COL Straw yellow, yellow white

L Silky to adamantine of fibres; dull of massive or earthy types

HD 1.0–2.0

SG 3.06–4.46

Optical properties

In this section, molybdite is pleochroic, with α and β colourless, and γ pale yellow. It has a very high relief (n_β = 1.808–1.827), extreme birefringence (δ = 0.199–0.206), and is biaxial positive with a small $2V$ (28°).

Tests Heated alone on charcoal, ferrimolybdite yields a white encrustation, which is yellow when hot and copper red round the assay. This encrustation becomes blue when touched with the reducing flame.

Occurrence Ferrimolybdite is an oxidation product of molybdenite, and is often associated with limonite, etc.

9

The silicate minerals

ABBREVIATIONS

Physical properties

CS	crystal system	**L**	lustre
F&H	form and habit	**TR**	transparency
TW	twinning (if present)	**T**	tenacity
COL	colour	**HD**	hardness (e.g. 2.5–3.0)
S	streak (if present)	**SG**	specific gravity (e.g. 3.55–3.72)
CL	cleavage (if present)	**DF**	distinguishing features (when
F	fracture (if present)		needed)

Other properties are rare and are written out in full (e.g. **FEEL**).

Optical properties

n_α, n_β, n_e or n_o, n_e or n are the mineral RIs
δ birefringence
$2V = X^\text{o} - Y^\text{o} +$ ve OAP is parallel to, e.g., (010)

COL	colour	**B**	birefringence
P	pleochroism (if present)	**IF**	interference figure
H	habit	**E**	extinction angle (if present)
CL	cleavage	**TW**	twinning (if present)
R	relief	**Z**	zoning (if present)
A	alteration (if present)	**DF**	distinguishing features (when needed)

The order within each section is as follows:

> MINERAL NAME, with alternatives
> C composition of mineral
> Physical properties (see above)
> Optical properties (see above)
> Varieties
> Tests
> Occurrence
> Uses (or Production and uses)

The above is the scheme for common minerals; for rare minerals the various subdivisions are condensed or not included.

9.1 Crystal chemistry of silicate minerals

All silicate minerals contain silicate oxyanions $[SiO_4]^{4-}$, and the classification of silicate minerals depends upon the degree of polymerization of
the $[SiO_4]$ tetrahedral units. The system of classification adopted here is
the usual one, which depends upon how many oxygens in each tetrahedron are shared with other similar tetrahedra. There are *six* main
divisions.

Nesosilicates (orthosilicates, island silicates). A group of silicate minerals contain discrete $[SiO_4]$ tetrahedra, and this is often revealed in
their formula; e.g. olivine, $(Mg,Fe)_2[SiO_4]$, and garnet, $X_3Y_2[SiO_4]_3$.
Important nesosilicates include the olivine group, the garnet group,
Al_2SiO_5 polymorphs, zircon, sphene, staurolite, chloritoid and topaz.

Cyclosilicates (ring silicates). In this division, tetrahedra sharing two
oxygens link together to form a ring of composition $[Si_nO_{3n}]^{2n-}$,
where *n* is a positive integer. A typical ring composition is
$[Si_6O_{18}]^{12-}$, and minerals of this group include tourmaline, beryl and
cordierite, although the last two are sometimes included with the
tektosilicates.

Sorosilicates. This division contains groups of two tetrahedra sharing a
common oxygen, with the composition $[Si_2O_7]^{6-}$, and minerals
include the epidote group, the melilite group, idocrase and pumpellyite.

Inosilicates (chain silicates). (a) When 2.5 oxygens are shared by adjacent
tetrahedra, *single-chain* silicates with a chain composition $[Si_2O_6]_n^{4-}$ are
formed, the single chains stacked parallel to the *c* axis. The minerals with
this composition have a prismatic habit and include the pyroxene group
and the pyroxenoids. (b) *Double-chain* silicates also occur, with the
composition of the double chains being $[Si_4O_{11}]_n^{6-}$, and minerals of this
type include the amphibole group.

Phyllosilicates (sheet silicates). When three oxygens are shared between
adjacent tetrahedra, sheet structures result with the composition
$[Si_4O_{10}]^{4-}$, and these sheets are 'stacked' along the *c* axis with other
'sheets' containing either magnesium and iron ions (**brucite sheet**) or
aluminium ions (**gibbsite sheet**). Combinations of these sheets produces
three different mineral types: (a) the **1 : 1 type**, a two-layer unit with one
silicate sheet and a brucite or gibbsite sheet (this type includes kaolinite
and serpentine); (b) the **2 : 1 type**, a three-layer unit with two silicate
sheets and a brucite or gibbsite sheet 'sandwiched' between them (this
type includes the mica group, montmorillonite and all smectite clays,

and also illite, talc and pyrophyllite); and (c) the **2 : 2 type**, a four-layer unit with two silicate sheets and two brucite or gibbsite sheets (this type includes the chlorite group).

Tektosilicates (framework silicates). In this division, the four oxygens in each tetrahedron are shared with other tetrahedra, and such a framework structure, if composed entirely of silicon and oxygens, would have the composition $[SiO_2]_n$, which, of course, is the composition of quartz. Important tektosilicates include quartz, the feldspar group, feldspathoids, zeolite group and scapolite.

The above classification has been used to separate the silicate minerals into their appropriate crystal chemical groups in the following sections.

9.2 Nesosilicates

Olivine group

OLIVINE, peridote, chrysolite
The olivine group consists of an isomorphous series with the general formula R_2SiO_4, in which R is Mg^{2+} or Fe^{2+}. The structure is described in Section 5.3, and is repeated in the introduction to this chapter. The magnesian end-member is **forsterite** (Fo), Mg_2SiO_4, and the iron end-member is **fayalite** (Fa), Fe_2SiO_4.
$C (Mg,Fe)_2SiO_4$, with Ni^{2+} and Co^{2+} substituting for Mg^{2+} and Fe^{2+}. Ca^{2+} and Mn^{2+} may also substitute for Mg and Fe in the lattice, but both are minor components.

Physical properties

CS Orthorhombic

F&H Stubby prismatic crystals with the {010} faces well developed, modified by terminal faces {021}; also as massive and compact grains

COL Shades of green, pale green, olive green, brownish; white or yellow on forsterite, brown or black in fayalite

CL Poor on {010}

F Cracks sub-parallel to {001} usually present

L Vitreous

TR Transparent to translucent

HD 6–7

SG 3.22(Fo) to 4.39(Fa)

Optical properties

$n_\alpha = 1.635$(Fo) $- 1.824$(Fa)
$n_\beta = 1.651$ $- 1.864$
$n_\gamma = 1.670$ $- 1.875$
$\delta = 0.035$ $- 0.051$
$2V = 82° - 134°$

Pure Fo is +ve and pure Fa is −ve, with the change at Mg : Fe = 15 : 85. OAP is parallel to (001)

COL Colourless (Fo) to faint greenish (Fa)

H Anhedral in plutonic rocks to euhedral (well shaped) in extrusive rocks

CL See previous page

R High to very high

A Olivine is very susceptible to hydrothermal alteration, weathering and the effects of low-grade metamorphism, with several secondary minerals being produced, particularly serpentine

B High, with maximum third order interference colours

IF Since $2V$ is large ($\pm 90°$), a single isogyre is needed to obtain a sign; remember that Fo is +ve and Fa is −ve

Varieties **Forsterite** is white, greenish or yellowish in crystalline limestones or in ultrabasic or ultramafic igneous rocks, especially dunites. **Fayalite** is brown or black in colour, fusible before the blowpipe, occurring in slags, in cavities, etc., in rhyolites as in Yellowstone Park, and in granite pegmatites as in the Mourne Mountains. **Peridot** is a gem variety of olivine, transparent and pale green, found in Egypt, Burma and Brazil.

Tests Most varieties are infusible before the blowpipe. The olivines are recognized by their colour and physical characteristics. They are decomposed by hydrochloric acid (HCl) with gelatinization.

Occurrence Mg-rich olivine is an essential mineral in most ultrabasic igneous rocks; for example, peridotite, picrites and dunites which are almost entirely composed of olivine. Mg-rich olivine occurs in basic igneous rocks such as gabbros, norites, dolerites and basalts; and more iron-rich olivines may occur in some intermediate rocks such as pitchstones, as well as in some other igneous rocks such as larvikites, teschenites and alkali-basalts, where the olivine frequently occurs in the presence of quartz. Pure forsterite is found in metamorphosed impure dolomites, but can be serpentinized, giving rise to **ophicalcite**, a rock consisting of white calcite and serpentine. Olivines in igneous rocks may show reaction rims (called kelyphytic rims or corona structures) against plagioclase crystals.

ZIRCON
C $ZrSiO_4$

Physical properties

CS Tetragonal

F&H Crystals usually small, prismatic with a square cross section, and with {111} pyramidal faces; also found as round, detrital grains

COL Colourless, grey, pale yellow, greenish, reddish brown

CL Imperfect {110} parallel to prism faces; poor parallel to {111}

F Conchoidal

L Adamantine

TR Transparent to opaque

H 7.5

SG 4.6–4.7

Optical properties

$n_o = 1.923 - 1.960$

$n_e = 1.968 - 2.015$

$\delta = 0.042 - 0.065$

Uniaxial +ve

COL Colourless

H Tiny euhedral prisms with square cross sections

CL See above

R Extremely high; the mineral has an intense black border

B Prismatic sections show third- or fourth order colours; basal sections are isotropic

IF Positive uniaxial, the crystal is length slow

DF Apatite is −ve with lower RIs; sphene is pale brown and usually diamond-shaped and biaxial +ve; monazite is similar to sphene. Cassiterite and rutile are both coloured minerals.

Varieties **Hyacinth** is a gem variety, red and transparent; **jargoon** is also a gem variety, colourless to smoky; **zirconite** is a grey or brown variety.

Tests Heated before the blowpipe alone, zircon is infusible. Coloured varieties become colourless and transparent when heated. If zircon is powdered and fused with sodium carbonate, and then dissolved in dilute hydrochloric acid (HCl), the resulting solution turns turmeric paper orange.

Occurrence Zircon occurs as an accessory mineral in igneous rocks, especially the more acidic types (granites, various syenites, etc.) and pegmatites. Decomposed pegmatites have been worked for zircon in Madagascar and Brazil. Zircon is also a constituent of crystalline limestones, gneisses and other metamorphic rocks, and detrital zircon-bearing deposits have been worked in Sri Lanka and Burma. Zircon is a resistant mineral and is commonly found in heavy residues from sandstones along with ilmenite, rutile and monazite, occurring in beach sands in Australia, at Travancore in India and in Florida and Brazil, and such deposits have been worked commercially.

THORITE

C $ThSiO_4$

Physical properties

CS Tetragonal

F&H Similar to zircon

COL Black; orange–yellow in the variety **orangite**

S Dark brown

CL Good {110} prismatic

F Conchoidal

L Vitreous on fresh surfaces

HD 4.5

SG 5.2–5.4

Tests Thorite is usually hydrated, yielding water on heating. It is soluble in HCl with gelatinization on concentration of the solution. The SG is noticeable.

Occurrence As large crystals in the syenite–pegmatites of the Langesund-fiord district of Norway, and in similar rocks elsewhere.

WILLEMITE, wilhelmite

C Zn_2SiO_4

Physical Properties

CS Trigonal

F&G Prismatic crystals with hexagonal cross section and rhombohedral terminations

COL Green, yellow or brown

CL {0001} and {11$\bar{2}$0} imperfect

L Vitreous to resinous

HD 5.5

SG 3.9–4.1

Varieties **Troostite** is an Mn-bearing variety occurring in large crystals.

Tests After heating on charcoal, moistening with cobalt nitrate and reheating, willemite gives a green mass. The solution gelatinizes in HCl when concentrated.

Occurrence Willemite occurs with **zincite** and **franklinite** in the deposit at Franklin Furnace, New Jersey (see under these minerals also); also at Moresnet and Vieille Montagne (Belgium) and at Raibl (Carinthia).

SPHENE

C $CaTiSiO_4(O,OH,F)$; Ca may be replaced in part by Na, and Ti by Al, etc.

Physical properties

CS Monoclinic

F&H Crystals usually wedge- or lozenge-shaped; also massive

TW Common on {100}

COL Brown, grey, green, yellow or black

Cl Good {110} prismatic

F Imperfect conchoidal

L Adamantine or resinous

TR Transparent to opaque

T Brittle

HD 5

SG 3.45–3.55

Optical properties

$n_\alpha = 1.843 - 1.950$
$n_\beta = 1.879 - 2.034$
$n_\gamma = 1.943 - 2.110$
$\delta = 0.100 - 0.192$
$2V = \sim 20° - 40°$ +ve
OAP is parallel to (010)

COL Variable; normally colourless to pale yellow, but greyish purple in soda-rich rocks, and brownish in soda-poor rocks

P Common in crystals from alkali-igneous rocks, with α colourless, β pink to yellowish brown, and γ pink to orange brown

H Occurs as four-sided lozenge-shaped crystals; also as irregular grains

CL {110} good

R Extremely high

A Sphene alters to leucoxene (a hydrated titanium dioxide). Leucoxene may also form from alteration of other Ti-bearing minerals such as ilmenite, where it is seen as three sets of lines or stripes making equal angles with one another.

B Extreme, but colours masked by the colour of the mineral

IF 2V small, but dispersion is high and a good figure is difficult to obtain

TW Crystals are often twinned, the {100} twin plane occurring along the long axis of the lozenge

Varieties **Greenovite** is a sphene containing a little MnO, and red or pink in colour.

Tests Heated before the blowpipe, the yellow varieties remain yellow, but darker types become yellow. Sphene is partly soluble in hot HCl. The solution becomes violet after concentration, when a little tin is added.

Occurrence Sphene occurs as a primary accessory mineral in calc-alkaline igneous rocks and alkali igneous rocks, but is more abundant in rocks in lime such as contact metamorphosed limestones, particularly skarns.

GARNET GROUP

Garnets are silicate minerals containing various divalent and trivalent cations, the general formula being $X_3Y_2Si_3O_{12}$, with X = divalent ions such as Ca^{2+}, Mg, Fe, Mn, etc., and Y = trivalent ions such as Al^{3+}, Fe, Cr and Ti.

C The principal minerals of this family are:

grossular	$Ca_3Al_2Si_3O_{12}$	RI : 1.734	SG : 3.594
pyrope	$Mg_3Al_2Si_3O_{12}$	1.714	3.582
almandine	$Fe_3Al_2Si_3O_{12}$	1.830	4.318
spessartine	$Mn_3Al_2Si_3O_{12}$	1.800	4.190
andradite	$Ca_3Fe_2Si_3O_{12}$	1.887	3.859
uvarovite	$Ca_3Cr_2Si_3O_{12}$	~ 1.86	~ 3.90

Rare earth elements (REE) may also substitute in the Y position. Solid solutions within the family are common and cover a wide range of possible compositions. A variety called **hydrogrossular**, $[Ca_3Al_2(SiO_4)_2(SiO_4)_{1-m}(OH)_{4m}]$, also exists which contains hydroxyl groups in the structure (RI = 1.734–1.675; SG = 3.13–3.59).

Physical properties

CS All garnets are cubic

F&H Forms include {110} dodecahedron (Gr, Sp, Uv, Hy), or combination of tetrahexahedron {210} and dodecahedron (Al, An)', or as angular fragments (Py)

COL Pale olive green, greenish white, yellow or pink (Gr); deep crimson (Py); deep red (Al); deep hyacinth or brownish red (Sp); dark brown, yellowish green or brownish green (An); emerald green (Uv); white (Hy)

CL Garnets do not possess any cleavage

F Imperfect conchoidal or uneven

L Vitreous

TR Transparent to translucent (Py); translucent (Gr, Al, Sp, Uv); opaque (An, Hy)

HD 7.0–7.5 (all garnets except hydrogrossular which is <7.0)

SG See previous page

Optical properties

COL Colourless to pale pinkish

H Subhedral, somewhat rounded crystals traversed by branching cracks

and with inclusions of quartz, micas, etc.

R High to very high

Garnets are isotropic under XP. Strain polarization and abnormal twinning may be seen in hydrogrossular

Varieties **Cinnamon-stone** (Ca–Al garnet) has a yellowish colour, a vitreous or poorer lustre and a flattish conchoidal fracture. It occurs in Wicklow and Aberdeenshire and is used in jewellery. **Common garnet** (Fe variety) is brownish red and translucent to opaque; **precious garnet** (Fe variety) is the transparent variety. Varieties of andradite include **colophonite**, a coarse, dark red, resinous type; and **pyreneite**, a black, opaque well formed variety from the Pyrenees. **Melanite** is also a black variety with a dull to vitreous lustre occurring in alkali igneous rocks; **topazolite** is a yellow, transparent, topaz-like variety; and **demantoid** is a bright green gem variety.

Tests Grossular and pyrope fuse easily, and are soluble in HCl after ignition, yielding gelatinous silica on concentration. Almandine and andradite fuse to magnetic globules. Spessartine gives manganese reactions before the blowpipe. Uvarovite is infusible heated alone before the blowpipe; it gives a clear chrome-green borax bead.

Occurrence Garnets occur in a variety of rock types. Almandine is common in medium-grade metamorphic rocks (schists and gneisses)

derived from pelitic sediments or from basic igneous rocks; and is sometimes found in granites, as in Connemara, and in detrital deposits in Sri Lanka together with pyrope. Grossular and andradite are common in metamorphosed impure limestones, and in skarns (An), often with hedenbergite and magnetite; andradite may occur as a result of metasomatism of some igneous rocks such as andesites. Uvarovite occurs in serpentinites which are rich in chromite, as in the Urals and in Unst, Shetlands. Pyrope occurs in ultrabasic igneous rocks such as garnet peridotites, etc., where it is associated with olivines, pyroxenes, etc., as in Saxony and Czechoslovakia. Hydrogrossular is probably more common than was first thought. It is found in metamorphosed marls, in altered gabbros and in **rodingites**, which are metasomatized calc-silicate rocks associated with serpentinites.

Uses The uses are twofold: as an abrasive and as a gemstone. Garnet is mined from metamorphic rocks in New York State and Idaho in the USA, and from alluvial deposits in Spain, Sri Lanka and elsewhere, and is used as an abrasive in the polishing of wood. The gem varieties include trade names such as Bohemian Garnet, Cape Ruby, carbuncle, cinnamon-stone, etc., and the gem varieties come mostly from Czechoslovakia, India, Sri Lanka, South Africa and Mozambique.

IDOCRASE, vesuvianite

C $Ca_{10}(Mg,Fe)_2Al_4Si_9O_{34}(OH,F)_4$; both $[SiO_4]$ and $[Si_2O_7]$ groups are present, as in the epidote group of minerals

Physical properties

CS Tetragonal

F&H Prismatic crystals with {100} and {110} forms, and terminal faces {111} and {001} present; also massive

COL Brown, green, yellowish

CL Poor {110} and {001} cleavages present

F Subconchoidal to uneven

L Vitreous

ſ Subtransparent to subtranslucent

HD 6–7

SG 3.33

Optical properties

$n_\alpha = 1.708$–1.752

$n_e = 1.700$–1.746

$\delta = 0.001$–0.008

Uniaxial −ve

COL Colourless, pale yellowish brown, with occasional beautiful zoning present

H Usually occurs as large, often shapeless grains

CL See above

R High

DF Similar to zoisite (biaxial +ve), but idocrase is usually found with grossular, diopside or wollastonite

Tests Idocrase fuses with intumescence before the blowpipe, forming a greenish or brownish glass.

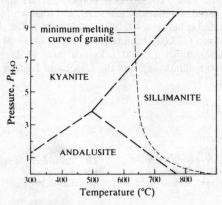

Figure 9.1 Stability relations of the three Al_2SiO_2 polymorphs, with the minimum melting curve of granite superposed.

Occurrence Idocrase occurs in thermally metamorphosed impure limestones, together with grossular, scapolite, wollastonite, etc., as in the limestone of Monte Somma, Italy; and also in regionally metamorphosed limestones, as in the Loch Tay limestone of Scotland. Idocrase is also common in rodingites (p. 356).

Al_2SiO_5 polymorphs
There are three polymorphs, each with the same composition but with different atomic structures; namely **andalusite, kyanite** and **sillimanite**. A related mineral, **mullite**, which is rare in nature, may form under very high temperatures from alumina-rich material.

The three minerals have closely related atomic lattices, containing independent [SiO_4] tetrahedra and chains of Al–O groups, with the Al^{3+} ion in different coordinations.

The Al_2SiO_5 minerals occur as index minerals in metamorphosed aluminous rocks. The pressure–temperature diagram (Fig. 9.1) shows the stability relations of the three polymorphs. The minimum melting curve of granite has been superimposed on to the diagram, so that melting in metamorphic rocks will take place to the right of this curve, and the rocks can be described as **migmatites**. The polymorph occurring under one set of *P–T* conditions may be unstable under a later set of conditions; for example, the andalusite schists of the Inchbae in Ross-shire, Scotland have been changed into kyanite schists by later regional metamorphism, and there are numerous examples of one polymorph **inverting** to another as the conditions change.

Production and uses The Al_2SiO_5 polymorphs are mined in several parts of the world as refractory materials, and total world production in 1983 is estimated at between 400 000 t and 500 000 t. The main producers are South Africa (103 000 t, andalusite), the USA (~90 000 t, kyanite and sillimanite), India (35 000 t kyanite and 14 000 t sillimanite), France (40 000 t sillimanite); and the USSR and Eastern bloc countries (~100 000 t).

ANDALUSITE

Physical properties

CS Orthorhombic	**L** Vitreous
COL Pearl grey, purplish red	**TR** Translucent to opaque
CL Poor {110} prismatic	**HD** 6.5–7.5
F Uneven, tough	**SG** 3.13–3.16

Optical properties

$n_\alpha = 1.629–1.649$

$n_\beta = 1.633–1.653$

$n_\gamma = 1.638–1.660$

$\delta = 0.009–0.011$

$2V = 78°–86°$ −ve

OAP is parallel to (010)

COL Colourless, but thick sections may be pleochroic in pinks and greens

P α pink, and β,γ greenish yellow

H Stumpy, often poikiloblastic; prisms with square cross sections seen in some metamorphic rocks; also as grains and aggregates

CL See above

R Moderate

A Andalusite can alter to sericite under hydrothermal conditions

B Greys and whites of first order common

IF $2V$ is large, therefore a single isogyre is needed to obtain a sign; crystals are length fast.

Varieties **Chiastolite** or **macle** is a variety found in some metamorphic rocks, and in certain slates, such as the Skiddaw slate of Cumberland, and the Killas of Cornwall, resulting from the thermal metamorphism of pelites. The basal section of these crystals exhibit definite cruciform lozenge-shaped or tesselated markings due to impurities enclosed in the crystals during their formation. These varieties are small in the UK, but similar crystals from foreign localities attain much larger sizes. In some crystals the corners may wear away, producing a form simulating that of a twinned crystal. **Manganandalusite**, or **viridine**, is a manganiferous variety.

Tests Heated before the blowpipe, andalusite is infusible. It is not acted on by acids. A blue colour is produced when it is heated with cobalt nitrate solution on charcoal.

Occurrence Andalusite occurs in thermally metamorphosed pelites, forming andalusite hornfelses in the inner zones of thermal aureoles; and in regional metamorphic rocks such as the Fyvie schists of Aberdeenshire, Scotland, where it occurs with cordierite. These rocks are part of the so-called 'Buchan' metamorphic rocks, representing a regional development under high heat flow and low pressure. Andalusite also occurs as an accessory mineral in certain granites, such as in Cornwall, where its presence is connected with argillaceous inclusions in the granite.

KYANITE, cyanite, disthene

Physical properties

CS Triclinic

F&H Found as long, thin-bladed crystals occurring as index minerals in schists and gneisses; sometimes in radiating rosettes embedded in quartz

COL Light blue, sometimes white, and sometimes zoned from blue in the core of the crystal to colourless at the margins; also grey green, and rarely black

S White

CL {100} and {010} good; a parting on {001} is usually present

L Pearly for cleavage faces

TR Transparent to subtranslucent

H 5.5–7.0

SG 3.58–3.65

Optical properties

$n_\alpha = 1.712–1.718$
$n_\beta = 1.721–1.723$
$n_\gamma = 1.727–1.734$
$\delta = 0.015–0.016$
$2V = 82°–ve$

OAP is perpendicular to the (100) face with *a* axis = Bxa

COL Colourless, rarely pale blue

P Thick sections may rarely show α colourless and β,γ pale blue

H Occurs as subhedral bladed prisms in metamorphic rocks

CL See above

R High – much higher than the other polymorphs

A Same as andalusite

B Low

IF A single isogyre should be used to obtain the sign since the 2V is so large

E Oblique on prism edge, with slow (γ) ^prism edge = ~30°. Note that andalusite and sillimanite *both* have straight extinction measured on a prism edge or face.

Occurrence Occurs in regionally metamorphosed pelites under moderate heat flow and moderate to high pressure, forming kyanite schists and gneisses; sometimes found in eclogites.

SILLIMANITE, fibrolite

Physical properties

CS   Orthorhombic

F&H Usually occurs as long acicular crystals with diamond-shaped cross sections; and also in wisp-like aggregates (fibrolite)

COL Shades of brown, grey and green

CL   Good {010} prismatic

F   Uneven

L   Vitreous

TR   Transparent to translucent

HD   6.5–7.5

SG   3.23–3.27

Optical properties

n_α = 1.654–1.651

n_β = 1.658–1.662

n_γ = 1.678–1.683

δ = 0.019–0.222

$2V$ = 21° – 30° +ve

OAP is parallel to (010)

COL Colourless

H   Occurs as long slender crystals with diamond-shaped cross sections, common in thermal aureoles; or as mats of fibrous crystals, common in regional metamorphic rocks

CL   {010} good; cuts the diamond-shaped cross section parallel to the long axis

R   Moderate

B   Low, up to first order red, although yellow is most common

IF   The low $2V$ (~30°) means that both isogyres can be seen at the same time in a basal section

DF   Sillimanite is distinguished from andalusite by its diamond-shaped cross section, small positive $2V$, and being length slow; kyanite possesses oblique extinction in certain prismatic sections and often displays two or three cleavages

Occurrence   Sillimanite occurs in hornfelses in the inner zones of thermal aureoles, resulting from the thermal metamorphism of pelitic rocks; and in high-grade regionally metamorphosed rocks of similar composition.

MULLITE

C $Al_6Si_2O_{13}$ or $[3Al_2O_3 . 2SiO_2]$

Properties and occurrence   Occurs as orthorhombic prisms that are similar to sillimanite in appearance. Found in shales fused by immersion in basic magma in Mull, Scotland, and in buchites formed by burning coal seams, it is formed by heating other aluminosilicates. Mullite is produced commercially by calcining bauxite at high temperatures. The product is mixed with bitumen and other cements or resins, and used commercially as a non-skid road surface in high-density traffic areas.

TOPAZ

C $Al_2SiO_4(OH,F)_2$

Physical properties

CS Orthorhombic

F&H Prismatic crystals displaying the forms {110}, {010} and {120} with terminal forms {020}, {111} and rarely {001}; also columnar and granular

COL Various shades of grey, white, yellow, occasionally blue or pink; the pink colour of many topaz crystals in jewellery being produced by artificial heat. The topaz is wrapped in **amadou** (a tinder), which is ignited and allowed to smoulder.

CL Basal {001} perfect

F Subconchoidal to uneven

L Vitreous

TR Transparent to translucent

HD 8

SG 3.49–3.57

Optical properties

n_α = 1.606–1.629
n_β = 1.609–1.631
n_γ = 1.616–1.638
δ = 0.008–0.011
$2V$ = 48°–68° +ve
OAP is parallel to (010)
COL Colourless
H Usually occurs as prismatic crystals
CL See above

R Moderate

A Topaz alters to clay minerals

B Low

IF $2V$ is moderate, but a single optic axis is still needed to obtain the sign

DF Both topaz and quartz contain inclusions (of water or some other liquid), but topaz is biaxial and has a perfect cleavage

Tests Heated alone topaz is infusible, but if moistened with cobalt nitrate and reheated, it assumes a blue colour, due to aluminium. When fused with microcosmic salt it gives off silicon fluoride which etches glass.

Occurrence Topaz is a late-stage mineral occurring in acid igneous rocks (granites, rhyolites, etc.) and good crystals are found in drusy cavities, as in the Mourne Mountains granite, and in the lithophysal cavities of rhyolites, as in Colorado. It is a mineral which also occurs in tin-bearing pegmatites and in tin veins generally, associated with other pneumatolytic minerals such as fluorite, cassiterite and tourmaline. It may occur in the contact-altered zone adjacent to granite intrusions, and is a constituent of **greisen**.

Uses Mainly used as a gemstone, the chief sources are the Urals (USSR), Brazil, Japan and Zimbabwe. The original 'Cairngorms' were topaz crystals, but these have since been replaced by crystals of citrine (brown quartz), often obtained from Brazil, although still called 'Cairngorms'.

ILVAITE, lievrite, yenite
C $Ca(Fe^{2+}Fe^{3+})O[Si_2O_7](OH)$

Physical properties

CS Orthorhombic	**F** Uneven
F&H Appears as prismatic crystals with the sides deeply striated lengthwise; also compact and massive	**L** Submetallic
	TR Opaque
	T Brittle
COL Black or brownish black	**HD** 5.5–6.0
S Black, brownish or greenish	**SG** 3.8–4.1
CL Perfect {001} and {010}	

Tests Ilvaite fuses to a black magnetic globule before the blowpipe. With borax it yields a dark green and almost opaque bead. It is soluble in HCl, forming a jelly.

Occurrence Ilvaite is associated with magnetite, zinc and copper ores, as in localities in Elba, the Harz, Tyrol, Saxony, Norway and Rhode Island.

STAUROLITE

C Probably $(Fe^{2+}Mg)_2(Al,Fe^{3+})_9O_6Si_4O_{16}(O,OH)_2$; Mn^{2+} may replace Fe^{2+} in the structure

Physical properties

CS Orthorhombic	**S** Colourless or greyish
F&H Good prismatic crystals with {110} and {010} forms present, and terminated by {001}	**CL** {010} imperfect
	F Conchoidal
TW Common on {232} and {032}, the latter of which produces the characteristic cruciform twinned crystal of staurolite	**L** Subvitreous to resinous; crystals often have a dull, rough surface
	TR Translucent to opaque, usually opaque
COL Reddish brown, brownish black; sometimes yellowish brown	**HD** 7.5
	SG 3.74–3.85

Optical properties

$n_\alpha = 1.739–1.747$	**CL** See above
$n_\beta = 1.745–1.753$	**R** High
$n_\gamma = 1.752–1.761$	**B** Low, but may be masked by mineral colour
$\delta = 0.013–0.014$	
$2V = 82°–90°$ +ve	**IF** Since $2V$ is very large, a single optic axis figure is needed
OAP is parallel to (100)	
COL Yellow or pale yellow	**TW** Twinning may sometimes be seen under XP
P Always pleochroic, with α colourless, β pale yellow and γ deep yellow	
H Staurolite occurs as large, squat prisms usually containing inclusions of quartz	

Tests Varieties containing manganese fuse easily to a black magnetic glass, but the other varieties are infusible.

Occurrence Staurolite occurs in regionally metamorphosed rocks, such as iron-rich pelites which may also have high $Fe^{3+} : Fe^{2+}$ ratios, at medium grades of metamorphism, and often in association with garnet and kyanite. It may have developed from chloritoid as the metamorphic grade increased.

CHLORITOID, ottrelite

C $(Fe^{2+}Mg)_2(Al,Fe^{3+})Al_3O_2[SiO_4]_2(OH)_4$; Mn^{2+} may replace Fe^{2+}

Physical properties

CS Monoclinic or triclinic; two polymorphs exist, crystallizing in different systems

F&H Crystals tabular, pseudohexagonal

COL Dark green, greenish black

CL {001} perfect
L Pearly on cleavage planes
T Brittle
HD 6.5
SG 3.51–3.80

Optical properties

$n_\alpha = 1.713–1.730$
$n_\beta = 1.719–1.734$
$n_\gamma = 1.723–1.740$
$\delta = 0.010$
$2V = 45°–68°$ +ve, but there can be a much greater spread of $2V$ values
OAP is parallel to (010)

COL Colourless, greens and blues
P Always present and very marked, with α pale green, β blue and γ colourless to pale yellow
H Occurs as lath-like crystals (similar to the micas)
CL {001} perfect; another poor prismatic one may be present which helps to distinguish chloritoid from the micas and chlorite

R High
B Low, but usually masked by the mineral colour
IF A (100) section of chloritoid will give a good Bxa figure, but in some cases a single optic axis (one isogyre) may be required
TW Lamellar twinning may be seen in some mineral sections
Z Hourglass zoning may occur in some prismatic sections

Occurrence Chloritoid occurs in regionally metamorphosed pelitic rocks with a high $Fe^{3+} : Fe^{2+}$ ratio, at fairly low metamorphic grades. Chloritoid develops at the same time as biotite, and changes to staurolite as the grade (P and T) increases. In non-stress environments, chloritoid is usually triclinic, e.g. in quartz carbonate veins and altered lava flows. Ottrelite, the Mn-rich variety, characterizes the ottrelite phyllites of the Ardennes.

9.3 Sorosilicates

Epidote group

The epidote group of minerals are complex silicates with the general formula $X_2Y_3Z_3O_{12}(OH,F)$, where $X =$ Ca, Ce^{3+}, La^{3+}, Y^{3+}, Fe^{2+}, Mn^{2+} and Mg^{2+}, $Y = Al^{3+}$, Fe^{3+}, Mn^{3+}, Fe^{2+} and Ti^{3+}, and $Z = Si^{4+}$. The formula can be written as $X_2Y_3[Si_2O_7][SiO_4]O(OH,F)$, showing that the atomic structure contains *both* $[Si_2O_7]$ and $[SiO_4]$ groups. The epidote minerals crystallize both in the orthorhombic and monoclinic crystal systems, and are conventionally described in that order.

ZOISITE (α- and β-zoisite)

C $Ca_2Al_3Si_3O_{12}(OH)$. Two orthorhombic forms exist, α-zoisite and β-zoisite, which can contain up to 5% of Fe_2O_3. **Thulite** is the manganese-rich zoisite.

Physical properties

CS   Orthorhombic

F&H   Crystals are elongated along the *a* crystallographic axis (originally thought to be the *b* axis), but the optic orientation varies in the α and β types; it is found in clusters of elongate prismatic crystals with rectangular cross sections, often with longitudinal striations; also massive.

COL   White, grey, greenish; thulite is rose red

CL   {100} usually perfect; {001} poor and not always present

L   Vitreous

TR   Transport to translucent

HD   6

SG   3.15–3.36

Optical properties

$n_\alpha = 1.696$
$n_\beta = 1.696$ } RIs may vary depending upon the amount of Fe^{3+}, Mn^{2+},
$n_\gamma = 1.702$ } etc. in the structure

$\delta = 0.006$

$2V = 0°–60°$ +ve

OAP is parallel to (010) in α-zoisite, and parallel to (001) in β-zoisite

COL   Colourless; thulite is pink and pleochroic in pinks

H   Zoisite appears as clusters of slender prismatic crystals, but sometimes as shapeless grains

CL   See above

R   High

B   Very low; α-zoisite shows low first-order greys and whites, and β-zoisite shows anomalous interference colours of a deep blue

IF   A (100) section gives a Bxa figure

Occurrence   Zoisite occurs in hydrothermally altered igneous rocks such as anorthosites and gabbros, developing from calcic plagioclase, where it is accompanied by amphiboles, the process being called **saussuritization**. It may also ocur in medium-grade schists in association with Na-plagioclase, amphibole, biotite and garnet, and also in metamorphosed impure limestones.

CLINOZOISITE

C $Ca_2Al_3Si_3O_{12}(OH)$, with some Al^{3+} replaced by Fe^{3+}

Physical properties

CS Monoclinic

F&H Small prismatic crystals elongated along the *b* axis, usually occurring in aggregates

COL Grey or greyish white

CL {001} perfect

HD 6.5

SG 3.12–3.28

Optical properties

$n_\alpha = 1.710$⎫ Based on a composition with ~1% Fe_2O_3;
$n_\beta = 1.715$⎬ RIs will increase with increasing iron,
$n_\gamma = 1.719$⎭ and decrease if iron amount is reduced

$\delta = 0.005$–0.015 (depending on iron amount)

$2V = 14°$–$90°$ +ve

OAP is parallel to (010)

COL Colourless or very pale yellow

H Appears as small columnar aggregates of crystals

CL See above

R High

B Very low, with anomalous interference colours of deep blue (Berlin blue) or greenish yellow seen

IF A (100) section gives a Bxa figure, but a single optic axis figure may be needed to obtain the sign

E Oblique on a (010) section with slow (γ) ^ cleavage = $20°\pm$; on a prismatic section, e.g. (100), extinction is straight on cleavage or edge

Occurrence Clinozoisite occurs in low-grade regionally metamorphosed rocks forming from micaceous minerals; it also forms, like zoisite, as a product of saussuritization.

EPIDOTE, pistacite

C $Ca_2(Al_2Fe^{3+})Si_3O_{12}(OH)$

Physical properties

CS Monoclinic

F&H Crystals elongated along the *b* axis and often in divergent aggregates of crystals; also as granular masses

COL Shades of green, pistachio green, blackish green, dark oily green; red in **withamite**

CL Perfect basal {001} appearing as a prismatic cleavage since epidote is elongated along the *b* axis

F Uneven

L Vitreous

TR Transport to opaque

HD 6

SG 3.38–3.49

Optical properties

$n_\alpha = 1.715$⎫
$n_\beta = 1.725$⎬ RIs increase with increasing iron content
$n_\gamma = 1.734$⎭

$\delta = 0.019$–0.049 (depending on iron content)

$2V = 64°$–$90°$ −ve

OAP is parallel to (010)

COL Colourless to pale green

P Slightly pleochroic, with c colourless, pale yellow, β greenish and γ yellowish; the colour intensifying with increasing iron content

H Prismatic, elongate crystals present with pseudo-hexagonal cross sections

CL See previous page

R High

B Moderate to high, up to upper third order colours; anomalous interference colours may occasionally appear

IF $2V$ is large, so a single optic figure is needed

E Oblique on an (010) section (i.e. a section which has a pseudo-hexagonal outline); a prismatic section has straight extinction

TW Some sections may show lamellar twinning

Varieties **Pistacite** is a pistachio green variety; **arendalite** is a variety from Arendal in Norway, occuring as very fine crystals, externally blackish green to dark green in colour on fractured surfaces; and **withamite** is a red variety occurring in andesites from Glencoe in Scotland, which is strongly pleochroic in thin section in reds and yellows.

Tests Heated before the blowpipe, epidote gives reactions for iron.

Occurrence Common in low-grade regional metamorphic rocks derived from either impure calcareous rocks or basic igneous rocks that are rich in calcic plagioclase. Secondary epidote forms in basic igneous rocks from saussuritization of plagioclase feldspar and from the breakdown of amphiboles, both changes resulting from late-stage hydrothermal alteration.

PIEMONTITE, PIEDMONTITE

C $Ca_2(Mn,Al,Fe^{3+})_2AlSi_3O_{12}(OH)$

HD 6

SG 3.45–3.52

Characters and occurrence

n_α = 1.732 ⎫
n_β = 1.750 ⎬ RIs increase with increasing Mn and Fe content
n_γ = 1.762 ⎭
δ = 0.030
$2V$ = 64°–85° −ve

OAP is parallel to (010)

Most properties are similar to those of epidote. This is a manganiferous epidote, dark reddish in colour, and with strong and spectacular pleochroism, with α yellow, β purplish blue and γ red. Birefringence is moderate to high, but masked by the strong mineral colour. Heated before the blowpipe it fuses readily to a black glass, a character distinguishing it from pistacite and zoisite, which fuse only on thin edges. Piemontite occurs as a constituent of certain manganiferous metamorphic rocks such as piemontite-bearing mica schists, and also as a secondary mineral in some porphyries as in the classical *porfido antico rosso* of Egypt.

ALLANITE, ORTHITE

C $(Ca,Ce)_2(Fe^{2+}Fe^{3+})Al_2Si_3O_{12}(OH)$
HD 5.0–6.5
SG 3.2–4.2

Characters and occurrence

n_α = 1.690 ⎫
n_β = 1.700 ⎬ Ris increase with increasing iron (Ti,REE) content
n_γ = 1.706 ⎭
δ = 0.013 – 0.036
$2V$ = 57°–90° +ve; 90°–40° −ve

Most properties are similar to those of epidote. This is a cerium-bearing epidote, brown or black in colour, occurring in tabular or prismatic crystals, or as grains. It is often metamict due to damage to its lattice from the decay of radioactive elements such as thorium. In thin section, allanite is strongly pleochroic, with α reddish brown or light brown, β yellowish brown and γ dark greenish brown. Birefringence is moderate but masked by mineral colour. Allanite inclusions cause pronounced pleochroic haloes in hornblende and biotite crystals. Allanite often forms a core to epidote crystals. It occurs as an accessory mineral in syenites in particular, and also in granites and diorites and their metamorphic derivatives.

LAWSONITE

C $CaAl_2(OH)_2[Si_2O_7].H_2O$
HD 6
SG 3.05–3.12

Characters and occurrence A rare orthorhombic mineral, with $2V$ large (±80°) positive, colourless or pale coloured in hand specimen, and colourless or pleochroic in pale yellows in thin section. Relief is moderate to high (RI~1.670±), and birefringence is moderate (low second order). It is found in low-grade – relatively high-pressure regionally metamorphosed rocks, and is particularly characteristic of glaucophane schists (**blue-schists**), where it is associated with pumpellyite, albite, aegirine and jadeite. It may also be secondary, being derived from calcic plagioclase by the hydrothermal alteration of basic igneous rocks.

PUMPELLYITE

C $Ca_2Al_2(Mg,Al,Fe^{2+}Fe^{3+})[SiO_4][Si_2O_7](OH)_2(H_2O.OH)$
HD 6
SG 3.18–3.23

Characters and occurrence

n_α = 1.674
n_β = 1.675 } RIs increase with increasing iron content
n_γ = 1.688
δ = 0.002–0.022
$2V$ = 10°–85° +ve
OAP is parallel to (010)

A monoclinic mineral which is greenish or yellowish in colour in the hand specimen. In thin section, it is pale yellow or green and pleochroic in yellows. It has a high relief and low to moderate birefringence. A section showing two cleavages has oblique extinction. It is a common mineral in low-grade regionally metamorphosed schists, particularly glaucophane schists where it frequently accompanies lawsonite. It may also appear as a secondary mineral in hydrothermally altered basic igneous rocks.

MELILITE GROUP (gehlenite to åkermanite)

C $(Ca,Na)_2(Mg,Al)[(Si,Al)_2O_7]$. The group comprises a series of minerals from the Ca–Al end-member **gehlenite**, to the Ca–Mg end-member **åkermanite**. The mineral **melilite** occupies an intermediate composition with Ca, Al and Mg present. Na^+ and Fe^{3+} can replace Ca and Al respectively.
H 5.0–6.0
SG 2.94–3.04

Characters and occurrence

n_o = 1.670 (geh)–1.632 (åk)
n_e = 1.658 (geh)–1.640 (åk)
δ = 0.000–0.012 (geh and åk)
Uniaxial: +ve (geh) −ve (åk)

The minerals in this group are tetragonal and usually appear as small tabular crystals or grains, white, yellowish or greenish in colour. In thin section, the minerals are colourless or yellowish, and show low to anomalous birefringence colours under XP. Inclusions are often present, peg-like in shape and running parallel to the c axis. They are decomposed by HCl, with gelatinization. Melilite occurs in basic lava flows, which are silica undersaturated and without feldspar, such as melilite basalts, nepheline and leucite basalts. Melilite may occur in some slags.

HEMIMORPHITE, calamine, electric calamine, galmei, silicate of zinc
C $Zn_4[Si_2O_7](OH)_2.H_2O$

Physical properties

CS Orthorhombic

F&H Usually occurs as prisms, the opposite ends of which are terminated by dissimilar faces; also massive, granular, fibrous, mammilated, encrusting, stalactitic or banded

COL White, yellowish brown; sometimes faintly greenish or bluish, and sometimes banded in blue and white

S Streak

CL {110} perfect; {101} poor

F Uneven

L Vitreous; subpearly, sometimes adamantine

TR Transparent to translucent

T Brittle

OTHERS Hemimorphite becomes electrically charged when heated; and phosphorescent when rubbed

HD 4.5–5.0

SG 3.40–3.50

Optical properties

$n_\alpha = 1.611$–1.617
$n_\beta = 1.614$–1.620
$n_\gamma = 1.632$–1.639
$\delta = 0.022$
$2V = 44°$–$47°$ +ve
OAP is parallel to (100)

COL Colourless

H Occurs as sheaf-like aggregates of thin, tabular, vertically striated crystals

R Moderate

A It may decompose to willemite plus water

B Moderate, low second order colours seen

IF A basal section will show a moderate Bxa figure

Tests Heated in the closed tube, hemimorpohite decrepitates, whitens and gives off water. Heated alone before the blowpipe it is almost infusible, but heated with sodium carbonate on charcoal it gives an incrustation which is yellow when hot, and white when cold. If this encrustation is moistened with cobalt nitrate and strongly reheated, it assumes a green colour. Hemimorphite gelatinizes with acids, and is decomposed even by acetic acid, with gelatinization occurring. It is soluble in a strong solution of caustic potash.

Occurrence Hemimorphite accompanies the sulphides of iron, zinc and lead and is found associated with smithsonite. It is a secondary ore found in the oxide zones of ore deposits.

9.4 Cyclosilicates

BERYL
C $Be_3Al_2Si_6O_{18}$

Physical properties

CS Hexagonal

F&H Hexagonal prisms, often large, occur. One from Albany, Maine was 5 m long and weighed 18 t. Several forms are usually present but prism faces are $\{10\bar{1}1\}$, and the basal form $\{0001\}$ is always present.

COL Various shades of green, blue, yellowish or white

CL Basal $\{0001\}$ imperfect

F Conchoidal or uneven

L Vitreous to resinous

TR Transparent to translucent, sometimes opaque

HD 7.5–8.0

SG 2.66–2.92

Optical properties

n_o = 1.560–1.602
n_e = 1.557–1.599
δ = 0.003–0.009
Uniaxial −ve

COL Colourless

H Prismatic crystals are most common

CL Basal cleavage is at right angles to the prism length

R Low to moderate

A Beryl is easily altered to clay minerals

B Low, first order greys

Varieties **Emerald** is an emerald green or pale green gemstone variety, the colour being due to a small amount of chromium present. **Aquamarine** is a pale blue gemstone variety. The term **beryl** is applied to all other types, which, on account of their opacity, are unfit for use as gemstones.

Tests Heated before the blowpipe alone, beryl and its varieties become clouded, but otherwise unaltered, except that, after protracted heating, the edges become rounded.

Occurrence Beryl occurs as an accessory mineral in acid igneous rocks, such as granites and pegmatites, in the druses or cavities of which large crystals often project, as in the granite of the Mourne Mountains. In these beryl is often associated with cassiterite. Beryl is also found in metamorphic rocks of various types – particularly schists. The gem varieties come mainly from Colombia, where the mineral is found in veins of calcite that cuts black Cretaceous shales; from Brazil, where it occurs in altered limestones; and in pegmatites from the Urals, where it occurs in mica schists within a thermal aureole; and from a few less important localities. The major producers for industry are India, Brazil, the USSR and Ruanda.

CORDIERITE, iolite, dichroite
C $Al_3(Mg,Fe)_2(Si_5Al)O_{18}$; the Mg:Fe ratio is variable and iron-rich cordierites have been reported

Physical properties

CS   Orthorhombic
(pseudo-hexagonal)
F&H   Short pseudo-hexagonal
crystals; usually granular or massive
TW   Cyclic twinning common on {130}
and {110}
COL   Blue of various shades

CL   {010} and {001} poor
F   Subconchoidal
L   Vitreous
TR   Transparent to translucent
T   Brittle
H   7
SG   2.53–2.78

Optical properties

n_α = 1.522–1.588
n_β = 1.524–1.574
n_γ = 1.527–1.578
δ = 0.005–0.020
$2V$= 65°–90° +ve; 90°–76° −ve
OAP is parallel to (010)
COL   Colourless, but iron-rich cordierites may be bluish in colour
P   Iron-rich varieties may show α colourless, β and γ violet
H   Crystals are large and full of inclusions in regional metamorphic rocks, but in thermal aureoles cordierite crystals are often hexagonal in outline
TW   Cyclic twinning is common, producing a pseudo-trigonal or pseudo-hexagonal pattern; lamellar twinning may also occur

CL   See above
R   Low, identical to that of most plagioclase feldspars
A   Cordierite alters to pinite (a mixture of muscovite and serpentine) which gives it a pale yellowish colour in thin section. Yellow pleochroic haloes may occur in cordierite crystals surrounding small crystals of zircon or monazite.
B   Low, similar to quartz or feldspar
IF $2V$ is large, so a single optic axis figure is needed to obtain the sign and size

Tests   Heated before the blowpipe, cordierite loses transparency and fuses with difficulty on the edges. The glassy or resinous appearance is characteristic, and fusibility on the edges distinguishes cordierite from quartz. It is much softer than sapphire.

Occurrence   Cordierite occurs mainly in metamorphic rocks; in regional metamorphic rocks, as in the cordierite gneisses or schists of high metamorphic grade, and in thermal metamorphic rocks as in cordierite horn-felses of pelitic composition. It may occur in cordierite norites, which have been shown to be crystallized products from partially melted pelitic material (pyrometamorphic products). It is also common in S-type granites and granodiorites and their extrusive equivalents.

Uses Sometimes used as a gemstone, the precious variety being called **water sapphire**.

TOURMALINE

C $Na(Mg,Fe^{2+},Mn,Li,Al)_3Al_6(BO_3)_3[Si_6O_{18}](OH,F)_4$. Fe^{3+} can replace Al in the lattice which contains $[Si_6O_{18}]$ rings. The colour of tourmaline can indicate its composition. The important members are:

dravite $NaMg_3Al_6(BO_3)_3[Si_6O_{18}](OH,F)_4$
schorl $Na(Fe^{2+}Mn)_3Al_6(BO_3)_3[Si_6O_{18}](OH,F)_4$
elbaite $Na(Li,Al)_3Al_6(BO_3)_3[Si_6O_{18}](OH,F)_4$

Physical properties

CS Trigonal

F&H Common as elongate prismatic crystals with a triangular cross section, and sometimes as acicular needles in radiating sheaves. The prismatic faces are often convex and striated parallel to the length. Tourmaline may also occur as massive, compact or columnar aggregates.

COL Commonly black or bluish black; more rarely blue, pink or green, and almost never colourless.

Colours are often zoned with banding occurring along the prism length, that is from top to bottom.

CL Prismatic $\{11\bar{2}0\}$ good, and rhombohedral $\{10\bar{1}0\}$ poor

F Subconchoidal to uneven

L Vitreous

TR Transparent to opaque

T Brittle

HD 7.0–7.5

SG 2.9–3.2

Optical properties

n_o = 1.610–1.630+
n_e = 1.635–1.655+
δ = 0.021–0.025+
Uniaxial −ve (all varieties)

COL Variable; elbaite is colourless, but others exhibit shades of blue, green and yellow

P The coloured varieties are pleochroic, with o shades of brown and e shades of yellow in dravite; and o dark green or blue and e pale green, blue or violet in schorl; a reddish tinge may occur if schorl is rich in Mn

H Tourmaline occurs as three- or six-sided elongate, or as acicular, prismatic crystals which sometimes occur in radiating clusters

CL See above

R Moderate

B Moderate, but colours may be masked in coloured varieties

Z Occasional coloured zoning may be present

Varieties **Rubellite** is a transparent red or pink variety which is sometimes cut as a gemstone. **Indicolite** is an indigo-blue variety, and **Brazilian sapphire** is a transparent Berlin blue variety which is commonly used as a gemstone; **Brazilian emerald** is a transparent green type. **Peridot of Ceylon** is a honey yellow variety. The common varieties, **elbaite**, **schorl** and **dravite**, are the varieties whose properties are described in this account.

Tests Heated before the blowpipe, dark-coloured varieties intumesce and fuse with difficulty. Red and green varieties become milk white and fuse slightly on the edges.

Occurrence Tourmaline ocurs in granite pegmatites, pneumatolytic veins and some granites, as the elbaite or schorl variety. In the pneumatolytic stage of alteration tourmaline may form after boron introduction has occurred, and the rock **luxullianite** forms in this environment. In pneumatolytic igneous assemblages tourmaline is frequently found in association with topaz, spodumene, cassiterite, fluorite and apatite. In metamorphic rocks, especially metamorphosed impure limestones, and metasomatic rocks the dravite variety occurs, and dravite has occasionally been reported from some basic igneous rocks. Tourmaline is a common detrial 'heavy' mineral in sedimentary rocks, and has been found as an authigenic mineral in some limestones.

Uses Some varieties are used as gemstones, mainly from Brazil, the USSR, Madagascar and the USA.

AXINITE
C $(Ca,Mn,Fe^{2+})_3Al_2(BO_3)Si_4O_{12}(OH)$

Physical properties

CS Triclinic

F&H Thin sharp-edged crystals common; rarely massive or lamellar

COL Clove brown, plum blue or pearly

CL {100} good; {001}, {110} and {011} poor

F Conchoidal

L Vitreous or bright

TR Transparent to subtranslucent

T Brittle

HD 6.5–7

SG 3.26–3.36

Optical properties

n_α = 1.674–1.693

n_β = 1.681–1.701

n_γ = 1.684–1.704

δ = 0.009–0.013

$2V$ = 63°–90° −ve

OAP is perpendicular to (111)

COL Colourless or pale violet brown

H Euhedral or subhedral wedge-shaped crystals most common

CL Distinct on {100}; other cleavages poor (see above)

R High

B Low, first order greys or yellows

IF $2V$ is large and a single optic axis figure is needed for sign

E Oblique to cleavage or face edges in *all* sections

DF Axinite is difficult to identify, but wedge-shaped sections, pale violet brown colour, oblique extinction in all sections and occurrence are characteristic features

Tests Heated before the blowpipe, axinite fuses readily, with intumescence, and colours the outer flame as pale green. It is unaffected by acids unless previously heated, whereupon it gelatinizes.

Occurrence Axinite occurs in thermal aureoles, at contact zones between granites and carbonate sediments, and in association with epidote group minerals, diopside, Ca-bearing garnets, idocrase, tourmaline and calcite. Axinite may occur in veins or cavities in igneous rocks, where it may be found with prehnite, epidote, hornblende, datolite, etc. It is a rare accessory mineral in some pegmatites.

9.5 Inosilicates (chain silicates)

Single-chain silicates

Pyroxene group

The pyroxene group includes both orthorhombic pyroxenes (orthopyroxenes, or opx) and monoclinic pyroxenes (clinopyroxenes, or cpx). They all possess $[SiO_3]_n$ chains in their atomic structure. A wide range of compositions is available in the pyroxenes, particularly the clinopyroxenes. The general formula is $X_{1-n}Y_{1+n}Z_2O_6$, where x = Ca or Na; Y = Mg, Fe^{2+}, Ni,Li, Fe^{3+}, Cr or Ti; and Z = Si or Al. In the orthopyroxenes, $n = \sim 1$ and thus $x = 0$, and the formula is reduced to $Y_2Z_2O_6$ or YZO_3, and virtually no monovalent or trivalent cations enter the structure, except under high pressures when Al^{3+} may enter. In the clinopyroxenes, n = 0–1, and the cations entering the structure must be such that the sum of charges in the $X + Y = Z$ groups balance the six O^{2-} anions.

The pyroxenes are characterized by two prismatic cleavages which intersect almost at right angles on the basal plane.

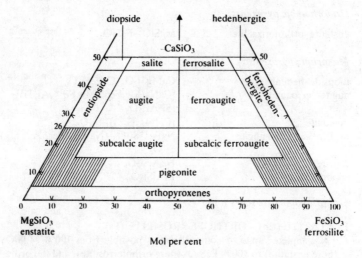

Figure 9.2 Compsoition diagram for pyroxenes.

A large number of the monoclinic pyroxenes can be considered as members of the ternary system $CaSiO_3$, or wollastonite (Wo), $MgSiO_3$, or (clino)enstatite (En), and $FeSiO_3$, or (clino)ferrosilite (Fs); and the nomenclature used to describe the pyroxenes in this ternary system can be seen in Figure 9.2. The orthorhombic pyroxenes occupy the field at the base of the ternary system, from $MgSiO_3$ to $FeSiO_3$, and with up to 5% $CaSiO_3$ included. The mid-points of two sides of the system, $(Ca_{0.5}Mg_{0.5})SiO_3$ and $(Ca_{0.5}Fe_{0.5})SiO_3$, represent the clinopyroxene minerals, diopside and hedenbergite respectively. The pyroxenes in Figure 9.2 represent 'normal' igneous pyroxenes.

Other monoclinic pyroxenes occur, which have appreciable amounts of Na in the X group, and Fe^{3+} or Al in the Y group. The sodium- and iron-bearing pyroxenes are found primarily in alkaline igneous rocks of various types, and the sodium- and aluminium-bearing pyroxenes occur primarily in igneous or metamorphic rocks formed under conditions of high pressure and temperature.

For our purposes the pyroxenes are divided into the main subsidiary groups given below:

Orthorhombic pyroxenes:

enstatite–orthoferrosilite $\quad$ $MgSiO_3$–$FeSiO_3$

Monoclinic pyroxenes:

diopside–hedenbergite	$CaMgSi_2O_6$–$CaFeSi_2O_6$
augite–ferroaugite	$[CaNa(Mg,Fe^{2+},Mn,Fe^{3+},Al,Ti)]_2$ $(Si,Al)_2O_6$
pigeonite	$[Ca[Mg,Fe^{2+}]]_2Si_2O_6$
aegirine–aegirine augite	$NaFe^{3+}Si_2O_6$–$(Na,Ca)(Fe^{2+}Fe^{3+}Mg)$ Si_2O_6
jadeite	$NaAlSi_2O_6$
spodumene	$LiAlSi_2O_6$

Orthorhombic pyroxenes (opx)

ENSTATITE (En) – ORTHOFERROSILITE (Fs)

C A complete range of compositions is possible from 100% $MgSiO_3$ (pure enstatite) to 100% $FeSiO_3$ (pure orthoferrosilite), and the orthopyroxenes may also contain up to 5% $CaSiO_3$, and up to 12% Al_2O_3 in high-grade metamorphic rocks. Pure enstatite is written $En_{100}Fs_0$ and pure orthoferrosilite En_0Fs_{100}, but these are usually shortened to either En_{100} for enstatite and En_0 for orthoferrosilite, or Fs_0 for enstatite and Fs_{100} for orthoferrosilite; that is, the composition is given in terms of the iron end-member. All other orthopyroxenes can be written in the same way; thus Fs_{26} means $En_{74}Fs_{26}$, or, to write the formula in full, $(Mg_{0.74}Fe_{0.26})SiO_3$. Pure orthoferrosilite is not found in nature because fayalite and quartz are stable to a depth of 45 km, but orthopyroxenes with up to 80% Fs have been recorded. Orthopyroxenes change their sign at approximately Fs_{12} (see below)

Physical properties

CS Orthorhombic

F&H Stout prisms common, showing {100}, {010} and {110} forms, terminated with faces of the form {101}

COL Mg-rich varieties grey, green-brown, yellow, colourless; Fe-rich varieties brownish green, brownish black

CL Two good {110} prismatic cleavages meeting at about 90° in a basal section; partings on {010} and {100} may be present

F Uneven

L Mg-rich varieties vitreous, fibrous looking on cleavage planes, Fe-rich varieties submetallic, with schillerization due to the presence of flakes of iron oxide

TR Subtranslucent to opaque

HD 5.0–6.0

SG 3.21 (En) to 3.96 (Fs)

Optical properties

$n_\alpha = En \quad Fs$
$n_\alpha = 1.650–1.768$
$n_\beta = 1.653–1.770$
$n_\gamma = 1.658–1.788$
$\delta = 0.008–0.020$
$2V = 60°–90° +ve (Fs_0–Fs_{13}); 90°–60° -ve (Fs_{13}–Fs_{87})$

Opx of composition $Fs_{87}–Fs_{100}$ are also positive, but are not found in naturally occurring rocks

OAP is parallel to (010); in some texts it is shown as parallel to (100)

COL Colourless usually, but varieties rich in Fe^{2+}, Al or Ti may be coloured in pale greens and pinks

P Some varieties are pleochroic, with α pink to pale brown, β yellow to brown and γ green

H Elongate prismatic crystals are most common

CL Two prismataic cleavages, {110}, meeting at 90° in the basal section – the two cleavages are parallel to each other. A prismatic section shows only *one* cleavage.

R Moderate (En) to high (Fs)

A Opx may alter to a fibrous aggregate of serpentine, called **bastite** or **schillerspar**. Sometimes opx alter to an amphibole, cummingtonite, with iron ores being released in this reaction.

B Low, first-order greys

IF A single isogyre is needed to get sign and size of 2V

E Straight on a prism face edge or cleavage trace

EXSOLUTION LAMELLAE In many slow-cooled pyroxenes, especially opx and augite, **exsolution lamellæ** occur which are parallel to certain crystallographic planes. An orthopyroxene may contain lamellæ parallel to the (100) plane, which represent Ca-rich cpx which were exsolved as the crystal cooled because of temperature-dependent limited solubility in the pyroxene system. In a crystallizing magma the orthopyroxene is replaced by a Ca-poor clinopyroxene (a pigeonite) when the Mg : Fe ratio of the magma is 70 : 30. The mineral pigeonite may likewise exsolve excess calcium, as Ca-rich cpx lamellae, as it cools but, in this case, the lamellae are parallel to the (001) basal plane of the *monoclinic* pigeonite. As the pigeonite cools still further, it itself inverts to an orthorhombic pyroxene (an opx) containing lamellae which are still parallel to the (001) plane of the original pigeonite. Thus an opx may have two sets of lamellae.

Ca-rich cpx (such as augite) may contain exsolution lamellae parallel to either (100) representing exsolved opx, or (001) representing exsolved pigeonite formed at high temperature. If the crystal of augite cools still further, the pigeonite lamellae may invert to opx, producing a second set of lamellae parallel to (100). Crystallographic orientations of these lamellae may be very complex, depending upon the compositions of the phases involved.

Varieties **Bronzite** and **hypersthene** are iron-bearing varieties, intermediate in composition between enstatite and orthoferrosilite. They have a bronze-like or pearly metallic lustre.

Tests Heated before the blowpipe, iron-rich varieties fuse to black enamel, and on charcoal to a black magnetic mass. Bronzite is the most infusible standard (6 on Von Kobel's Scale of Fusibility) and can only be rounded on the edges of fine splinters in the blowpipe flame.

Occurrence Orthopyroxenes occur in basic and ultrabasic igneous rocks of all types, and are the dominant pyroxene type present in norites. Mg-rich opx occur in ultrabasic and ultramafic igneous rocks such as pyroxenites, harzburgites and lherzolites, in association with forsterite, Mg-rich augite and Mg-rich spinel or garnet. Orthopyroxenes occur in rocks affected by high-temperature regional metamorphism, particularly charnockites and granulites, and also may occur in some high-grade hornfelses formed from pelitic rocks in the innermost zone of thermal aureoles. Orthopyroxenes commonly occur in some chondritic meteorites.

Monoclinic pyroxenes (cpx)
Several mineral pairs occur in the clinopyroxenes and some of the more common are dealt with here. The calcium-bearing cpx are discussed first, and then the alkali ones.

DIOPSIDE, (Di) – HEDENBERGITE (Hed)
(Also called diopside solid solution series, and written di_{ss} (or Di_{ss}).
C A complete range of compositions is possible from $CaMgSi_2O_6$(di) to $CaFe^{2+}Si_2O_6$ (hed); up to 10% Al_2O_3 may be present in the structure

Physical properties
CS Monoclinic

F&H Short prismatic crystals common; but may occasionally be granular

TW Simple and multiple twins common on {100} and {001}

COL Colourless, white, greenish or dark greenish (di); black (hed)

CL Two good {110} prismatic cleavages meeting at 87° on the basal plane. Several partings ({100}, {010} and {001}) may be present; a diopside with the {100} parting is called **diallage**, and one with the {001} parting is called **sahlite**.

L Vitreous

TR Transparent to opaque (hedenbergite is always opaque)

HD 5.5–6.5

SG 3.22 (di) to 3.56 (hed)

Optical properties

	Di	Hed
n_α =	1.664–	1.726
n_β =	1.672–	1.730
n_γ =	1.694–	1.751

δ = 0.030– 0.025

$2V$ = 50°–62° +ve

OAP is parallel to (010)

COL Colourless (di) to pale brownish green (hed)

P Hedenbergite is weakly pleochroic in greens, but this is not a diagnostic property

H Prismatic crystals common

CL Two prismataic cleavages meet at 87° in the basal plane

R Moderate (di) to high (hed)

B Moderate, with middle second-order colours

IF A (100) prismatic section shows a moderate $2V$

E Extinction angle is at a maximum on an (010) section; $\gamma \char"5E cl$ varies from ~38° (di) to ~48° (hed)

The RI and extinction angle ($\gamma \char"5E cl$) increase from diopside to hedenbergite; the birefringence decreases and the $2V$ remains more or less the same.

Varieties **Malacolite** is a white, pale green, yellow or colourless, translucent variety. **Sahlite** (or salite) has a more dull green colour than diopside, has less lustre and often shows a striation parallel to (001) – the direction of a parting it possesses. **Coccolite** is a granular variety white or green in colour, and **chrome diopside** is a bright green variety containing a few per cent of Cr_2O_3; it is an essential constituent of kimberlites. **Mansjöite** is a fluorine-bearing member of the series.

Occurrence Diopside occurs in a wide variety of metamorphic rocks, particularly metamorphosed dolomitic limestones and calcareous sedimentary rocks. Hedenbergite occurs in metamorphosed iron-rich sediments, and is found in skarns and in rocks called eulysites. Diopside may occur in some basic igneous extrusive rocks, and rarely in some lime-rich pegmatites; hedenbergite occurs in some acid igneous rocks such as ferrogabbros and granophyres, together with fayalite.

AUGITE–FERROAUGITE

C $CaNa(Mg,Fe^{2+}Mn,Fe^{3+}Al,Ti)_2(Si,Al)_2O_6$. Ferroaugite is an iron-rich augite. The optical properties of augites vary depending upon the Mg : Fe ratio and the amount, and lattice positions, of the cations Al, Fe^{3+} and Ti. Those ions in *tetrahedral* coordination (i.e. occupying Si^{4+} sites) will increase $2V$ and decrease the RI, whereas those ions in *octahedral* coordination (i.e. occupying Mg and Fe^{2+} sites) will decrease $2V$ and increase the RI. The determination of composition from optical properties can, however, be very difficult and also produce quite inaccurate results.

Physical properties

CS Monoclinic

F&H Prismatic crystals with forms {100}, {010} and {110} common, and also with {001} and {111} forms present

TW Common on {100} and {001}

COL Greenish black, brownish black or black

CL Similar to diopside with good {110} prismatic cleavages and poor {100} and {010} partings present

L Vitreous to resinous

TR Opaque

HD 5.5–6.5

SG 3.22–3.56

Optical properties

n_α = 1.662–1.735
n_β = 1.670–1.741
n_γ = 1.688–1.761

RIs depend upon both the Mg : Fe ratio and also the amount of minor constituents (Al, Ti and Fe^{3+}) present

δ = 0.018–0.033

$2V$ = 50°–55° +ve

OAP is parallel to (010)

COL Colourless to pale brown; titaniferous varieties are pale purple

H Subhedral prismatic crystals common in basic plutonic rocks, and euhedral crystals often present in basic extrusive rocks

CL As for diopside – two prismatic {110} cleavages meeting at 87° in the basal plane

R Moderate to high

A Similar to that of diopside

B Moderate to low second order colours

IF Good Bxa figure seen on a section approximately parallel to (101); a single optic axis figure may be needed to obtain the sign and size of $2V$

E An extinction angle of $\gamma \wedge cl = 38°–48°$ is at a maximum on an (010) face

Z 'Hourglass' zoning may be present in some augites, especially if titaniferous

DF Augite is very similar to diopside, but its mineral associations and the rock types in which it occurs will help to distinguish it

Variety **Diallage** is a variety of diopside or augite which is distinguished under the microscope by the presence of a parting parallel to {100}. It has been recommended that the name 'diallage' be dropped from use. It originally represented a lamellar or foliaceous mineral, occasionally fibrous, and with a metallic lustre. The mineral is green, brown or grey in colour, translucent, and is a common constituent of gabbros.

Occurrence Augites occur in igneous rocks, and are essential constituents of gabbros, dolerites and basalts. In gabbros, augites usually occur with orthopyroxenes. If cpx : opx < 1, the rock is termed a norite. Ti-augites typically occur in alkaline basalts. Augites may occur in some very high-grade metamorphic rocks such as granulites and charnockites.

PIGEONITE
C $Ca(Mg,Fe)Si_2O_6$

Characters A monoclinic mineral very similar to diopside and augite, it is distinguished from them by possessing a P2/c type of unit cell, compared with a C2/c for all other cpx minerals. Pigeonite is rare in nature. It is colourless in thin section and with RIs between 1.68 and 1.75. Birefringence is moderate with second order colours. It is biaxial positive with a *small* 2V ($2V = 0°$–$30°$) The OAP is different from diopside and augite being perpendicular to (010).

Occurrence Pigeonite occurs in rapidly chilled igneous rocks. In basic igneous rocks which cool slowly, a pigeonite may first crystallize from the magma, but will eventually invert to an orthopyroxene as cooling proceeds.

AEGIRINE (Ae) – AEGIRINE AUGITE, acmite, aegirite
C aegirine$Na Fe^{3+}Si_2O_6$ **aegirine augite** $Na,Ca (Fe^{2+}Fe^{3+} Mg)Si_2O_6$
Both aegirine and aegirine augite can accept a certain amount of Al^{3+} in their compositions; analyses of naturally occurring aegirines and aegirine augites can contain up to approximately 20% of the jadeite molecule.

Physical properties
CS Monoclinic
F&H Elongate prismatic crystals with either sharp terminations (aegirine), or blunt ends (aegirine augite). Acicular crystals may also occur.
COL Brown or dark green

CL Good {110} cleavages present; poor {100} parting usually present
L Vitreous
TR Subtransparent to opaque
HD 6
SG 3.55–3.60(ae) to 3.40–3.55 (aegirine augite)

Optical properties

	Aegirine	Aegirine augite
n_α	1.750–1.776	1.700–1.750
n_β	1.780–1.820	1.710–1.780
n_γ	1.800–1.836	1.730–1.800
δ	0.040–0.060	0.030–0.050
2V	40°–60° −ve	70°–90° −ve
2V		90°–70° +ve

OAP is parallel to (010) in both minerals
COL Greenish (ae) or brownish green (aegirine augite)
P Aegirine is strongly pleochroic, with α emerald green, β deep green and γ brownish green. Aegirine augite has a similar pleochroic scheme but the colours are less intense and each is more yellowish.

CL Similar to other pyroxenes
R High to very high
B High third-order colours are usually masked by the colour of aegirine
IF 2V variable but usually large, so that a single isogyre is needed to determine sign and size. Aegirine is negative, but aegirine augite can be positive or negative.
E Both minerals have small extinction angles seen in an (010) section, with α ^ cl = 0° to ~20°

Occurrence Aegirine and aegirine augite occur in late-crystallizing products of alkali magmas, especially syenites and nepheline syenites, often in association with alkali amphiboles. Other host rocks include alkali granites and some sodium-rich schists containing glaucophane

JADEITE
C $NaAlSi_2O_6$

Physical properties

CS Monoclinic	**CL** Good {110} prismatic cleavages
F&H Granular crystals or massive	**L** Vitreous to subvitreous
TW Simple or lamellar on {100} and {001}	**TR** Translucent
COL Jade green	**HD** 6
	SG 3.24–3.43

Optical properties

$n_\alpha = 1.640–1.658$
$n_\beta = 1.645–1.653$
$n_\gamma = 1.652–1.673$
$\delta = 0.012–0.015$
$2V = 67°–70°$ +ve
OAP is parallel to (010)
COL Colourless
H Occurs as granular aggregates of crystals
CL Similar to the other pyroxenes
R Moderate

A Jadeite can alter to amphiboles and dissociate to a mixture of nepheline and albite
B Moderate, showing second order colours
IF $2V$ is quite large so that a single optic axis figure is needed for sign and size
E On an (010) section the maximum extinction angle ($\gamma \char94 cl$) is 33° to 40°

Occurrence Jadeite is a rare pyroxene which can occur with albite in some regional metamorphic schists (blueschists), especially glaucophane schists which form under conditions of low T and high P.

Omphacite is a member of the clinopyroxenes intermediate between jadeite and augire, and is found mainly in eclogites, where it occurs with a pyrope-almandine garnet. Eclogite forms at high P, being stable between 500°C and 1000°C and of pressures above 10 kbar (= 1GPa), and is the high-pressure equivalent of gabbro.

Uses Jadeite is an ornamental stone, constituting one variety of **jade**. Another example of this beautiful material is the amphibole **nephrite**.

SPODUMENE
C $LiAlSi_2O_6$

Physical properties

CS Monoclinic

F&H Usually occurs as large prismatic crystals, comprising the forms {100}, {010} and {110}, with the terminations {021}, {221} and {001}

COL Grey or greenish

CL Perfect {110} prismatic

L Pearly, but vitreous on fracture surfaces

TR Translucent to subtransparent

HD 6

SG' 3.24–3.33

Optical properties Similar to those of diopside, with RIs $n_\alpha = 1.648$–1.663, $n_\beta = 1.655$–1.669 and $n_\gamma = 1.662$–1.679. Birefringence (δ) = 0.014–0.027, so that low second order colours are seen. $2V = 58°$–$68°$ +ve, and the OAP is parallel to (010)

Varieties **Hiddenite**, an emerald green colour, and **kunzite**, a lilac colour, are both transparent varieties of gem quality.

Tests Heated before the blowpipe, spodumene swells up, becomes opaque and finally fuses to a colourless glass. It colours the flame red due to lithium. It is unaffected by acids.

Occurrence Spodumene is a rare mineral occurring, often as large crystals, in lithium-rich, acid pegmatites, where it is associated with quartz, albite, lepidolite (Li-rich mica), beryl and tourmaline. In pegmatite dykes in the Black Hills of South Dakota, spodumene occurs as huge crystals measuring up to 12 m × 2 m × 1 m in size, which are estimated to weigh about 70 t, and which have been worked for their lithium content. Other similar deposits occur at King's Mountain (North Carolina) and in Brazil, Zimbabwe, South Africa and Western Australia.

Uses Spodumene is a principal source of the raw materials from which lithium salts are manufactured. The gem varieties, hiddenite and kunzite, are cut as gemstones.

Pyroxenoid group

The pyroxenoid minerals are not structurally related to the pyroxenes, but the three pyroxenoid minerals are chain silicates with more complex chains, and are described in this section.

WOLLASTONITE, tabular spar

C $CaSiO_3$

Physical properties

CS Triclinic

F&H Crystals columnar or fibrous, elongated along the *b* axis

COL White, grey, yellowish brown or brownish red

CL {100} perfect; {001} and {102} good

L Vitreous but pearly on cleavage planes

TR Subtransparent to translucent

HD 4.5–5.0

SG 2.87–3.09

Optical properties

n_α = 1.616–1.640

n_β = 1.628–1.650

n_γ = 1.631–1.650

δ = 0.013–0.014

$2V$ = 38°–60° −ve

OAP is almost parallel to (010)

COL Colourless

H Columnar or fibrous crystals usually occur

CL Two or three cleavage traces usually are present in any section

R Moderate

B Low; first order yellow is maximum colour

IF $2V$ is moderate and may be observed on a section approximately parallel to the basal plane; a single optic axis figure may be required

E Extinction angle measured against the crystal length on an (010) section is $\alpha \,\hat{}\, c$ axis (crystal edge) = 30°–40°

DF Wollastonite is similar to diopside, but optically negative

Tests Heated before the blowpipe, wollastonite fuses easily on its edges; it also gelatinizes with HCl

Occurrence Wollastonite is a product of the high-grade thermal metamorphism of impure limestones, often being associated with grossular and diopside. It has also been found in some alkaline igneous rocks. **Parawollastonite** (the monoclinic form of wollastonite) has a similar paragenesis, being reported from limestone blocks ejected during a volcanic eruption.

PECTOLITE

C $Ca_2NaH(SiO_3)_3$ with Mn^{2+} replacing Ca

Physical properties

CS Triclinic

F&H Clumps of fibrous crystals, each elongated along the *b* axis

COL White or greyish

CL {100} and {001} perfect

L Silky or subvitreous when fibrous; dull when massive

TR Subtranslucent to opaque

HD 4.5–5.0

SG 2.86–2.90

Optical properties

n_α = 1.595–1.610
n_β = 1.605–1.615
n_γ = 1.632–1.645
δ = 0.030–0.038
$2V$ = 50°–60° +ve
OAP is almost parallel to (100)
COL Colourless
H Fibrous crystals common
CL Prismatic cleavages present but crystals very small

R Moderate
B Moderate to high with low third order colours
IF Size of crystals make obtaining an interference figure impossible; crystals are length slow
E Oblique, with $\alpha \;\hat{}\; $ cleavage = 5°–11° measured on an (010) section

Tests Heated in a closed tube pectolite gives off water. Heated before the blowpipe it fuses to a glass. The yellow flame of sodium is seen. It gelatinizes with HCl.

Occurrence Deposited by hydrothermal solutions in veins, in cavities, and an amygdales in basalts where it is associated with zeolites, prehnite, calcite, etc. It is occasionally found in igneous rocks containing feldspathoidal minerals.

RHODONITE
C $(Mn,Ca)SiO_3$; Mn-rich mineral with some Fe^{2+} replacing Mn

Physical properties

CS Triclinic
F&H Large crystals tabular on the basal plane; occasionally massive
COL Flesh red or light brownish red when Mn-rich; greenish or yellowish when Fe-rich. It is often black on exposed surfaces, the colour darkening on exposure to the atmosphere due to oxidation of the manganese.

CL {110} and {1̄10} perfect; {001} good
F Uneven, sometimes conchoidal; tough when massive
L Vitreous
TR Transparent to opaque
HD 5.5–6.5
SG 3.57–3.76

Optical properties

n_α = 1.711–1.738
n_β = 1.717–1.741
n_γ = 1.724–1.751
δ = 0.011–0.014
$2V$ = 61°–76° +ve
OAP is almost parallel to (010)
COL Usually colourless in thin section, but large Mn-rich crystals may be pleochroic in pinks

H Large or massive crystals common
CL As above
R High
B Low, usually first order white
IF $2V$ is large, so that a single optic axis figure is needed to obtain sign and size

Varieties **Fowlerite** is a variety containing zinc, and **bustamite** contains calcium.

Tests Heated before the blowpipe, rhodonite blackens and fuses to a black glass, with slight intuminescence. In the oxidizing flame, it gives the manganese reactions with borax and the microcosmic salt bead.

Occurrence Rhodonite occurs in manganese ore bodies and is associated with metasomatic activity. It is found as a veinstone in lead and silver–lead veins, as at Broken Hill, NSW, where it is associated with dialogite and quartz.

When cut and polished, rhodonite is used for ornamental work. It can be used to impart a violet colour to glass. When it is mixed with the common salt-glazing, a black or deep blue glaze is formed.

Double chain silicates:
Amphibole group

The amphiboles are a group of hydroxylated chain silicates with some substitution of F and Cl for (OH). They include both orthorhombic and monoclinic members. The double-chain silicate structure allows a large number of elemental substitutions. A double chain has a composition of $[Si_4O_{11}]_n$, with some Al^{3+} replacing Si^{4+} by a coupled substitution of the type $(Al^{3+})^{iv} + (Al^{3+})^{vi} = (Y^{2+})^{vi} + (Si^{4+})^{iv}$. The chains are stacked parallel to the c crystallographic axis, and joined together by ions (called A, X and Y types) occupying particular lattice sites. The Y ions comprise Mg^{2+} and Fe^{2+} and include other ions such as Al^{3+}, Fe^{3+}, Mn^{2+} and Ti^{4+}. There are four distinct $X + Y$ sites in the amphiboles compared with two in the pyroxenes. The X ions include Ca^{2+} or Ca^{2+} and Na^+, although in the orthohombic amphiboles Mg^{2+} and Fe^{2+} ions occupy the X ion sites as well as the Y ion sites. The A ion sites are always occupied by Na^+ and sometimes also by K^+, although in calcium-poor and calcium-rich amphiboles the A ion sites remain unoccupied.

There are three main groups: (a) the Ca-poor amphiboles; (b) the Ca-rich amphiboles; and (c) the alkali amphiboles.

All amphiboles are prismatic, with two good prismatic cleavages meeting at 124° in the basal section.

The main physical and optical differences between the amphiboles and the pyroxenes (two of the most important and abundant rock-forming silicates) are given below:

Amphiboles	Pyroxenes
crystals elongate prisms, often acicular or bladed	crystals squat prisms
basal sections are six-sided and 'diamond' shaped	basal sections are eight-sided and square shaped
two prismatic cleavages meet at 124° in the basal section	two prismatic cleavages meet at 88° in the basal section
usually pleochroic	non-pleochroic except for the Na–Fe^{3+} bearing varieties
maximum extinction angle (measured to cleavage) on an (010) section = ~ 20°	maximum extinction angle (measured to cleavage) on an (010) section = ~ 48°
biaxial negative (rarely positive) with large $2V$ ($> 70°$)	biaxial positive (rarely negative) with moderate $2V$ (50°–60°)
exsolution lamallae very rare or not present	exsolution lamellae common in cpx and opx from basic igneous rocks
twins common, with no re-entrant angle	twins rare with a re-entrant angle

The calcium-poor amphiboles

The Ca-poor amphiboles, in which Ca + Na is about equal to zero, include the orthorhombic amphiboles and the Ca-poor monoclinic amphiboles. The general formula is $X_2Y_5Z_8O_{22}(OH,F)_2$, where X = Mg or Fe^{2+}, Y = Mg, Fe^{2+}, Fe^{3+}, Al, etc., and Z = Si or Al

Orthorhombic amphiboles

ANTHOPHYLLITE–GEDRITE

C anthophyllite $(Mg,Fe^{2+})_2(Mg,Fe^{2+})_5[Si_8O_{22}]OH,F)_2$, with Mg > Fe^{2+}
gedrite $(Mg,Fe^{2+})_2(Mg,Fe^{2+})_3 Al_2 [Si_6Al_2O_{22}](OH,F)_2$, with Fe^{2+} > Mg
Gedrite may contain 1–3% Na_2O.

Physical properties

CS Orthorhombic
F&H Elongate to acicular crystals common, often in clusters or in radiating aggregates
COL Shades of brown, with gedrite darker than anthophyllite

CL Perfect {110} prismatic cleavages; very poor cleavages on {100} and {010}
L Vitreous
TR Transparent to subtranslucent
HD 5.5–6.0
SG 2.85–3.57

Optical properties

$n_\alpha = 1.596-1.694$
$n_\beta = 1.605-1.710$
$n_\gamma = 1.615-1.722$
$\delta = 0.013-0.028$
$2V = 69°-90°$ −ve anthophyllite
78°–90° +ve gedrite
OAP is parallel to (010) in both minerals
Both minerals are length slow
COL Colourless (anthophyllite) to shades of brown (gedrite)
P Gedrite is pleochroic, with α and β pale brown, and γ dark brown

H Crystals are acicular or elongate prismatic
CL Two prismatic cleavages intersecting at 124° in the basal section
R Moderate (anth) to high (ged)
A Common to chlorite or talc in the presence of water
B Low to moderate, showing second order colours
IF $2V$ is large in both minerals so that a single optic axis figure is needed to obtain sign and size

Occurrence The orthorhombic amphiboles are found only in metamorphic rocks, where they form during high-grade regional metamorphism of ultrabasic igneous rocks or by Mg- and Fe-metasomatism of pelitic sediments. They are found in amphibolites, gneisses and hornfelses, often in association with cordierite.

Monoclinic amphiboles

CUMMINGTONITE–GRUNERITE

C $(Mg, Fe^{2+})_2 (Mg, Fe^{2+})_5[Si_8O_{22}](OH,F)_2$. Anthophyllites have Mg > Fe, whereas grunerites have Fe > Mg. This group of minerals is very similar to the orthorhombic amphiboles but has *no* aluminium in the structure.

Physical properties

CS Monoclinic
F&H Columnar to acicular, with fibrous crystals in radiating aggregates
COL Greyish brown to brown

CL Perfect {110} prismatic
L Vitreous
TR Transparent to subtranslucent
HD 5.0–6.0
SG 3.10–3.60

Optical properties

$n_\alpha = 1.635-1.696$
$n_\beta = 1.644-1.709$
$n_\gamma = 1.655-1.729$
$\delta = 0.020-0.045$
$2V = 65°-90°$ +ve (cumm)
90°–84° −ve (grun)
The $2V$ of 90° is for a mineral of composition Mg : Mg + Fe + Mn = 30 : 70
OAP is parallel to (010)

COL Colourless to pale green or brown
P Iron-rich cummingtonite is weakly pleochroic, with α and β colourless, and γ pale green; grunerite is weakly pleochroic with α and β very pale yellow or brown, and γ pale brown
H Elongate to fibrous crystals found singly or in clusters
CL Usual amphibole cleavages present

R Moderate to high
A Common to chlorite, etc. in the presence of water
B Moderate with interference colours from low second order (cumm) to low third order (grun)

IF Large, so a single optic axis figure is needed for sign and size
E In an {010} section, $\gamma \hat{} \ cl = 21°$

Varieties **Amosite** or **montasite** is the asbestiform variety of the cummingtonite–grunerite series.

Occurrence Cummingtonite occurs in metamorphosed basic igneous rocks where it is associated with the common hornblendes. Grunerite occurs in metamorphosed iron-rich sediments in association with magnetite and quartz, or garnet and fayalitic olivine, the latter minerals being found in eulysite bands.

The calcium-rich amphiboles

The Ca-rich amphiboles (Ca > Na) are monoclinic, and include tremolite–ferroactinolite, and the 'common hornblendes' (the hornblende group of minerals). The general formula is $AX_2Y_5Z_8O_{22}(OH,F)_2$, where A = Na (often no sodium is present and $A = 0$), X = Ca, Y = Mg, Fe, Al, etc., and Z = Si or Al.

TREMOLITE–FERROACTINOLITE

C $Ca_2 (Mg, Fe^{2+})_5[Si_8O_{22}](OH, F)_2$ with $A = 0$ (Na is not present). In actinolite Mg > Fe; whereas in ferroactinolite Mg < Fe.

Physical properties
CS Monoclinic
F&H Elongate, prismatic or acicular crystals common; often fibrous or columnar in radiating clusters; occasionally granular
COL Tremolite is white or greyish; ferroactinolite is green or brownish

CL typical {110} prismatic cleavages occur; a poor {100} parting may also be present
L Vitreous
TR Transparent to translucent
HD 5.0–6.0
SG 3.02–3.44

Optical properties
n_α = 1.599–1.688
n_β = 1.612–1.697
n_γ = 1.622–1.705
δ = 0.027–0.017
$2V$ = 86° (trem)–65° (ferroact)
OAP is parallel to (010)
COL Colourless (trem) to brownish green (ferroact)
P Actinolitic (Fe) varieties are pleochroic, with α pale yellow, β yellowish green and γ greenish blue
H Crystals usually elongate, acicular or fibrous; often in radiating aggregates

CL See previous section
R Moderate to high, increasing with increasing iron content
A Common to chlorite, etc. in the presence of water
B Moderate with second-order green maximum colour seen.
IF $2V$ large, so that a single optic axis figure is needed to obtain sign and size
E Maximum in an (010) section with $\gamma \hat{} \ cl = 21°$ (trem) to 11° (actinolitic types)

Varieties **Nephrite** is the asbestiform variety (see also under Asbestos, p. 382). Precious **jade** may be composed of either nephrite or jadeite (see under Pyroxenes, p. 395). **Uralite** is an actinolite which has replaced and pseudomorphed a pyroxene, often in a basic igneous plutonic rock. Where such a basic rock shows extensive alteration of this type it can be said to be **uralitized** (suffering **uralitization**); but nowadays the term amphibol(it)ized, abbreviated to **amphibolized**, is preferred.

Occurrence Tremolite and actinolite are metamorphic minerals, formed during thermal or regional metamorphism, especially involving impure dolomitic limestones. They occur in relatively low-grade metamorphic rocks, and actinolite is a characteristic mineral of **greenschist facies**, occurring with the 'common hornblendes' (see next section). Actinolite may also occur in **blueschists**, along with glaucophane, epidote and albite.

THE HORNBLENDE SERIES, 'common hornblendes'
C The general composition of the 'common hornblendes' can be given as:

$$Na_{0-1} (Mg_{3-5} Al_{2-0}) [(Si_{6-7} Al_{2-1}) O_{22}](OH,F)_2$$

The 'hornblende series' or common hornblendes' is the name given to any amphibole which has a composition occurring in a field of composition which is itself defined by four end-members, as follows:

hastingsite	$Ca_2 (Mg_4Al) [Si_7Al O_{22}] (OH,F)_2$
tschermakite	$Ca_2 (Mg_3Al_2) [Si_6Al_2 O_{22}] (OH,F)_2$
edenite	$NaCa_2 Mg_5 [Si_7Al O_{22}] (OH,F)_2$
pargasite	$NaCa_2 (Mg_4Al) [Si_6Al_2 O_{22}] (OH,F)_2$

Iron II (Fe^{2+}) may replace Mg in the above minerals, but all substitutions of this type have beem omitted from the above formulae for simplicity.

Physical properties

CS Monoclinic

F&H Prismatic crystals common, showing {110} and {010} prism forms and {011} terminations

TW Simple twinning common on {100}

COL Black or greenish black

CL Perfect on {110}, intersecting at 124° on the basal plane

L Vitreous

TR Translucent to opaque

HD 5.0–6.0

SG 3.02–3.50

Optical properties

$n_\alpha = 1.615–1.705$
$n_\beta = 1.618–1.729$
$n_\gamma = 1.632–1.730$

The large variations in RI are due to variations in the Mg : Fe ratio, and also the amount of Fe^{3+} and Al present

$\delta = 0.014–0.028$
$2V = 15°–90°$ −ve
(some Mg-hornblendes may have $2V = \sim 90°$ +ve)
OAP is parallel to (010)

COL Green or yellowish brown

P Variable with Mg-hornblendes, having α pale green or brown, and β and γ brownish green; and Fe-hornblendes having α yellow brown or green, β deep green or bluish green, and γ very dark green

H Usually prismatic, elongate crystals common

CL See above

R Moderate to high

B Variable with maximum interference colours low second order blue, often masked by the 'body' colour

IF Fe-hornblendes possess very low 2Vs, so that a complete OA figure can be seen in a (100) prismatic section. Other hornblendes, particularly Mg-rich varieties, have large 2Vs and a single optic axis figure is needed to determine sign and size

E Variable with a maximum γ ˆ cl = 34° in an (010) section

Tests Their physical properties are distinctive. Heated before the blowpipe, hornblendes fuse easily, forming a magnetic globule.

Occurrence Hornblendes are primary minerals in intermediate, and some acid, plutonic igneous rocks. In intermediate igneous rocks, hornblendes have an Mg : Fe ratio of ~ 1 : 1, which increases in hornblendes from basic igneous rocks and decreases in hornblendes from acid igneous rocks. Hornblendes are essential constituents of the amphibolite facies of regional metamorphism, and became more aluminous with increasing metamorphic grade. Tschermakite occurs in high-grade metamorphic rocks, and pargasite in metamorphosed impure dolomitic limestones. Secondary amphiboles in igneous rocks are usually tremolites or cummingtonites, rarely hornblendes.

The alkali amphiboles

In the alkali amphiboles Na is always greater in amount than Ca (i.e. Na > Ca), and the general formula is $AX_2Y_5Z_8O_{22}(OH,F)_2$, where A = Na or K, X = Na (or Na and Ca), Y = Mg, Fe, Al, etc., and Z = Si or Al.

GLAUCOPHANE–RIEBECKITE

C   **glaucophane** $Na_2(Mg_3Al_2)[Si_8O_{22}](OH,F)_2$
       riebeckite $Na_2(Fe_3^{2+}Fe_2^{3+})[Si_8O_{22}](OH,F)_2$

Physical properties

CS Monoclinic

F&H Glaucophane usually occurs as tiny prismatic crystals, which may be fibrous, massive or granular. Riebeckite occurs either as large, subhedral, prismatic (often poikilitic) crystals, or as tiny prismatic crystals in the ground mass of some alkali igneous rock types.

COL Glaucophane is blue, bluish black or bluish grey; riebeckite is dark bluish green or black

CL Perfect {110} prismatic; rare partings on {010} and {001} may occur

L Vitreous

TR Translucent

HD 6.0 (glaucophane); 5.0 (riebeckite)

SG 3.02 (glaucophane) to 3.43 (riebeckite)

Optical properties

	glauc	rieb
n_α	1.606–	1.701
n_β	1.622	–1.711
n_γ	1.627	–1.717
δ	0.008	–0.022

$2V = 0°–50°$ −ve(glauc); $0°–90°$ −ve (rieb)

OAP is parallel to (010) in both minerals. A mineral, **crossite**, intermediate between glaucophane and riebeckite, occurs in which the OAP is *perpendicular* to (010).

COL Glaucophane is pale coloured or colourless, whereas riebeckite is strongly coloured in blues or greens

P Both are pleochroic, with glaucophane having α colourless, β pale blue and γ blue; riebeckite has α blue, β deep blue and γ yellowish green

H Usually found as small prismatic crystals occurring in the ground mass, but riebeckite may also occur as large poikilitic crystals

CL See above

R Moderate to high

A Rare in glaucophane, but riebeckite may alter to **crocidolite**, a fibrous asbestiform mineral

B Low to moderate; glaucophane shows middle first order colours, but the interference colours of riebeckite are masked by the mineral colour

IF $2V$s vary considerably in size, but the strong dispersion and body colour of riebeckite makes a figure very difficult to obtain

E Glaucophane is length slow with $\gamma \hat{} $ cl = 6°–9°, whereas riebeckite is length fast with $\alpha \hat{} $ cl = 6°–8°, both extinction angles seen in an (010) section

TW Occasional simple or repeated twins on {100} seen

Varieties **Crocidolite** (or **blue asbestos**) is probably a fibrous asbestiform variety of riebeckite, indigo blue in colour, found in Griqualand, South Africa and elsewhere. When altered it assumes a golden yellowish brown colour and, when infiltrated with silica, it constitutes the semi-precious **cat's eye** or **tiger's eye**. **Crossite** is a variety intermediate between glaucophane and riebeckite in composition.

Occurrence Glaucophane is the essential amphibole in **blueschists** which form under conditions of high P and low T in metamorphosed sediments at destructive plate margins. Blueschists are commonly found in association with **ophiolites**. Riebeckite occurs in alkali igneous rocks, especially alkali granites, in association with aegirine. Fibrous riebeckite (crocidolite or blue asbestos) is formed from the metamorphism of massive ironstone deposits under conditions of moderate T and P.

RICHTERITE

C $Na,K(Na,Ca)(Mg, Fe^{2+}Fe^{3+}Mn,etc.)_5[Si_8O_{22}](OH,F)_2$

Physical properties

CS Monoclinic
F&H Prismatic crystals common
COL Brown, yellowish or brownish red; occasionally green

CL   Perfect {110} prismatic cleavages; good {100} and {001} partings
L   Subvitreous
HD   5.5
SG   2.97–3.45

Optical properties

n_α = 1.605–1.685
n_β = 1.618–1.700
n_γ = 1.627–1.712
δ = 0.022–0.027
$2V$ = 66°–90° −ve
OAP is parallel to (010)
COL Colourless or pale coloured
P   Slightly pleochroic, with α pale yellow or pale blue, β yellow, orange or blue, and γ (= α) pale yellow or pale blue

H   Prismatic crystals common
CL   See above
R   Moderate to high
B   Moderate, with low second order colours shown
IF $2V$ is large, so that a single optic axis figure is needed to obtain sign and size
E $\gamma \char94 cl = 15°–40°$ measured in an (010) section

Occurrence   A rare amphibole, richterite is found in skarns, and in thermally metamorphosed limestones.

KATOPHORITE, OXYHORNBLENDE (basaltic hornblende), KAERSUTITE

The above minerals are rare, brown amphiboles, and are treated together.

C   **katophorite** $Na(Na,Ca)(Mg,Fe^{2+})_4Fe^{3+}[Si_7Al\ O_{22}](OH)_2$
    oxyhornblende $NaCa_2(Mg,Fe^{2+}Fe^{3+}Ti,Al)_5[Si_6Al_2O_{22}](OH,O)_2$
    kaersutite $(Na,K)Ca_2(Mg,Fe^{2+})_4Ti[Si_6Al_2O_{22}](OH)_2$

Physical properties

CS Monoclinic

F&H All minerals occur as small euhedral crystals

COL Dark reddish brown or reddish (katophorite); dark brown or black (oxyhornblende and kaersutite)

CL Perfect {110} prismatic cleavages present; partings on {100} and {001} occur in kaersutite, and on {010} in katophorite

L Vitreous to subvitreous

HD 5.0 (katophorite); 5.0–6.0 (oxy-hornblende and kaersutite)

SG 3.20–3.50 (kat); 3.19–3.30 (oxy); 3.20–3.28 (kaer)

Optical properties

$n_{\alpha-\gamma}$ = 1.64–1.69(kat); 1.66–1.76(oxy); 1.67–1.77(kaer)

δ = 0.01–0.02(kat); 0.02–0.075 (oxy and kaer)

$2V$ = 0°–50° −ve (kat); $2V$ = 60°–80° −ve (oxy and kaer)

OAP is parallel to (010) (oxy and kaer); or perpendicular to (010) (kat).

COL All minerals are yellowish or brown

P Katophorite has α yellor or pale brown, β darkish brown, and γ brown Oxyhornblende has α yellow, and β,8 dark brown, Kaersulite has α yellow, β reddish brown, and γ brown. In iron-rich varieties, β and γ become more greenish

H Oxyhornblende and kaersutite occurs as phenocrysts in some alkali extrusives; katophorite occurs in some intrusive alkali igneous rocks

CL See above

R Moderate to high

B Low with upper first-order colours (kat); moderate to high with maximum interference colours of third or fourth order (oxy and kaer)

IF Variable or large, but strong mineral colour and dispersion makes interference figures difficult to obtain

E In (010) sections β ˆ cl = 20°–54° (kat); γ ˆcl = 0°–19° (oxy and kaer)

Occurrence All are rare amphiboles. Katophorite occurs in dark-coloured alkali intrusive igneous rocks; kaersutite in alkali (K-rich) extrusive rocks; and oxyhornblende in some intermediate hypabyssal and extrusive igneous rocks.

ECKERMANNITE–ARFVEDSONITE

C $NaNa_2(Mg,Fe^{2+})_4Al[Si_8O_{22}](OH)_2$; eckermannite has Mg > Fe, whereas arfvedsonite has Fe > Mg

Physical properties

CS Monoclinic

F&H Large prismatic crystals common; often poikilitic

COL Black, dark brown or green

CL Perfect {110} prismatic cleavages with common {010} parting

L Vitreous

TR Opaque

HD 5.5

ST 3.00–3.16 (eck); 3.30–3.50 (arfv)

Optical properties

	Eckermannite	Arfvedsonite
n_α	1.612–1.638	1.674–1.700
n_β	1.625–1.652	1.679–1.709
n_γ	1.630–1.654	1.686–1.710
δ	0.009 0.020	0.005–0.012
2V	80°–15° −ve	variable, probably −ve
OAP	parallel to (010)	perpendicular to (010)

COL Various shades of green

P Eckermannite has α blue green, β light green, and γ pale yellow green; arfvedsonite has α greenish blue to indigo, β lavender blue to brownish yellow, and γ blue grey to greenish yellow

H Large poikilitic crystals with corroded edges common

CL See above

R Moderate (eck) to high (arfv)

A Common to chloritic minerals

B Both minerals low, but interference colours usually masked by strong colours of each mineral

IF Interference figures are virtually impossible to obtain, especially with arfvedsonite, because of strong mineral colour and strong dispersion

E Also difficult to obtain, but $\alpha \,\hat{}\, cl = 0°$–50° in both minerals

Occurrence These minerals occur in alkali (Na-rich) plutonic igneous rocks, often in association with aegirine and apatite.

AENIGMATITE, cossyrite

C $Na_2Fe_5^{2+}Ti^{4+}Si_6O_{20}$

Characters and occurrence

HD 5.5

SG 3.74–3.85

$n_\alpha = 1.81$ $n_\beta = 1.82$ $n_\gamma = 1.88$ $\delta \approx 0.07$ $2V \approx 32°$ +ve

Aenigmatite is a triclinic mineral with a structure containing both *single* and *double* chains of Si–O tetrahedra. Aenigmatite occurs as elongate, black prismatic crystals which exhibit good {100} and {010} cleavages, in thin section, crystals are prismatic and dark red brown, in colour, often almost opaque and pleochroic with α reddish or yellowish brown, β brown, and γ dark brown. Birefringence is low, and is usually completely masked by the strong mineral colour. Biaxial positive with a small $2V$ and an OAP approximately parallel to (110). The extinction angle $\gamma \,\hat{}\, cl \approx 45°$, measured on an (010) section.

Aenigmatite occurs as small phenocrysts in silica-poor, alkali extrusive igneous rocks such as phonolites and trachytes.

ASBESTOS

Asbestos includes the fibrous forms of amphibole. The fibres in general are very long, thin, flexible and easily separated by the fingers. The colour may vary from white to greenish and brownish. The ancients called similar material *amianthus*, undefiled, alluding to the ease with which cloth woven from it was cleaned by throwing it into a fire; but the name amianthus is now restricted to the more silky kinds. **Mountain cork, mountain leather and mountain wood** are varieties of asbestos which vary in compactness and the matting of their fibres. In the strictest sense, the term 'asbestos' is confined to the fibrous forms of actinolite, but common asbestos includes fibrous varieties of a number of different silicates which are now considered.

The following minerals are now included as 'commercial asbestos':

chrysotile (fibrous serpentine) $Mg_6[Si_4O_{10}](OH)_8$
actinolite (asbestos proper) $Ca_2(Mg,Fe^{2+})_5[Si_8O_{22}](OH,F)_2$
amosite (fibrous anthophyllite) $(Mg, Fe)_2 (Mg, Fe)_5 [Si_8O_{22}](OH,F)_2$
crocidolite (fibrous riebeckite) $Na_2 (Fe_3^+Fe_2^{3+})[Si_8O_{22}](OH)_2$

All the above minerals occur as long fibrous crystals. The commercial value of the minerals depends almost wholly upon its property of being spun, and good asbestos yields long silky fibres when rubbed between the fingers. The heat-resisting value of all the mineral varieties is about the same. Chrysotile is of most general use, although certain other types, in particular crocidolite and amosite, are sometimes preferred for their acid-resisting qualities. Chrysotile is decomposed by HCl, but the other asbestos minerals are not.

The usefulness of asbestos depends upon its resistance to heat and to its property of being spun into yarn. The better grades, that is those with long fibres, are woven into fireproof fabrics and were recently used in brake linings, but this use is now banned in the UK and other countries. Those with shorter fibres are utilized in the manufacture of asbestos sheets, boards, roofing tiles, felt, boiler coverings, fireproof paints, insulating cements, etc.

Asbestos, in the commercial sense, occurs in three forms: (1) *cross-fibre*, when the fibres are at right angles to the vein walls; (2) *slip-fibre*, when the fibres are parallel with the walls and are formed along planes of movement; and (3) *mass fibre*, when the fibres occur in a confused fashion as in the anthophyllitic types.

The main producers of commercial asbestos in 1985 were the USSR (2.5 Mt), Canada (0.744 Mt), South Africa (0.220 Mt), Brazil (0.148 Mt) and Italy (0.136 Mt), with Swaziland (30 000 t) and Zimbabwe as minor producers. The world total is not available, but must be about 4 Mt.

Inhalation of small quantities of asbestos dust, specifically the minute, needle-like fibres, can cause serious lung diseases (**pneumoconiosis**) many years later; and the specific lung disease associated with asbestos fibres, **asbestosis**, will certainly reduce the amount of asbestos being mined and used commercially in the future. It is interesting to note that the previous edition of this book quoted about 3.5 Mt mined in 1965, which is almost the same as was mined in 1985. Further details of the various minerals that comprise commercial asbestos will be found under their respective descriptions.

9.6 Phyllosilicates (sheet silicates)

Mica group

The atomic structure of the micas consists of a sheet of cations (Fe^{2+}, Fe^{3+}, Mg, Al) and hydroxyl anions $(OH)^-$, called either a **brucite sheet** if the ions are Mg, etc. and (OH), or a **gibbsite sheet** if the ions are Al and (OH). These octahedral sheets are linked to sheets of linked $[SiO_4]$ tetrahedra, which have the general composition $[Si_4O_{10}]_n$. The complete 'sandwich' unit therefore possesses two tetrahedral sheets (Si–O) and one octahedral one, and is termed a 2:1 structure. Each 2:1 sandwich unit is linked to another similar unit by means of weakly bonded monovalent cations such as K^+ and Na^+, the perfect cleavage possessed by the micas occurring along this plane. F^- or Cl^- ions may occasionally replace the (OH) ions in the lattice, and the general formula is:

$$X_2Y_{4-6}Z_8O_{20}(OH,F)_4 \text{ with } X = \text{K or Na}, \ Y = \text{Mg, } Fe^{2+}, Fe^{3+} \text{ or Al and}$$
$$Z = \text{Si or Al}$$

When a blunt steel punch is placed on a cleaved plate of mica and lightly struck, a small six-rayed star, the **percussion figure**, is produced. The three cracks which constitute these stars have a constant relation to the form of the crystal from which the plate is cleaved, and one of these cracks is always in the direction of the plane of symmetry of the crystal.

Cleavage plates (basal sections) of micas show biaxial interference figures which can be orientated with respect to the plane of symmetry, as revealed by the percussion star figure. *Muscovite* and *paragonite* have the OAP perpendicular to the plane of symmetry, whereas all other micas have the OAP parallel to, that is lying in, the plane of symmetry.

The specific gravity values of the mica group range from 2.7 to 3.3, and the hardness values occur in the range 2.0–4.0.

The micas differ from the chlorites and other micaceous minerals in several ways; (1) their alkali content; (2) the elastic properties of cleavage flakes of micas; and (3) some of their optical properties.

Muscovite is used in electrical and visual applications, and phlogopite is used in high-risk environments.

MUSCOVITE, common mica, potash mica, muscovy glass

C $K_2Al_4[Si_6Al_2O_{20}](OH,F)_4$; Al substitutes for Si in the Z group to the extent of about one atom in every four

Physical properties

CS Monoclinic, pseudohexagonal

F&H Occurs as six-sided hexagonal plates, or as disseminated scales or massive

COL Colourless, white or pale yellow

CL One perfect cleavage parallel to the basal plane with large and thin laminae being easily separated

L Pearly

TR Transparent to translucent; when held up to a bright light, thin cleavage flakes may exhibit **asterism**, star-like rays of light being transmitted

T Thin cleavage flakes are both flexible and elastic

HD 2.5–3.0

SG 2.77–2.88

Optical properties

$n_\alpha = 1.552–1.574$
$n_\beta = 1.582–1.610$
$n_\gamma = 1.587–1.616$
$\delta = 0.036–0.049$
$2V = 30°–47° -ve$
OAP is perpendicular to (010)

COL Colourless

H Usually occurs as shapeless plates or in aggregates of crystals; when the crystals are very fine grained the mineral is called **sericite**

CL One perfect basal {001} cleavage is seen on a prismatic section; basal sections show no cleavage

R Low to moderate (if Fe enters the structure)

A Absent

B High; upper third order colours are seen on prismatic sections (sections showing cleavage). Basal sections show first order grey which has a speckled appearance.

IF A basal section gives an excellent Bxa (−ve) figure, all of which can be seen using a ×40 or ×45 objective lens

E Straight on cleavage

Varieties **Sericite, damourite** and **gilbertite** are secondary micas of muscovitic composition, resulting from the hydrothermal alteration of many rock-forming silicate minerals including feldspars, Al_2SiO_5 polymorphs, etc., which usually occur as fine grains of fibres.

Tests Before the blowpipe, muscovite whitens and fuses only on thin edges. It is not decomposed by acids, and it yields water when heated in the closed tube.

Occurrence Muscovite occurs in low-grade metamorphic rocks where it forms from pyrophyllite or illite. Muscovite remains in these rocks as the grade increases, and is a common constituent of schists and gneisses. At very high temperatures (above 600°C) muscovite becomes unstable, breaking down in the presence of quartz to give K-feldspar and sillimanite:

$$KAl_2[Si_3AlO_{10}](OH)_2 + \quad SiO_2 = \quad KAl\,Si_3O_8 + \quad Al_2SiO_5 + \quad H_2O$$

muscovite $\qquad\qquad$ quartz $\quad$ K-feldspar $\quad$ sillimanite $\quad$ water

Muscovite occurs as a primary crystallizing mineral in acid igneous plutonic rocks, such as granites and pegmatites. Muscovite is a common constituent of detrital sedimentary rocks, especially the arenites.

Production and uses In 1986, commercial mica was produced from a number of countries, such as the USSR (50 000 t), the USA (13 000 t), India (produced 6200 t of crude mica, but exported 22 000 t), Madagascar (1000 t) and Brazil, and with small quantities from Zimbabwe and Tanzania. Although production figures are not always available, the total world production is probably about 80 000 t. It is interesting to note that the USSR, easily the largest producer, also is India's biggest customer, taking about 75% of its mica exports. Most mica mined is muscovite and phlogopite.

Mica **splittings** (the trade name for mica split into flakes) is used in electrical equipment which needs material with good insulating and bending properties. The mica sheets formed by the splitting are called **micanite**, and are molded into shapes with a bonding of shellac, or some other heat-resisting material. These are used in 'V' rings for commutators and also as sleeves to enclose piping. Major applications are in the very large commutators in electrical generating stations. Mica is also used in the industrial and domestic heating field, for example as mica windows in heating stoves. High-quality mica is used in the manufacture of gauges, compass cards, quarter-wave plates and in geiger counters. Ground-up mica is used in roofing materials and in the manufacture of lubricants and wall finishes, but these markets are diminishing, although some mica is now being used in the manufacture of non-stick coatings for various kitchen utensils. Mica powder is used to give the 'frost' effect on Christmas cards, and also for Christmas tree 'snow'.

PARAGONITE
C $\quad Na_2Al_4[Si_6Al_2O_{20}](OH)_4$

Characters and occurrence A pale yellow or pale green mica, colourless in thin section, and resembling muscovite in general properties. It occurs in massive, scaly aggregates and has been reported from phyllites, schists and gneisses, quartz veins and sedimentary rocks.

GLAUCONITE

C $(KH_3O)_2(Fe^{3+} Al,Mg,Fe^{2+})_4[Si_{7.0-7.5}Al_{1.0-0.5}O_{20}](OH)_4$. Glauconite is the ferric equivalent of illite. Interlayer cations are mainly K^+ with appreciable amounts of water molecules or $(H_3O)^+$ ions.

Varieties Aluminous glauconites with Fe^{3+} : Al less than 3 : 1 are called **skolites**. A variety enriched in Mg and Fe^{2+} and depleted in Al and Fe^{3+} is called **celadonite**.

Characters and occurrence
HD 2 **SG** 2.4–3.0
Glauconite is amorphous, granular or earthy, and olive green, yellowish, greyish or blackish green in colour. It has a dull or glistening lustre, and is opaque.

In thin section, glauconite is pale yellow or green, the colour increasing with increasing iron content. It usually occurs as rounded green granules or pellets which reveal a fine aggregate structure under XP.

Heated before the blowpipe, it fuses easily to a dark magnetic glass, and gives off water.

It occurs extensively disseminated in small grains in the chalk marl, chloritic marl and greensands of the Cretaceous of England, and also in rocks much older than these as for example, the Cambrian Comely Sandstone. In addition, it occurs in oceanic sediments now in the actual process of formation. Its presence in a sediment indicates a shallow-water, marine origin for the deposit, and a slow rate of deposition. It may also form by the alteration of ferromagnesian silicates, especially biotite. Celadonite occurs as radiating aggregates in vesicles in basaltic rocks.

PHLOGOPITE

C $K_2 (Mg, Fe^{2+})_6 [Si_6Al_2O_{20}](OH,F)_4$. In phlogopite Mg : Fe is more than 2 : 1, whereas in biotite it is less than 2 : 1. Mn and Ti may replace Mg and Fe^{2+}

Physical properties

CS Monoclinic

F&H Six-sided prismatic crystals common; sometimes occurs as scales

COL White, colourless, brown, copper red

CL Perfect {001} basal cleavage giving thin laminae

L Pearly, often submetallic on cleavage planes

TR Transparent to subtransparent; asterism often seen on cleavage flakes

T Cleavage flakes are elastic and tough

HD 2.0–2.5

SG 2.76–2.90

Optical properties

n_α = 1.530–1.590 ⎫
n_β = 1.557–1.637 ⎬ RIs increase with increasing iron content
n_γ = 1.558–1.637 ⎭
δ = 0.028–0.049
$2V$ = 0°–15° − ve

OAP is parallel to (010)

COL Pale brown, nearly colourless

P Weak with pale colours; α; yellow, β and γ deep yellow or brownish red

H Phlogopite occurs as small, subhedral, tabular crystals

CL The perfect basal {001} cleavage is seen in prismatic sections

R Low to moderate

B High, with third order colours

IF Small negative $2V$ seen in a basal section

E Straight, but may show an extinction angle of a few degrees

OTHER Phlogopites in kimberlites often show dark reaction rims

Occurrence   Phlogopite is a mineral found in metamorphosed impure magnesian limestones, where it forms by reactions between the dolomite and either K-feldspar or muscovite. It is a common constituent of kimberlites, and is a minor constituent of ultramafic rocks. It is a primary mineral in some leucite-rich rocks

Uses

See under Muscovite (p. 399).

BIOTITE

C $K_2 (Mg, Fe^{2+})_{6-4} (Fe^{3+} Al, Ti)_{0-2}[Si_{6-5} Al_{2-3}O_{20}]OH,F)_4$ Mn may also be present. The type of mica depends upon its $Mg : Fe^{2+}$ ratio ; that is, upon the percentage of $Mg : (Mg + Fe)$ present.

phlogopite $Mg : (Mg + Fe)$ from 100% to 70%

biotite $Mg : (Mg + Fe)$ from 69% to 20%, and with increasing R^{3+}

siderophyllite $Mg : (Mg + Fe)$ less than 10%, and with $R^{3+} = Al^{3+}$

lepidomelane $Mg : (Mg + Fe)$ less than 10% and with $R^{3+} = Fe^{3+}$

annite $Mg : (Mg + Fe)$ approx. equal to 0, and with no R^{3+}

Physical properties

CS Monoclinic, pseudo-hexagonal

F&H Six-sided prismatic crystals common, tabular parallel to (001)

COL Black or dark green in thick crystals, but in transmitted light thin laminæ appear brown green or blood red

CL {001} perfect

L Splendent, and pearly on the cleavage

TR Transparent to opaque

T Thin cleavage laminæ are flexible and elastic

HD 2.5–3.0

SG 2.70–3.30

Optical properties

n_α = 1.525–1.625

n_β = 1.605–1.696

n_γ = 1.605–1.696

δ = 0.040–0.080

$2V$ = 0°–25° −ve

OAP is parallel to (010)

COL Brownish or greenish

P Strong with α yellow, and β and γ dark brown or brownish green

H Usually occurs as platy crystals, lath-like in cross section, where the basal cleavage also appears

CL See previous section

R Moderate

A Biotite alters to chlorite under hydrothermal action. The alteration is noted by the presence of areas of pale green chlorite, often interdigitated with areas of fresh biotite.

B High to very high, but colours are always masked by the mineral colour

IF A non-pleochroic basal section will give a Bxa figure with a very small negative $2V$

Varieties See under composition: **siderophyllite** is alumina-rich; **lepidomelane** and **haughtonite** are rich in ferric iron (although some ferrous iron is also present); and **annite** is rich in ferrous iron.

Tests Heated with fluxes, biotite gives a strong iron reaction. It decomposes in strong sulphuric acid, leaving a residue of siliceous scales.

Occurrence Biotite is a common mineral in a variety of rocks. It forms from chlorite in metamorphosed pelitic rocks, and exists over a wide range of regional metamorphic conditions, its composition changing with increased grade, becoming more Mg- (and Ti-) rich. Biotites are primary minerals in acid and intermediate plutonic igneous rocks and in some basic plutonic rocks, but it is not a common mineral in acid and intermediate extrusive and hypabyssal rocks. Biotite is a common mineral in clastic sedimentary rocks, particularly arenaceous rocks, but it is prone to oxidation and degradation.

LEPIDOLITE

C $K_2 (Li, Al)_{5-6}[Si_{6-7}Al_{2-1}O_{20}](OH,F)_4$

Physical properties

CS Monoclinic

F&H Occurs in plates similar to muscovite, but mostly as small scales or granules

COL Lilac, rose red, violet grey, sometimes white

CL Perfect {001} basal cleavage

L Pearly

TR Translucent

HD 2.5–4.0

SG 2.8–2.9

Optical properties Similar to those of muscovite. Biaxial negative with $2V = 0°–58°$, and with the OAP parallel to (010).

Tests Heated before the blowpipe, lepidolite gives the red lithium flame. It usually gives the reaction for fluorine.

Occurrence Lepidolite occurs in late-stage pegmatites, associated with tourmaline, topaz, and other minerals of pneumatolytic origin, as in the eastern USA, Elba, Madagascar, etc. It occurs as a gangue mineral in tin veins in Saxony.

ZINNWALDITE

C $K_2 (Fe^{2+}_{2-1}, Li_{2-3}, Al_2) [Si_{6-7} Al_{2-1}O_{20}] (F, OH)_4$

Characters and occurrence Zinwaldite is a mica, similar to biotite but with lithium in the structure. It is violet, pale yellow or brown in colour, and occurs in modifications of the Zinnwald–Erzgebirge granite, associated with cassiterite, etc., and in all occurrences it is pneumatolytic in origin.

Brittle micas

Brittle micas resemble normal micas closely but are harder, and their cleavage sheets are much less elastic, which gives this group their name.

MARGARITE

C $Ca_2Al_4 [Si_4Al_4O_{20}](OH)_4$

Calcium ions occupy the K sites of the normal micas, and the Si : Al ratio is increased in the Z group.

Physical properties

CS Monoclinic

F&H Usually as fine aggregates

COL Pink, white or grey

CL Perfect basal {001} cleavage present

T Cleavage flakes are brittle, not elastic like the micas

HD 3.5–4.5

SG 3.0–3.1

Optical properties Margarite is colourless in thin section, and is distinguished from talc and muscovite by higher RIs (1.630–1.650), lower birefringence (first order yellow), and larger $2V$ (40°–67° −ve). The OAP is perpendicular to (010).

Occurrence It is associated with corundum in emery deposits, and in mica schists in association with tourmaline and staurolite.

Minerals closely related to margarite include **clintonite** and **xanthophyllite**, both of which resemble phlogopite in colour and properties, except for their much lower birefringence (first order yellow).

STILPNOMELANE
$C (K,Na,Ca)_{0-1.4} (Fe^{3+} Fe^{2+} MG,Al,Mn)_{5.9-8.2} [Si_8O_{20}](OH)_4(OH,F)_{3.6-8.5}$

Characters and occurrence
HD 3.0–4.0 **SG** 2.59–2.96

A monoclinic crystal with similar physical and optical properties to biotite. It possesses a second cleavage parallel to {010} in addition to the perfect {001} basal cleavage. It is biaxial negative, but $2V \approx 0°$ so the crystal is virtually uniaxial. Stilpnomelane is darker in colour than biotite, and with a more striking pleochroism, with α golden yellow, and β and γ deep brown to black. RIs are variable, with $n_\alpha = 1.543-1.634$, and $n_\beta = 1.576-1.745$. Birefringence is moderate to very high.

Stilpnomelane occurs in metamorphosed iron- and manganese-rich sedimentary deposits. It is found in the Lake Superior ironstones, and in some metamorphic glaucophane schists.

PYROPHYLLITE
$C \quad Al_4[Si_8O_{20}](OH)_4$

Characters and occurrence
HD 1.0–2.0 **SG** 2.65–2.90

A monoclinic mineral with properties very similar to talc and muscovite. It has a low relief, with RIs in the range 1.530–1.600, and with a high birefringence. It is biaxial with $2V = 53°–62°$ −ve, and the OAP is perpendicular to (010).

When heated with cobalt nitrate on charcoal, it gives the blue aluminium reaction.

Pyrophyllite is an uncommon mineral, occurring as a secondary product from the hydrothermal alteration of feldspar. It occurs as foliated masses in crystalline schists. It has similar uses as talc and is mined for the same purpose in North Carolina and South Korea.

TALC
C $Mg_6 [Si_8O_{20}] (OH)_4$

Physical properties
CS Monoclinic
F&H Rare crystals are tabular. Usually massive with a foliaceous structure; also granular, compact and cryptocrystalline.
COL White, silvery white, apple green, greenish grey, dark green
CL Perfect {001} basal cleavage

L Pearly
TR Subtransparent to translucent
T Thin cleavage plates are flexible, but not elastic
FEEL Greasy
HD 1.0 (softest mineral on Mohs' scale of hardness)
SG 2.58–2.83

Optical properties
$n_\alpha = 1.539–1.550$
$n_\beta = 1.589–1.594$
$n_\gamma = 1.589–1.600$
$\delta = \approx 0.05$
$2V = 0°–30° -ve$
OAP is perpendicular to (010)
COL Colourless
H Usually occurs as laths, wisps or plates in schistose rocks

CL Perfect basal cleavage present
R Low
B High, with third order colours shown
IF A basal section (without cleavage trace) gives good Bxa figure
E Straight on the cleavage

Varieties   **Steatite** or **soapstone** is a massive variety of talc, mostly white or various shades of grey but sometimes greenish or reddish, and having a greasy or soapy feel. **Potstone** is an impure massive talc or soapstone, greyish green, dark green, iron grey or brownish black in colour. It is easily turned on a lathe and, since it stands heat well, it can be used as cooking vessels. **Rensselaerite** is a variety of soapstone, pseudomorphous after pyroxene, which occurs in Jefferson County, New York, and also in Canada. It is white, yellow or black in colour and is harder than normal talc. It takes a high polish, and was formerly made into inkstands and ornamental articles. **French chalk** is a steatite used by tailors for marking cloth. **Indurated talc** is an impure slaty variety which is harder than French chalk.

Tests   Heated alone before the blowpipe, talc whitens and exfoliates, fusing to an enamel on the edges only. It is not decomposed by acids, except for the rensselaerite variety.

Occurrence   Talk is formed during low-grade metamorphism of siliceous dolomites, and also by the hydrothermal alteration of ultrabasic rocks, where talc may occur along faults and shear planes. It is often associated with serpentinization, with the serpentine changing to talc plus magnetite by the addition of CO_2

Talc is mined in many countries, with a world production figure in 1985 estimated at ~3 Mt. The main producers are the USA (1.1 Mt), the USSR (520 000 t), China (357 000 t), France (311 000 t), South Korea (194 000 t), West Germany (134 000 t) and Italy (130 000 t), with talc also produced by Japan, India and the UK, where it occurs in a serpentine at Unst in the Shetland Islands.

Uses Talc is used as a filler in paints, paper, rubber, etc., and in plasters, foundry facings and lubricants. It is also used to remove grease from cloth, in leather-making, for crayons, toilet (talcum) powders, etc., and as an absorbent for nitroglycerine. Soapstone slabs are used as hearthstones, sinks, laboratory table-tops and so on. Harder varieties are carved into ornaments.

CHLORITE GROUP

Included with chlorite are many allied minerals, all of which are related in composition to the micas but do not contain any alkalis. Each mineral is a sheet silicate with the typical $[Si_4O_{10}]$ tetrahedral unit present, and they may be considered as hydrous silicates of aluminium, iron and magnesium. The different types of mineral depend upon the iron : magnesium ratio which occurs.

In general, the chlorite minerals are monoclinic, and often pseudo-hexagonal. Their colour is green and they all possess a perfect {001} basal cleavage, which gives cleavage flakes which are flexible but not elastic, and the average hardness is ≈ 2.

The most important chlorites are the minerals:

clinochlore   a tabular, monoclinic chlorite; biaxial with a positive $2V$
penninite   a monoclinic, pseudo-hexagonal chlorite; biaxial with $2V \approx 0°$
ripidolite   a mineral with a granular, radiating or tubular habit, biaxial with $2V \approx 0°$

Four minerals – amesite, chamosite, greenalite and cronstedtite – are chemically similar to the chlorites, but are termed **septechlorites** because they contain some serpentine-like layers.

The description which follows is a general one of a typical chlorite.

CHLORITE

C $(Mg, Al, Fe^{2+})_{12} [(SI, Al)_8 O_{20}] (OH)_8$

Physical properties

CS Monolithic, pseudo-hexagonal

F&H Tabular crystals common; also occurs as granular masses, and disseminated scales and foliæ in metamorphic rocks.

CL Perfect {001} basal cleavage

L More or less pearly

TR Subtransparent to opaque

T Cleavage flakes are flexible, not elastic

FEEL Cleavage flakes are greasy to the touch

HD 2.0–3.0

SG 2.6–3.3

Optical properties

$n_\alpha = 1.57–1.66$

$n_\beta = 1.57–1.67$

$n_\gamma = 1.57–1.67$

$\delta = 0.00–0.01$

$2V = 20°–60°$ +ve or −ve

OAP is parallel to (010)

COL Colourless or green

Green chlorites have α pale green to colourless, and β and γ darker green

H Chlorite appears either as radiating aggregates filling cavities, or as an alteration product of primary crystallizing minerals in igneous rocks, such as biotite, or as small irregular flakes and laths

CL Perfect {001} cleavage seen in prismatic (lath-like) sections

B Very low with anomalous 'Berlin blue' colours usually seen

IF A basal section gives an excellent Bxa figure, which can be positive or negative with a huge range of $2V$ values, so that in some cases a single OA figure may be required to obtain sign and size of $2V$

E Usually straight, but a small angle $\alpha \char"005E cl$ or $\gamma \char"005E cl$ may exist

Tests Distinguished by colour and physical properties. In a closed tube chlorite gives water when heated; it fuses with difficulty on thin edges.

Occurrence Chlorite is a common primary mineral in low-grade regional metamorphic rocks such as **greenschists**, where it changes to biotite with increasing grade, muscovite also being involved in the reaction. These metamorphic changes usually occur in pelitic rocks – original argillaceous sediments – but basic igneous rocks and tuffs will also produce chlorite during regional metamorphism. In igneous rocks, chlorite is mostly a secondary mineral, forming from the hydrothermal breakdown of pyroxene, amphibole and biotite. It may occur as an infilling in lava flows, and as a primary mineral in some low-temperature veins. Chlorites are common in argillaceous sedimentary rocks, where they occur with clay minerals.

SERPENTINE; there are three main varieties of serpentine namely **chrysotile, lizardite, antigorite**

C $Mg_6 [Si_4 O_{10}] (OH)_8$

Physical properties

CS Monoclinic

F&H Chrysotile is fibrous, elongated parallel to the *a* crystallographic axis. Lizardite and antigorite both occur as flat, tabular crystals, but may also appear massive or granular.

COL Various shades of green to almost black; sometimes red, yellow or brown. Massive varieties are often veined and spotted with red, white or green. The white veins are often steatite (talc) which in many cases, envelops crushed fragments of the darker serpentine, producing very ornamental-looking patches of brecciated serpentine.

CL Fibrous varieties have a prismatic cleavage; tabular varieties have a basal one

F Conchoidal

L Subresinous to oily

TR Transparent to opaque

T Tough

FEEL Slightly soapy at times

HD 2.5 (chrysotile and lizardite); 2.0–3.5 (antigorite). Can be cut with a knife, and is frequently turned into vases, etc.

SG 2.55–2.60

Optical properties

n_α = 1.53–1.57

n_β = 1.57

n_γ = 1.55–1.57

δ = 0.015± (chrysotile); 0.006± (lizardite and antigorite)

2V variable −ve

OAP is parallel to (010) in chrysotile and antigorite

COL Colourless or pale green or yellow

H,CL See above

R Low

B Low or very low, often with anomalous colours showing

IF Variable negative, but difficult to obtain

E Straight on fibres; chrysotile is length slow

Serpentine often appears as a complete or partial pseudomorph after olivine, or occasionally pyroxene; magnetite usually accompanies this alteration.

Varieties **Precious** or **noble serpentine** is a greenish, translucent variety. **Picrolite** is a columnar variety, occurring along shear planes. **Bastite** (schillerspar) forms from the alteration of orthopyroxene. It is olive green, blackish green or brown in colour, and with a metallic lustre on cleavage surfaces. Bastite has a hardness of 3.5–4.0 and an SG of 2.5–2.7. In thin section, bastite appears as a fibrous mineral pseudomorphing orthopyroxene, the fibres lying parallel to the *c* crystallographic axis of the pyroxene. It is pale green, weakly pleochroic, and with low to anomalous birefringence. Bastite occurs within serpentine at Baste in the Harz, in the Lizard in Cornwall, and in most occurrences of serpentine, where it tends to enhance the appearance of the stone. **Ophicalcite** is a rock composed of green serpentine in white calcite, and results from the dedolomitization after the metamorphism of a siliceous dolomite. Mg-olivine (forsterite) and calcite are first formed, and then the olivine is serpentinized.

Occurrence Serpentine results from the alteration, either during metamorphism or by late-stage hydrothermal action, at temperatures below 400°C, of ultramafic rocks rich in magnesium, containing olivine, pyroxene or amphibole, such as picrites, peridotites, etc., and in serpentinites in ophiolite complexes, found in many places throughout the world.

Uses Serpentine is used mainly as a building stone, and for ornamental work in general, primarily for interior cladding panels. Chrysotile, the fibrous variety, is one of the most important varieties of commercial asbestos.

GARNIERITE, noumeite
C   (Ni, Mg)$_6$ [Si$_4$O$_{10}$](OH)$_8$; a nickeliferous serpentine

Physical properties
F&H Amorphous; a soft and friable	**L**	Dull
mineral	**HD**	3.0–4.0
COL Apple green to off-white	**SG**	2.2–2.8

Varieties **Genthite** is a hydrated magnesium silicate related to gernierite.

Tests Garnierite adheres to the tongue. When heated in the closed tube, it yields water and blackens. In the borax bead it gives nickel reactions; in the microcosmic bead it gives a nickel reaction and leaves behind an insoluble skeleton of silica, indicating a silicate mineral.

Occurrence Garnierite occurs in serpentine near Noumea, New Caledonia, in veins associated with chromite and talc. A residual deposit rich in nickel is formed by the lateritic decay of the nickeliferous serpentine. It is also found at Riddle in Oregon, Webster in North Carolina, and Revda in the Urals.

Uses Garnierite is an important source of nickel, and the New Caledonia deposits, before the development of the nickeliferous pyrrhotite deposits of Sudbury (Canada), were the chief source of this metal. The two working mines in New Caledonia produced 1.7 Mt of ore in 1985, which yielded 45 000 t of nickel.

Clay minerals

Clay minerals are important products from the weathering of rocks. In particular, feldspars give rise to clays, with K-feldspar reacting in the presence of water to give illite, and plagioclase feldspar reacting in a similar manner to give montmorillonite. If excess water is present, both reactions will eventually produce kaolinite which is the final product. The weathered material either remains where it is and gives rise to residual clays, or is transported by various agencies (water, wind and ice) and deposited as beds of clay in the sea or in lakes, as a superficial deposit of boulder clay, or as loess or adobe deposits. Clays have the properties of absorbing water and becoming either plastic (specifically defined by a *plastic limit*) or liquid (defined by the *liquid limit*). These values are called the *Atterberg limits*. Clays also become hard when heated to a suitable temperature. Certain other substances are also considered briefly, such as fuller's earth and bentonite, which do not become plastic when wetted. All clay minerals contain [Si_4O_{10}] tetrahedral sheets, similar to the micas, and tend to occur as minute flaky crystals. The minerals dealt with here include the following:

kaolinite, nacrite, dickite
illite
montmorillonite, beidellite, nontronite
vermiculite
sepiolite (meerschaum)
allophane

Related minerals include **apophyllite** and **prehnite**. All clay related minerals have a very small grain size in the natural state, and identification by optical techniques is virtually impossible. Serious study and precise identification of clay minerals are carried out either by X-ray diffraction techniques (XRD; see section 4.15), by using a scanning electron microscope (SEM), or by using an electron microprobe.

Kaolinite group (kandites)

KAOLINITE, nacrite, dickite, China clay, kaolin
C $Al_4[Si_4O_{10}](OH)_8$
Nacrite and dickite have identical chemical compositions to kaolinite, but their atomic layer stackings along the *c* crystallographic axis are different from that of kaolinite.

Physical properties
CS Triclinic or monoclinic
F&G Small pseudohexagonal tabular crystals; usually massive
COL White when pure, grey or yellowish
CL Perfect {001} basal cleavage
L Dull and earthy

FEEL Greasy feel; often very soft material, crumbling to powder when pressed between the fingers
SMELL Clayey smell
HD 2.0–2.5
SG 2.61–2.68

Optical properties
n_α = 1.55–1.56
n_β, n_γ = 1.56–1.57
δ = 0.006
$2V$ = 24°–50° −ve

A detailed discussion of the optics of kaolinite is beyond the scope of this book, but its optical properties are similar to those of muscovite.

Varieties **Kaolinite, nacrite** and **dickite** have already been mentioned. **Kaolin** or **China clay** consists partly of crystalline and partly of amorphous material. **Lithomarge** is a white, yellow or reddish clay, consisting of kaolinite and halloysite, often speckled and mottled, adhering strongly to the tongue, and having a greasy feel; when scratched with a fingernail it shows a shining streak. It is infusible before the blowpipe. It is found in Cornwall, Saxony and elsewhere.

Tests Heated on charcoal with cobalt nitrate, kaolinite gives a blue mass due to aluminium. It yields water on heating in a closed tube, and is insoluble in alcohol.

Occurrence Nacrite and dickite are relatively rare minerals occurring in association with metallic ores. Kaolinite forms from the alteration of feldspars in granites. This alteration may be caused either by the weathering process, as described in the introduction to the clay minerals, or by pneumatolytic action of gases on feldspars. This latter type of alteration may account for the Cornwall occurrences of kaolinite, since it is found in association with cassiterite, tourmaline, and other minerals of undoubted pneumatolytic origin.

Production and uses Total world production in 1985 was in excess of 12 Mt, and the main producers are the USA (7.22 Mt), the UK (3.60 Mt), the USSR (2.90 Mt), South Korea (0.66 Mt), China (0.60 Mt), Austria (0.50

Mt), West Germany (0.41 Mt) and Thailand (0.11 Mt), with Spain, Malaysia and Italy each producing significant amounts.

Kaolinite is used for the manufacture of fine porcelain and china, porcelain fittings, etc., and as fillers in paper, rubber and paint manufacture.

HALLOYSITE

C $Al_4[Si_4O_{10}](OH)_8$. $8H_2O$; similar to kaolinite but containing interlayer water molecules.

Characters and occurrence   Similar to kaolinite, it occurs with kaolinite in deposits of China clay. Tests are similar to those for kaolinite. The weathered surface of reddish material often found on lava flows, called **red bole**, is a clay which includes impure halloysite. Bole gives a shining streak when scratched with a fingernail, has a conchoidal fracture, and falls to pieces with a crackling noise when placed in water. Heated before the blowpipe it fuses easily to a yellow or green enamel. It is found primarily in Italy, Silesia and Asia Minor.

Illite group

ILLITE, HYDROMUSCOVITE

C $K_{1-1.5} Al_4 [Si_{7-6.5} Al_{1-1.5}O_{20}] (OH)_4$. Hydromuscovite has $(H_3O)^+$ ions replacing the K^+ ions, and some molecules present in the structure.

Characters and occurrence   Properties are similar to those of kaolinite, although the birefringence is slightly higher, and $2V$ is very small ($-$ve). Illite is formed by the weathering of K-feldspar or muscovite, and is a dominant mineral in any shales and argillaceous rocks.

Montmorillonite group (smectites)

MONTMORILLONITE, BENTONITE

C   Essentially $\{Al_4[Si_6Al_2O_{20}] (OH)_4\}^{2-} .nH_2O$

Al may be replaced by Fe and Mg, and K^+, Na^+ or Ca^{2+} ions may partly fill the interlayer sites. The positive cations (Na, Ca etc.) are exchangeable bases, and their presence accounts for the high base exchange capacity of the mineral. An overall residual negative charge often remains, and in deep marine muds the smectite clays can absorb radiogenic ions (caesium, strontium-90, etc.). A large amount of water can be accommodated in these interlayer gaps, the amount of water often varying depending upon the type of alkali ion present; thus an Na-montmorillonite will absorb more water molecules than a K-montmorillonite. This water may be driven off on heating, or during normal drying-out processes, and the mineral then swells greatly on absorbing water again; some Na-montmorillonites have been recorded as absorbing water to 700% of their initial volume. In civil engineering these are called *expansive clays*.

Characters and occurrence   A soft clay mineral coloured white, greyish or greenish, with properties similar to those of illite. Montmorillonite, together with beidellite, constitutes a large proportion of the mineral bentonite, which occurs as thin beds in Cretaceous and Tertiary rocks of the western USA. Bentonite is believed to result from the decomposition of volcanic ash, and is employed for various purposes, such as decolorizing of oils, in water softening, as a filler for thickening *drilling muds* (in which barite is also a constituent) in the oil industry, and as an absorbent in many processes.

Although the production figures for bentonite are not given, the main producers in 1985 were Italy (300 000 t), Hungary (60 000 t) and Cyprus (52 000 t), with the Philippines also producing. Although figures are not available, the USA and the USSR also produce significant quantities.

BEIDELLITE
C   Essentially $Al_4[(SI,Al)_8O_{20}][OH]_4.nH_2O$, with some replacement of Si by Al in the Z group.

Characters and occurrence   Beidellite occurs as thin plates, white or reddish in colour, forming from the alteration of micas, etc.

VERMICULITE
C $(Mg, Ca)(Mg, Fe^{2+})_5 (Al, Fe^{3+})[(Si, Al)_8O_{20}](OH)_4.8H_2O$, with Al replacing Si in the Z group, and with Al and Fe^{3+} in the Y group. The X group contains Mg and Ca. Vermiculite is closely related to the smectites, but its special characteristics merit its description as a separate mineral.

Physical properties

CX	Monoclinic	**CL**	Perfect basal parallel to {001}
F&H	Small fibrous-looking mineral, often found pseudomorphing biotite	**HD**	≈1.5
		SG	≈2.3
COL	Yellow, brown or green		

Tests   Expands on heating, with the sheets of atoms being forced apart by the escape of steam.

Occurrence and uses It forms from the hydrothermal alteration of biotite. Vermiculite is used in the building industry for the production of insulating materials, lightweight cements and plasters, and for refractory purposes, these uses depending upon its expansion on heating. In 1985 the USA produced 230 000 t from Montana, and South Africa produced 184 000 t.

Palygorskite–sepiolite series

This group includes the minerals **palygorskite, attapulgite** and **sepiolite**. These minerals are rare but the more common 'mineral' **meerschaum** is a variety of sepiolite, or a mixture of mainly sepiolite with some other clays present. Meerschaum is the only mineral of this series described here.

MEERSCHAUM, sepiolite
C $(Mg, Al, Fe^{3+})_8 [Si_4O_{10}]_3 (OH)_4 . 12H_2O$; this is the composition of sepiolite

Physical properties

CS Not known; appears to be a mixture of amorphous material and fibrous biaxial material (which is probably sepiolite)

F&H Appears as earthy clay-like amorphous masses; sometimes fibrous

COL White, greyish-white, sometimes with a faint pink or yellow tint

L Dull and earthy

HD 2.0–2.5; can be easily scratched with a fingernail

SG 2.0; floats on water when dry

Tests Decomposed in hydrochloric acid, with gelatinization. When heated it gives off water. Heated on charcoal with cobalt nitrate, it gives a pink mass.

Occurrence Found in beds or irregular masses in alluvial deposits derived from serpentinite masses, as in Asia Minor; in veins with silica, possibly derived from dolomitic rocks, as in New Mexico; and also within serpentinite masses. It comes from Turkey, Samos, Negropont in the Grecian Archipelago, Morocco, Spain, and from Kenya and Tanzania.

Uses Meerschaum is mainly used to make pipe and pipe bowls, because of its absorbent nature and its lightness. It is coloured by smoke after a time, and displays additional beauty in the eyes of connoisseurs. Before being made into pipes, the meerschaum is first soaked in tallow and afterwards in wax, and it then takes a good polish. It was formerly used in North Africa as a substitute for soap. Very rarely, it has been used as a building stone.

ALLOPHANE
C $Al_4[Si_4O_{10}](OH)_4$

The name allophane is now used to denote the amorphous clay that is soluble in dilute hydrochloric acid. It probably has affinities with kaolinite, and is found as crusts in coal beds.

FULLER'S EARTH
C Fuller's earth is a clay which is mainly composed of montmorillonite It is a greenish brown, greenish grey, bluish or yellowish material, soft and earthy in texture, and with a soapy feel. It yields to the fingernail with a shining streak, and adheres to the tongue. It has a high absorptive capacity, and when placed in water it falls to a powder but does not form a paste. It fuses to a porous slag and ultimately to a white blebby glass.

It forms in Europe from the weathering of basic igneous rocks or pyroclastic deposits, and may occur in deposits formed from them.

Fuller's earth was used for cleaning (fulling) woollen fabrics and cloth, its absorbent properties allowing it to remove natural oils and greases. It is now used in the refining of oils and fats. In England, fuller's earth is found at Nutfield near Reigate, Detling near Maidstone, Bletchingley in Surrey, Woburn in Bedfordshire, and at Bath; in total the UK produces around 300 000 t per annum. It is worked in Florida, Georgia, and Arkansas in the USA, and also in Germany

APOPHYLLITE
C $KFCa_4[Si_8O_{20}].8H_2O$

Physical properties

CS Tetragonal

F&H Crystals of two habits commonly occur: first, combinations of {100}, {001} and minor forms give a crystal of cube-like aspect; and second, prismatic crystals of {100} with {111} terminations. It also occurs massive or foliaceous.

COL Milk white to colourless, greyish; sometimes greenish, yellowish or reddish

S Colourless

CL Perfect {001} basal

F Uneven

L Vitreous, pearly on some faces

TR Translucent to transparent

T Brittle

HD 4.5–5.0

SG 2.3–2.4

Optical properties

In this section apophyllite is colourless with RIs ≈ 1.53–1.54 and uniaxial +ve.

Tests Heated in a closed tube, apophyllite gives off water, exfoliates and whitens. Heated before the blowpipe it exfoliates and fuses to a white, vesicular enamel. It colours the flame violet due to potassium, and is soluble in HCl with separation of silica.

Occurrence Apopyhllite is found infilling amygdales in basic lava flows, often accompanied by zeolites.

PREHNITE
C $Ca_2Al[Si_3AlO_{10}](OH)_2$

Physical properties

CS	Orthorhombic	**F**	Uneven
F&H	Usually in botryoidal masses with a radiating crystalline structure	**L**	Vitreous
		TR	Subtransparent to translucent
COL	Pale green to colourless	**T**	Rather brittle
S	Colourless	**HD**	6.0–6.5
CL	Basal {001} good, {110} poor	**SG**	2.90–2.95

Optical properties

Prehnite is colourless in thin section, with moderate relief (RIs 1.61–1.66), and biaxial +ve with a $2V \approx 65°–70°$.

Tests Heated in a closed tube, prehnite gives off a little water. Heated before the blowpipe, it fuses with intumescence to a bubbly, enamel-like glass. After fusion, it gelatinizes with HCl. Water is given off prehnite only at red heat.

Occurrence Prehnite occurs in certain amphibolites, crystalline limestones, etc., and in association with zeolites in the amygdales of basalts. In New Zealand, prehnite-bearing rocks occur on a regional scale, in a zone of low-grade incipient metamorphism. The prehnite is found in association with pumpellyite.

9.7   Tektosilicates (framework silicates)

Feldspar group

The feldspars are the most important group of rock-forming silicate minerals occurring in igneous, sedimentary and metamorphic rocks. Their range of compositions has led to them being used as a means of classifying igneous rocks, since they are absence only from certain ultramafic and ultra-alkaline igneous rock types and carbonatites. In metamorphic rocks, feldspars are absent only from some low-grade pelites, pure marbles, pure quartzites and most eclogites. Feldspars are common in arenaceous

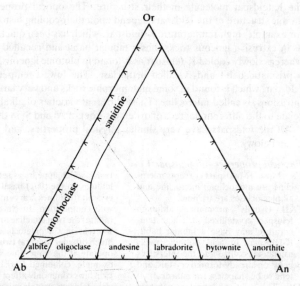

Figure 9.3 Composition diagram for feldspars.

sedimentary rocks, but are less common in argillaceous types. In these fine-grained rocks feldspars are difficult to recognize because of their very small size, but X-ray techniques will reveal their presence.

There are two main groups of feldspars:

(1) *Alkali feldspars*, which occupy a range of compositions between albite, $NaAlSi_3O_8$, and K-feldspar, $KAlSi_3O_8$.
(2) *Plagioclase feldspars*, which occupy a range of compositions between albite, $NaAlSi_3O_8$, and anorthite, $CaAl_2Si_2O_8$. If pure anorthite is written Ab_0An_{100}, or more commonly An_{100}, then pure albite is written An_0. A complete range of plagioclase feldspars can be described with **albite** (An_0–An_{10}), **oligoclase** (An_{10}–An_{30}), **andesine** (An_{30}–An_{50}), **labradorite** (An_{50}–An_{70}), **bytownite** (An_{70}–An_{90}), and **anorthite** (An_{90}–An_{100}).

The feldspars are usually represented by means of a ternary diagram with K-feldspar (K fs), albite (Ab) and anorthite (An) as the apices (see Fig. 9.3). The fields occupied by the two feldspar groups show that the alkali feldspars may contain up to 10% of the anorthite molecule in their structure and, likewise, the plagioclase feldspars may contain up to 10% of the K-feldspar molecule in their structure. The optical properties and atomic structure of the feldspars depend upon their cooling history. Thus, for example, high-temperature K-feldspar, which has been quickly cooled as in extrusive igneous rocks, has a tabular habit and is called **sanidine**, whereas slowly cooled K-feldspar as is found in plutonic igneous rocks has a prismatic habit and is called **orthoclase**. The lowest-temperature K-feldspar, which is found in some metamorphic rocks and very large granite intrusions, is called **microcline**. These different varieties of alkali feldspar relate to the different degrees of order–disorder of Al and Si in the lattice.

All the feldspars have very similar physical properties, and these are given below.

Physical properties (all feldspars)

CS Most K-feldspars except microcline are monoclinic; microcline and all plagioclases are triclinic

F&H Usually prismatic, although feldspars crystallized at a high temperature may have a tabular habit. K-feldspars in acid plutonic rocks usually crystallize into the interstices between already-formed crystals, and tend to be shapeless (or **anhedral**).

TW This is a very important property in the feldspars. K-feldspars, such as sanidine and orthoclase, are usually simply twinned on three laws: (1) the *Carlsbad Law*, with (100) as the twin plane; (2) the *Baveno Law*, with (021) as the twin plane; and (3) the *Manebach Law*, with (001) as the twin plane. The plagioclase feldspars and microcline may show simple twins such as those just described, but invariably also exhibit repeated or lamellar twinning which may be on two laws: (1) *Albite Law*, with (010) as the twin plane; and (2) *Pericline Law*, with the *b* crystallographic axis as the twin axis. In a crystal of microcline, both twin laws usually are in operation so that the twinning appears as fine, intersecting lines (called **cross-hatching**) on the (001) basal plane of the crystal. Cross-hatching does not occur in plagioclase feldspars to any great extent, and the most common twins seen are albite twins, which appears as sets of parallel lines on the (001) basal plane of a crystal; the lines vary in thickness with composition. In Na plagioclase (albite) the twin lamellae are very fine, whereas in Ca plagioclase they are much broader. Several twin laws may operate at the same time so that, for example, combined Carlsbad–Albite twin laws commonly affect the plagioclase feldspar crystals in basic igneous rocks.

COL Variable, usually white or light-coloured, but orthoclase is often pink or reddish, and Ca-rich plagioclase is often dark grey with a bluish tinge. Sometimes greenish or bright green (as in **amazonstone**, a variety of microcline), colourless, etc.

CL All feldspars possess two cleavages, a prismatic parallel to {010} and a basal parallel to {001}, which meet at ~90° on the (100) face; several partings may also occur

F Conchoidal to uneven and splintery

L Vitreous to pearly on cleavage planes; iridescence may be seen on cleavage faces in labradorite plagioclase feldspars; sometimes dull

TR Subtransparent to translucent and opaque; the higher temperature members are more transparent than those formed at lower temperatures

HD 6.0–6.5

SG   Alkali feldspars   2.56 (pure K-feldspar) to 2.63 (albite)
     Plagioclase feldspars 2.63 (albite) to 2.76 (anorthite)

Perthites

Feldspars formed at high temperatures that have cooled slowly, such as in plutonic rocks, show unmixing with perthite development. Most alkali feldspars in such rocks occur as perthites, consisting mainly of an alkali feldspar host with an exsolved Na-rich plagioclase feldspar phase which segregated from the host during cooling. Alkali feldspars crystallize with disordered Al and Si atoms. At high temperatures the sanidine – high albite pair is thermodynamically stable and, if quickly cooled, as for example in a lava flow, the resulting alkali feldspars are homogeneous with no perthite development. Ordering is sluggish and depends upon diffusion and other processes. Slow cooling from high temperatures results in unmixing as the solvus curve is intersected, with the segregation of Na and Ca atoms to form *perthitic intergrowths*. Such intergrowths have textures varying in size from *macroperthites* which are visible to the naked eye, through *microperthites* which are visible under the microscope, to *cryptoperthites* which are detected only by X-ray and other techniques. The nature of the perthite depends upon the ordering of the Si and Al atoms achieved by the lattice as it cools. Some perthites consist of roughly equal amounts of intergrown alkali feldspar and plagioclase, called *mesoperthites*, which occur in granulites and charnockites. In plutonic rocks the low-temperature alkali feldspar series, all of which are perthites, is orthoclase – low albite, and in low-grade metamorphic rocks, where the temperatures of crystallization are still lower, the alkali feldspar series is microcline – low albite, although microcline may occur in some granites and aplites.

High-temperature plagioclase feldspars, such as the end-members albite and anorthite, do not show unmixing when quickly cooled, and crystallize as non-perthitic plagioclase feldspars. This series is called high albite–high anorthite. At lower temperatures unmixing takes place as the plagioclase feldspars cool, with the development of perthites, or *antiperthites* as they are called in the plagioclase group. Complex intergrowth often form, of which three are important: (1) *peristerites*, containing equal amounts of alkali feldspar (albite) and a more Ca-rich plagioclase; (2) *Bøggild intergrowths* (An_{40}–An_{60}), which can often be recognized by their labradorite iridescence; and (3) *Huttenlocher intergrowths* (An_{70}–An_{85}). Bøggild and Huttenlocher intergrowths are submicroscopic and can only be detected by X-ray techniques. Plagioclase feldspars showing unmixing are found either in slowly cooled plutonic rocks or in metamorphic rocks.

ALKALI FELDSPARS

C A continuous series exists at high temperatures from K-feldspar, $KAlSi_3O_8$, to albite, $NaAlSi_3O_8$, for high alkali feldspars. Three series of alkali feldspars exist depending upon their mode of formation. These are:

sanidine – high albite series	$Ab_0 - Ab_{63}$	sanidine
	$Ab_{63} - Ab_{90}$	anorthoclase
	$Ab_{90} - Ab_{100}$	high albite
orthoclase – low albite series	$Or_{100} - Or_{85}$	orthoclase
	$Or_{85} - Or_{20}$	orthoclase cryptoperthite
	$Or_{20} - Or_0$	low albite
microcline – low albite series	$Or_{100} - Or_{92}$	microcline
	$Or_{92} - Or_{20}$	microcline cryptoperthite
	$Or_{20} - Or_0$	low albite

Optical properties

	Albite	Orthoclase
n_α	1.527–1.518	
n_β	1.531–1.522	
n_γ	1.539–1.524	
δ	0.012–0.006	

$2V$ is variable in size and sign depending upon composition, with the following values:

15°–40°	sandidine	42°–52°	anorthoclase
52°–54°	high albite	35°–50°	orthoclase
66°–90°	microcline	78°–84°	low albite

All are negative (−ve), except low albite which is positive (+ve).

OAP also varies; in sanidine it is parallel to (010). In anorthoclase, orthoclase and microcline it is roughly perpendicular to the c axis.

COL Colourless, but may be cloudy if altered

H Euhedral and prismatic in most porphyritic rocks to anhedral in acid plutonic rocks

CL Two cleavages occur, {010} and {001} intersecting at ~90° on the (100) face

R Low, just less than 1.54

A Alteration to clay minerals is common in K-feldspars during hydro-thermal action or weathering, with illite or kaolinite clays, including dickite, being formed. The clay minerals occur as tiny, discrete grains within the feldspar crystal and, as alteration increases, the clays increase in size to a point at which they are termed **sericite**. The sodium-rich alkali feldspars (albite) also alter in the same way, with montmorillonite being formed. The presence of these alteration products makes the feldspar crystals cloudy under normal light.

B Low first order colours are seen

IF See above for 2*V* signs and sizes; a single isogyre is always required for this study

E Not used in K-feldspars (see extinction in plagioclase feldspar for details regarding albite)

TW See the introduction; orthoclase shows simple twins, albite exhibits repeated twinning on the albite law, with black and white lamellae being seen, and microcline (and anorthoclase) always exhibit cross-hatching.

PERTHITES See discussion under this heading on p. 419

DF It is difficult to identify which type of alkali feldspar is present. Low albite is +ve; sanidine has a small negative 2*V*, and anorthoclase and microcline usually show cross-hatching. Although quartz has a similar birefringence, it is uniaxial +ve, with an RI>1.54.

Varieties

(1) *K-feldspar types*. **Common orthoclase** includes the subtranslucent or dull varieties. **Adularia** is a low-temperature variety, colourless and of prismatic habit. **Moonstone** is a pearly to opalescent variety, and the name also applies to some varieties of albite. **Sunstone** and **aventurine feldspar** are types of adularia spangled with minute crystals of hematite, limonite, etc. **Murchisonite** is a red orthoclase with a peculiar golden yellow lustre.

(2) *Na-feldspar types*. **Aventurine** and **moonstone** are similar to the K-feldspar varieties already discussed. **Pericline** is a white, semi-opaque variety, prismatic and elongated along the *b* axis. **Cleavelandite** is a white, lamellar variety.

Tests Orthoclase fuses only on the edges of thin splinters. Heated with borax it forms a transparent glass. It is insoluble in the microcosmic salt bead, unaffected by acids, and only gives the potassium flame with difficulty. Albite fuses with difficulty, colouring the flame yellow.

Occurrence The alkali feldspars are essential minerals in alkali and acid igneous rocks, particularly syenites, granites and granodiorites, syenites, felsites, trachytes, rhyolites, dacites, and various acid porphyries. Alkali feldspars are common in pegmatites, hydrothermal veins and high grade gneisses. Extrusive rocks may contain sanidine, high albite and 'high' phases, whereas plutonic rocks contain 'low' phases such as orthoclase, low albite and occasionally microcline. Perthitic types can be correlated with temperature depending upon the size of the exsolved phase; the finer the perthite the higher the temperature; thus cryptoperthites (<5 nm) are found in extrusive rocks, and macroperthites (>0.1 mm) are found in plutonic rocks.

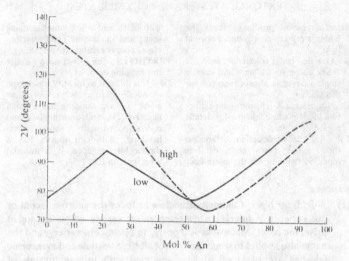

Figure 9.4 The variation of $2V$ in plagioclase feldspars. The 'high' curve is for feldspars from extrusive rocks, and the 'low' curve for feldspars in plutonic rocks.

Some pegmatites contain intergrowths of alkali feldspar and quartz (called *graphic intergrowth*), and some orbicular granites, including the Rapakivi granites contain large crystals of alkali feldspar mantled by plagioclase feldspar, or vice versa.

K-feldspars are common in metamorphic rocks, and microcline is a common variety present. In high-grade metamorphic rocks such as charnockites, sillimanite gneisses, migmatites and granulites, orthoclase is found in association with sillimanite, forming from the breakdown of micas and quartz.

In sedimentary rocks orthoclase and microcline are common constituents of arenaceous rocks such as feldspathic arenites, greywackes and arkoses. Authigenic, untwinned alkali feldspars may form as primary feldspars within some sedimentary rocks.

PLAGIOCLASE FELDSPARS

C   A continuous series exists at high temperatures from high albite, $NaAlSi_3O_8$, to high anorthite, $CaAl_2Si_2O_8$. A complex set of unmixed perthitic phases exists at moderate temperatures, below various solvus boundaries, until at low temperatures ($< 400°C$), low albite and low

anorthite occur as coexistent phases. The two main series are high albite –
high anorthite and low albite – low anorthite.

Optical properties

	Albite	Anorthite
n_α	1.527 –	1.577
n_β	1.531 –	1.585
n_γ	1.539 –	1.590
δ	0.012 –	0.013

$2V$ and the OAP are very variable in plagioclase feldspars (see Fig. 9.4).

COL Colourless, with cloudy patches
occurring if altered to clays

H Subhedral prismatic in hypabysal
and plutonic igneous rocks, to euhedral
prismatic or tabular in extrusive rocks,
where it may occur as phenocrysts; anhedral in metamorphic rocks.

CL Similar to the alkali feldspars; two
cleavages {010} and {001} meeting on an
(100) face

R Low, but apart from albite, all plagioclase feldspars have RIs > 1.54

A Plagioclase feldspars alter to montmorillonite when limited water is available, but alteration proceeds with production of kaolinite when excess water is
available. This alteration may be the
result either of late-stage hydrothermal
activity or of chemical weathering. As
with alkali feldspars, alteration to clays
give the feldspars a cloudy or opaque
appearance under the microscope.

B Low first-order colours, increasing
with increasing Ca content; anorthite
shows whites or very pale yellow

IF Not easy to obtain, but a single optic
axis figure is always needed. Figure 9.4
gives details of $2V$ size and sign. Make
sure that the correct curve is used (high
for extrusive rock feldspars, low or hypabyssal and plutonic).

E In albite the extinction angle is
measured using albite twin lamellae. The
twinned crystal is rotated first to the left
until one set of lamellae is in extinction
and then to the other side until the other
set is in extinction; the extinction angle is
this angle divided by 2. This angle is
plotted into the appropriate curve in
Figure 9.5, and the composition of the
feldspar obtained. Note that the

maximum angle from a number of grains.
is used and not the average value. For best
results no twin lamellae should be seen in
the parallel position, that is when the
twin planes are parallel to the N–S crosswires. Note that two curves occur for high
and low types, and thus, for example,
albite twins in extrusive rocks should be
plotted into the high curve. In some basic
and ultrabasic igneous rocks, combined
Carlsbad–Albite twins may be present.
In these, each half of the Carlsbad twin is
treated separately, and the extinction
angle for that albite twin obtained. Thus
two albite twin extinction angles are
obtained; (1) for the left-hand half, and
(2) for the right-hand half. These results
are then plotted into Figure 9.6, the
smaller of the two angles being plotted up
the ordinate (vertical axis), and the larger
of the two angles being plotted along the
horizontal at that point, into the 'nest' of
curves until the exact angle is reached.
The composition is read off along the
abscissa. Negative values on the ordinate
are needed for feldspars with an RI of less
than 1.54. Note that in this type of twin,
no twin lamellae should be observed in
the vertical position, that is parallel to the
N–S crosswire.

TW See under physical properties of
feldspars. Note that in albite twins the
lamellae become broader as the plagioclase becomes more Ca-rich.

Z Zoning is an important property in
plagioclase feldspars, and is particularly
common in feldspars in extrusive rocks.
Normal zoning occurs where there is a
continuous change from a Ca-rich core to
a Na-rich margin, whereas reversed
zoning is the opposite; that is, from an
Na-rich core to a Ca-rich margin. Oscilla-
tory zoning may also occur with separated zones of equal extinction occurring
in a random manner. In every case, the
range of zoning should be given, from,
for example, An_x (core) to an An_y
(margin)

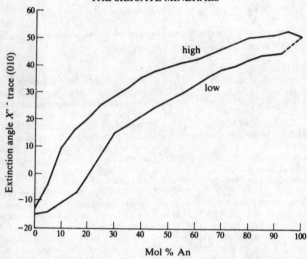

Figure 9.5 The maximum extinction angles for albite twins in high and low plagioclase feldspars

Tests Na-rich plagioclase feldspars fuse to a colourless glass more easily than K-feldspar and albite. Oligoclase is insoluble in acids. The Ca-rich types also fuse to a colourless glass, but rather more easily. They are also soluble in acids; labradorite in hot HCl, and anorthite in HCl with some separation of gelatinous silica.

Occurrence Plagioclase feldspars are present in all igneous rocks, with the exception of some ultramafic and ultra-alkaline types. Plagioclase feldspars often comprise more than 50% of the rock's volume, and the composition of the plagioclase is used to classify the type of igneous rock in which it is found. Na-rich plagioclase feldspars (oligoclase) occurs in acid igneous rocks, whereas Ca-rich types occur in basic and ultrabasic igneous rocks. In differentiated igneous intrusions the plagioclase feldspar is frequently zoned, and shows a large compositional range from the early formed, high-temperature, basic varieties, to the later formed, lower-temperature, more acidic varieties. Antiperthites occur in most plutonic rock types, with Bøggild or Huttonlocher intergrowths occurring in basic varieties, although these are rare and difficult to identify, and peristerites occurring in acidic varieties.

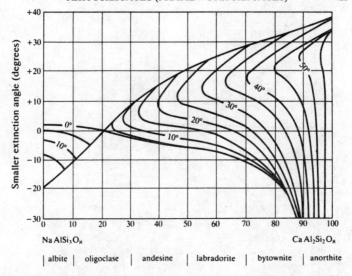

Figure 9.6 Extinction angles of Carlsbad–Albite twins. The extinction angle for an albite twin is measured in each half of a Carlsbad twin. The smaller angle is plotted along the ordinate and the larger angle into the nest of curves. Thus, for example, a Carlsbad–Albite twin with angles of extinction of 10° (the smaller) and 30° (the larger) has a composition of An_{60}. The negative ordinate values (below the horizontal line representing 0°) are needed for feldspars which have refractive indices of less than 1.54.

Anorthosite intrusions contain plagioclase feldspars (An_{40-60}) as the major constituent, often occupying over 80% of the volume of the rock; anorthosite layers in stratiform intrusions (Bushveld, Skaergaard) contain a more Ca-rich plagioclase feldspar (An_{80+}). Pure albite is found in spilites, although it often occurs mantling a more Ca-rich variety. In metamorphic rocks, the plagioclase feldspar becomes increasingly Ca-rich as the metamorphic grade increases; thus albite is the common feldspar in low-grade rocks (greenschists, etc.), whereas andesine and labradorite are common in high-grade rocks (granulites and high-grade gneisses); this is caused by the progressive breakdown of the epidote group minerals. Pure anorthite may occur thermally metamorphosed calcareous rocks.

Plagioclase feldspars occur as detrital grains in many terrigenous sedimentary rocks, and authigenic untwinned albite may form in some arenites during sedimentation.

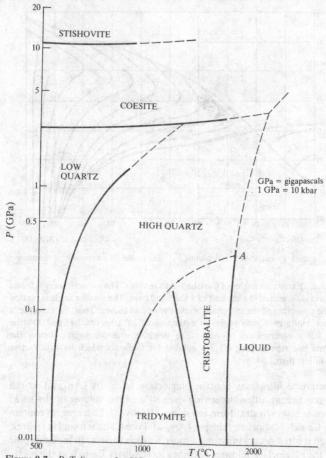

Figure 9.7 *P–T* diagram for SiO₂. GPa = gigapascals; 1 GPa = 10 kbar.

BARIUM FELDSPARS celsian, hyalophane
C BaAl₂Si₂O₈ (celsian)
(Ka,Na,Ba) [(Si,Al)₄O₈], with K > Ba (hyalophane)

Characters and occurrence
HD 6.0–6.5 (both) **SG** 3.10–3.39 (celsian), 2.58–2.82 (hyalophane)
RI = 1.580–1.600 (celsian); 1.520–1.547 (hyalophane)

Similar to orthoclase in most properties, except for RIs, hardness and specific gravity (see above).

Barium feldspars are rare. They occur in dolomitic limestones as at Jakobsberg, Sweden, and Binnental, Switzerland. Celsian is found associated with manganese deposits and stratiform barite deposits.

Silica group

Silica occurs in a number of forms:

crystalline   quartz, tridymite, crystobalite, coesite, stishovite
cryptocrystalline chalcedony, jasper, flint, opal, agate, diatomite, etc.

Crystalline silica

The various forms of crystalline silica can be represented on a simple $P-T$ diagram. The phase relationships show that the lowest T form of quartz (α-**quartz**) *inverts* to high quartz (β-**quartz**) at 573°C, at normal atmospheric pressure, but the temperature of this inversion increases with increasing pressure (to about 670°C at 3kbar, see Fig. 9.7). β-quartz inverts to **tridymite** at 867°C, and tridymite inverts to **crystobalite** at 1470°C. At 1713°C crystobalite melts and the *liquidus* boundary is reached. At very high pressures two other structural phases are known, namely **coesite** and **stishovite**, both of which are recorded from meteor impact sites, but these minerals very rarely occur in normal terrestrial rocks, although other minerals possessing the same type of atomic lattices may exist in the upper mantle. Coesite has been recorded in the Alps as inclusions in pyrope garnet in gneisses, which must have formed at pressures of 30 kbar (i.e. at depths of >85 km).

QUARTZ
C SiO_2

Physical properties
CS   Trigonal

F&H   Crystals occur as hexagonal prisms terminated by rhombohedral faces, which, when equally developed, resemble a hexagonal bipyramid. Quartz crystals often contain inclusions of other minerals, particularly rutile, as thin, reddish brown, needle-like crystals, and cavities containing liquids are also common. The prismatic faces are always striated perpendicular to their length; that is, perpendicular to the *c* axis. Quartz also occurs massive granular and sometimes stalactitic.

TW   Quartz is twinned on a number of laws, of which the most common are the *Dauphiné Law*, which has the *c* axis as the twin plane; the *Brazil Law*, which has $\{11\bar{2}0\}$ as the twin plane; and the *Japanese Law*, which has $\{11\bar{2}2\}$ as the twin plane. Other twin planes include {10Ī1}, {10Ī2} and {11Ī1}. The enantiomorphism of quaratz gives rise to left- and right-handed crystals, both of which are relatively common. Twinning is purely a physical feature, and cannot be seen under the microscope.

COL Usually colourless or white, but some varieties may be coloured

CL Quartz does *not* possess a cleavage
F Conchoidal
L Vitreous, occasionally resinous
TR Transparent when clear, to translucent when white
HD 7 (one of the minerals on Mohs' scale)
SG 2.65

Optical properties
$n_\alpha = 1.544$
$n_\beta = 1.553$
$\delta = 0.009$
Uniaxial +ve; quartz is length slow
COL Colourless
H Euhedral quartz crystals may occur as phenocrysts in some porphyritic igneous rocks, but quartz usually occurs as shapeless (anhedral) interstitial grains in igneous and metamorphic rocks, and as rounded grains in sedimentary clastic rocks

CL None
R Low, just greater than 1.54
A None
B Low, with first order greys and whites
IF Good +ve uniaxial figure seen on a basal section
TW Not seen in thin section

Varieties **Rock crystal** is the purest and most transparent variety of quartz. It is used in jewellery and for windows in deep-sea diving vessels. **Amethyst** is a transparent, purple, semi-precious variety, with impurities of ferric iron giving the colour. The colour fades on exposure but may be partly restored by moistening the crystal. **Smoky quartz** and **citrine** are varieties of pale brown or pale yellow quartz. This variety is sometimes called **Cairngorm** (from the Cairngorm Mountains of Scotland), but the original Cairngorms were crystals of topaz. **Rose quartz** is a pale pink variety which owes its colour to impurities of Li, Na and Ti in the structure. **Morion** is a almost black variety. **Milky quartz** is a common, white variety containing numerous microscopic air cavities which give the quartz its white colour. The milkiness may be superficial and such crystals are called *quartz en chemise*. **Cat's eye** is quartz with a fibrous structure which imparts a peculiar opalescent lustre to the cut face; this crystal is often a pseudomorph after some fibrous mineral. **Aventurine quartz** is a quartz crystal contining spangles of mica, hematite, etc. **Ferruginous quartz** contains various iron oxides, giving the crystal a reddish or brownish appearance.

Tests Heated alone before the blowpipe, quartz is unaltered. It is soluble in borax and sodium carbonate beads, but insoluble in the microcosmic salt bead test.

Occurrence   Quartz is an essential constituent of acid igneous plutonic rocks such as granites, granodiorites and pegmatites. It may also be present in some diorites and gabbros, always occurring as shapeless interstitial grains. In extrusive and hypabyssal rocks such as rhyolites, dacites, pitchstones and various porphyries, quartz often occurs as phenocrysts, often with corroded edges. Quartz is a common gangue mineral in hydrothermal and other veins, accompanying the economic ore minerals. Quartz is a common detrital mineral because of its hardness, lack of cleavage and stability. It is an essential mineral in the coarser types of terrigenous rocks such as conglomerates, arenites, etc., and also occurs in fine-grained varieties such as siltstones and mudstones, although its identification may be difficult. Authigenic quartz may form during sedimentary diagenesis, often growing around pre-existing quartz grains. Quartz occurs in many metamorphic rocks, especially pelites and psammites, and remains until the very highest grades when it enters into the reaction:

$$\text{muscovite} + \text{quartz} = \text{K-feldspar} + \text{sillimanite} + \text{water}$$

TRIDYMITE AND CRYSTOBALITE
C SiO_2

Physical properties
CS   Orthorhombic (tridymite) tetragonal (crystobalite)

F&H   Tridymite occurs as small, overlapping scales or plates, whereas crystobalite occurs as small octahedra or cubes. They both are colourless, have no cleavage and possess a vitreous lustre.

HD   7 (tridymite); 6.0–7.0 (crystobalite)

SG   2.26 (tridymite); 2.38 (crystobalite)

Optical properties
n_α = 1.469–1.479 (tridymite)
n_β = 1.470–1.480
n_γ = 1.473–1.483
δ = 0.004
$2V$ = 40°–90° +ve
OAP is parallel to (100)

n_o = 1.478 (crystobalite)
n_e = 1.483

δ = 0.003
Uniaxial −ve

Both minerals are colourless with similar optical properties to quartz. Note that their RIs are considerably less than 1.54. Tridymite sometimes shows sector twins in thin section. When present both minerals usually occur as very minute crystals.

Occurrence Tridymite is rare in rocks but can occur in quickly chilled igneous rocks such as rhyolites, pitchstones, etc., sometimes in association with sanidine, augite and fayalitic olivine. It has been recorded from high-temperature thermally metamorphosed impure limestones.

Crystobalite may occur in cavities in some volcanic rocks, and has been recorded from some thermally metamorphosed sandstones.

CRYPTOCRYSTALLINE SILICA

C $SiO_2.nH_2O$; most varieties are mixtures of cryptocrystalline silica and hydrous silica, from SiO_2 (chalcedonic silica) to opal (hydrous silica)

Physical properties

CS None

F&H The different varieties may occur either as fibrous structures, or veinstones, or as nodules in sedimentary or extrusive rocks, the surface of these nodules being reniform, stalactitic or botryoidal. Opal occurs infilling cracks or cavities in igneous rocks, and as veins and nodules in sandstones and shales in Australia and Mexico.

COL Variable; jasper is often red, agate banded in concentric coloured rings, onyx black-and-white striped, flint and chalcedony greyish black, and opal variable, with a play of colours on the surface. The opalescence of opal is probably due to a layer of water molecules, trapped in the silica atoms of the structure near the surface.

F   All possess a conchoidal fracture

L   Most are waxy, some subvitreous, and opal is opalescent (iridescent)

TR Transparent (some opal) to translucent or opaque

HD 6.5–7.0 (chalcedonic types); 5.5–6.5 (opal)

SG 2.50–2.67 (chalcedonic types); 1.99–2.25 (opal)

Optical properties

n = 1.526–1.553 (chalcedony)   1.435–1.460 (opal)

Uniaxial +ve (sometimes biaxial)   Isotropic

In chalcedony, RIs and birefringence decrease with increasing water content. All types are colourless, some with banding or zoning present.

Varieties

(1)   **Chalcedony** includes a number of subvarieties based mainly on colour. **Carnelian** is a reddish or yellowish red, subtranslucent variety, and **sard** is brownish; both are used as stones in signet rings and for similar work. **Prase** is a translucent, dull, green variety, and **plasma** is a subtranslucent, bright green variety, speckled with white; **bloodstone** or **heliotrope** is similar, but speckled with red. **Crysoprase** is apple green. **Agate** is a variegated chalcedony composed of different coloured concentric bands, with sharp or diffuse boundaries. Used in brooches, snuff boxes and pendants, agates come from

Saxony, Bavaria, Arabia, India, and Perthshire in Scotland, where they are known as Scotch pebbles. **Moss agate** or **mocha stone** is a chalcedony containing small dendrites (tree-like growths), which consist of iron oxide or an iron-rich chlorite. **Onyx** or **sardonyx** are flat banded varieties, onyx having white and grey, brown or black bands, and sardonyx having white, or bluish white and red, or brownish red bands.

(2) **Flint** is usually black or shades of grey in colour, and is found as nodules, occurring as bands in the Upper Chalk of England. Flint breaks with a conchoidal fracture, giving sharp edges which early man used for cutting in tools and weapons. Flints were used for igniting tinder since they produce sparks when struck with steel – the steel particles being raised to a state of incandescence by the blow – and were used in early flint-lock pistols and guns. Flint is used in tube-mills, and calcined flint in the pottery industry. In South-east England, flint is used in road-making and for building. **Hornstone** and **chert** are grey to black opaque varieties which resemble flints, but break with a more or less flat fracture. Nodules and beds occur in the Carboniferous limestone of North Wales, and chert is often associated with black shales and spilites.

(3) **Jasper** is an impure, opaque variety, red, brown or yellow in colour and opaque, even on thin edges. **Egyptian** or **ribbon jasper** is beautifully banded in shades of brown 'Porcelain jasper' is thermally metamorphosed clay or shale, which is distinguished from true jasper by being fusible on its edges before the blowpipe.

(4) **Precious opal** is the gem variety, exhibiting opalescence and a brilliant play of colours. **Hydrophane** is a white or yellowish variety which, when immersed in water, becomes translucent and opalescent. **Hyalite** is a colourless, transparent, glassy type occurring in small botryoidal or stalactitic habits. **Menilite** or **liver opal** is an opaque, brownish variety found in flattened or rounded concretions with pale exteriors, as at Menil Montant (Paris). **Wood opal** is wood which has been replaced by opal. **Siliceous sinter** consists either of hydrous or anhydrous silica. It has a loose, porous texture, and is deposited from the waters of hot springs, as at Taupo (New Zealand), and is common at the geysers of Iceland and USA. In these occurrences it is called **geyserite**, and lines the tubes or bores of geysers; it is deposited in cauliflower-like encrustations on the surface of the neighbouring ground. **Float stone** is a porous variety which floats on water, and occurs in chalk deposits at Menil Montant (Paris). **Diatomite, diatomaceous earth** or *kieselguhr* is a deposit from the tests or skeletons of

siliceous organisms, such as algae and diatoms, forming beds in lakes, and thick deposits where siliceous volcanic emanations have supplied abundant material for diatom growth, as in the Miocene beds of California. Diatomite is dried and used as an absorbent, a filtering medium, for high-temperature insulation, and also in cement, glazes, pigments, etc. World production is ~1.5 Mt per annum, with the main producers being the USA, the USSR and Denmark, together accounting for over 1 Mt.

Feldspathoid group

The group of minerals termed **feldspathoids** include those minerals which have certain similarities with the feldspars, particularly in their chemistry and structure. The main members considered here are:

nepheline and **kalsilite**
leucite
the **sodalite group** (sodalite, nosean, haüyne and lazurite)
cancrinite – **vishnevite**

The feldspathoids are all silica-deficient compared to the feldspars, and their occurrence is restricted to undersaturated alkali igneous rocks, except for lazurite which occurs in contact metamorphosed limestone

NEPHELINE–KALSILITE

C $Na_3(Na,K)[Al_4Si_4O_{16}]$ (nepheline); $KAlSiO_4$ (kalsilite). Note that up to 25% of the kalsilite molecule can be accommodated in nepheline.

Physical properties

CS Hexagonal

F&H Squat prisms with basal faces common; also anhedral and massive

COL Colourless, white, purplish brown, yellowish, dark green

CL Perfect $\{10\bar{1}0\}$ prismatic, and poor $\{0001\}$ basal. The cleavages are difficult to see in thin section unless the nepheline is highly altered.

F Subconchoidal

L Vitreous or greasy

TR Transparent to opaque

HD 5.5–6.0

SG 2.56–2.66

Optical properties

n_o = 1.529–1.546
n_e = 1.526–1.542
δ = 0.003–0.005
Uniaxial −ve; crystals are length fast

COL Colourless

H Usually anhedral, occurring in the interstices between earlier-formed minerals. Occasionally found as small, irregular, exsolved blebs within feldspars, especially K-feldspars. Euhedral nepheline crystals have a hexagonal outline.

CL Prismatic perfect and basal poor; see above

R Low; about equal to, or just less than, 1.54

A Nepheline may alter to either natrolite or analcime (both zeolites) or sodalite or cancrinite (other feldspathoids) by the addition of water, silica or other volatiles (Cl, etc.)

B Low, showing first order greys. Small inclusions occur within nepheline, giving the crystal a clear 'night sky' effect.

Varieties **Eleolite** is a dark-coloured variety, with a greasy lustre, often shapeless, and found in syenites.

Tests Heated before the blowpipe, nepheline fuses to a colourless glass. It gelatinizes with acids.

Occurrence Nepheline is a characteristic primary crystallizing mineral of alkaline igneous rocks. It is an essential constituent of silica-deficient nepheline syenites and may occur in some volcanic rocks in association with high-temperature feldspars. Nepheline may be of metasomatic origin, and may occur in basic rocks near their contact with carbonate rocks. Some alkali dolerites (teschenites) may contain nepheline. **Kalsilite** forms a limited solid solution with nepheline, with up to 25% of the nepheline molecule being replaced by kalsilite, although this can increase with increasing temperatures. Kalsilite has been reported from the ground mass of some K-rich lavas.

LEUCITE
C $KAlSi_2O_6$

Physical properties

CS Tetragonal, pseudo-cubic; cubic at temperatures of 500–600°C

F&H {211} trapezohedron form common; less common as cube {100} and rhombic dodecahedron {110}; also found as disseminated grains

COL White or pale grey

CL Very poor on {110}

F Conchoidal

L Vitreous on fracture surfaces

TR Translucent to opaque

HD 5.5–6.0

SG 2.47–2.50

Optical properties

n = 1.508–1.511

δ = 0.001

Mostly isotropic, but may be uniaxial −ve. Sign impossible to obtain because of twinning.

COL Colourless

H Usually appears as euhedral crystals with eight-sided sections

CL Very poor on {110}

R Low

A A mixture of nepheline and feldspar, called **pseudo-leucite**, can replace leucite in some rocks.

B Isotropic to low

TW Repeated twinning on {110} is common as a type of cross-hatching under XP

Tests Leucite is infusible before the blowpipe. It gives a blue aluminium colour when heated with cobalt nitrate in the oxidizing flame. It is soluble in HCl, without gelatinization.

Occurrence Leucite occurs in K-rich basic extrusive rocks such as leucite basanite, leucite tephrite, lamproite and leucitophyre, which may also be silica deficient. Pseudo-leucite may occur in some alkali, basic extrusive rocks, and also in some alkali, basic igneous plutonic rocks where leucite breaks down at temperatures below the solidus.

Sodalite group

SODALITE

C $Na_8[Al_6Si_6O_{24}]Cl_2$

Physical properties

CS Cubic

F&H Rhombic dodecahedra as crystals; or massive

COL Grey, bluish, yellowish

CL Poor on {110}

F Conchoidal or uneven

L Vitreous

TR Subtransparent to translucent

HD 5.5–6.0

SG 2.27–2.88

Optical properties

n = 1.483–1.487

COL Colourless

H Usually anhedral crystals filling interstices

CL Poor {110} dodecahedral cleavage occurs

R Moderate, less than 1.54

DF Sodalite and analcime are virtually indistinguishable in thin section, but may be identified by the associated minerals and which type of rock they are found in

Tests Sodalite is soluble in nitric acid, the solution reacting for a chloride. It is decomposed by HCl, with separation of gelatinous silica. Heated before the blowpipe, it fuses with intumescence to a colourless glass.

Occurrence   Sodalite occurs in nepheline syenites in association with nepheline and fluorite. It occurs in metasomatized calcareous rocks near alkaline igneous intrusions.

NOSEAN, HAÜYNE, CANCRINITE

C   nosean $Na_8 [Al_6Si_6O_{24}] SO_4$
haüyne $(Na,Ca)_{4-8} [Al_6Si_6O_{24}] (SO_4, S)_{1-2}$
cancrinite $(Na,Ca,K)_{6-8} [Al_6Si_6O_{24}] (CO_3,SO_4,Cl).1-5H_2O$

Physical properties

CS Nosean and haüyne are cubic; cancrinite is hexagonal

F&H Nosean and haüyne occur as small octahedra or dodecahedra, often rounded. Cancrinite usually occurs massive.

COL Nosean is greyish or brown; haüyne is bright blue or greenish blue; and cancrinite is yellowish or white

CL {110} poor in nosean and hauyne; perfect {10$\bar{1}$0} prismatic in carnite

L Subvitreous to dull

HD 5.5–6.0

SG 2.30–2.40 (nosean); 2.44–2.50 (haüyne); 2.32–2.51 (cancrinite)

Optical properties

All the minerals of this group, except cancrinite–vishnevite, are isotropic with the following RIs:

$n = 1.495$ (nosean)
$n = 1.496–1.505$ (haüyne)

	Cancrinite		Vishnevite
n_o	1.503	–	1.488
n_e	1.528	–	1.490
δ	0.025	–	0.002

Uniaxial −ve (both end-members)

Cancrinite is carbonate-rich and Vishnevite is sulphur-rich

COL All are colourless in thin section; haüyne may be pale blue

H Nosean occurs as rounded crystals usually with a dark border around each one. Haüyne is characterized by an abundance of dark inclusions, arranged either as a darkish border or in a symmetrical pattern.

CL See above

T Low, all less than 1.54

A Cancrinite may be an alteration product from nepheline

B Nosean and haüyne are isotropic, but cancrinite has a high birefringence with second and third order colours shown

Tests   Nosean and haüyne are decomposed by HCl with the separation of gelatinous silica, the solutions giving the reaction for sulphate with barium chloride solution.

Occurrence   Nosean occurs in leucitophyres and phonolites. Haüyne also occurs in phonolites and other undersaturated igneous rocks, but is usually accompanied by sulphide minerals such as pyrite. Cancrinite is a common mineral in nepheline-bearing rocks.

LAZURITE, LAPIS LAZULI
C $(Na,Ca)_8 [(Al,Si)_{12}O_{24}] (S,SO_4)$

Characters and occurrence
HD   5.0–5.5; **SG**   2.38–2.45; $n = {\sim}1.50$

Lapis lazuli is dark blue, violet or greenish blue in the hand specimen, with similar physical and optical properties to the other sodalite group minerals.

It fuses, with intuminescence, to a white glass, and is decomposed by HCl, with the evolution of hydrogen sulphide, leaving behind a gelatinous deposit of silica.

It occurs in contact limestone deposits with calcite and pyrite, as at Lake Baikal.

Uses   Lapis lazuli is cut and polished for ornamental purposes, but is too soft for most jewellery. Ancient Egyptian amulets were commonly made of lapis. Powdered lazurite was used for ultramarine paint, but this is now artificially made by heating together clay, sodium carbonate and sulphur.

SCAPOLITE
C $(Na,Ca,K)_4 [Al_3(Al,Si)_3Si_6O_{24}](Cl,CO_3,SO_4,OH)$; two end-members exist, **marialite** (Na and Cl) and **meionite** (Ca and CO_3)

Physical properties
CS   Tetragonal

T&H Tetragonal or ditetragonal prisms terminated with bipyramids; it also occurs massive, granular and occasionally columnar

COL White, or pale shades of green or blue

CL   Prismatic cleavages {100} and {110} good

F   Subconchoidal

L   Vitreous to pearly or resinous

TR   Transparent to nearly opaque

T   Brittle

HD   5.0–6.0

SG   2.50 (marialite) to 2.74 (meionite)

Optical properties

	Marialite	Meionite
n_o	1.540–1.541	1.556–1.564
n_e	1.546–1.550	1.590–1.600
δ	0.005–0.009	0.034–0.038

Uniaxial −ve (all scapolites)

COL Colourless

H Large spongy,, prismatic crystals are common in metamorphosed carbonate rocks. Granular and fibrous habits are found in garnet-bearing metamorphic rocks.

CL Good {100} and {110} prismatic cleavages present

R Low

A Scapolite may alter to a fine aggregate of chlorite, sericite, calcite, plagioclase, clays, etc.

B Low (marialite) to high (meionite)

IF Uniaxial negative if large crystal available, but aggregates of crystals are too small for size determination

Z Compositional zoning is common

Tests Heated before the blowpipe, scapolite fuses, with intuminescence, to a white glass. It is imperfectly decomposed by HCl; meionite easily, marialite less easily.

Occurrence Scapolite occurs in some pegmatites, replacing quartz or plagioclase, but it mainly occurs in metamorphic or metasomatic rocks. It occurs as a primary mineral in calcareous rocks subjected to moderate to high-grade regional metamorphism. Scapolite is found in association with sphene, calcite, diposide, plagioclase, epidote and garnet. In contact metamorphism, scapolite forms in carbonate rocks by introduction of Na and Cl from the igneous intrusion; grossular, wollastonite and fluorite are associated minerals.

Zeolite group

C (general) (Na_2, K_2, Ca, Ba) $[(Al, Si)O_2]n.yH_2O$

The zeolites are hydrated aluminosilicates of sodium, potassium and calcium, with other alkali and alkali earths also incorporated, such as Ba and Sr, and each zeolite has the atomic ratio $O : (Al + Si) = 2$. Dehydration of the zeolite minerals is continuous and in part reversible, and they possess important base exchange properties.

Analcime has close affinities with both the zeolites and the feldspathoids, but is here included with the zeolites.

All zeolites possess an aluminosilicate framework composed of $[(Al, Si)O_4]^{4-}$ to $^{6-}$, each oxygen of which is shared between two tetrahedra. The negative charge(s) on the framework are balanced by Na^+, K^+ or Ca^{2+} cations situated in cavities in the structure, similar to the lattices of the feldspars and the feldspathoids. In the zeolites the cavities are large and there are many open channels, giving the framework a somewhat loose structure, and the zeolites lowish specific gravities (2.0–2.3). Water molecules are also present in the structural channels, but these are lightly bonded and can be removed and partly replaced with little disruption to the overall structural framework.

Zeolites typically occur in amygdales and cavities in basic volcanic rocks and in other late-stage environments.

Zeolites are widely used as indicator minerals in thick lava piles, such as *ocean-floor basalts*, to determine temperature and depth of burial. A typical sequence from a recent Icelandic lava pile is:

> *top*
> zeolite-free zone
> chabazite–thomsonite
>
> analcime (+ natrolite)
>
> mesolite–scolecite
> *bottom*

The most important zeolites are discussed in this text beginning with analcime, and followed by those zeolites with a fibrous habit, and finally the remainder, as follows:

analcime

natrolite ⎫
mesolite   ⎬ fibrous zeolites
scolecite
thomsonite ⎭

phillipsite
harmotome
chabazite
heulandite
stilbite
laumontite

Optical properties (all zeolites except analcime) All the zeolite minerals have similar optical properties, the only differences being in their RIs, interference figures and extinction. The optical properties of all the zeolites dealt with in this text are listed together, so that quick comparisons can be made. The habit, crystal system, cleavage, etc., can be obtained from their physical properties, which are given separately for each mineral:

Mineral	RIs		Birefringence (δ)	Sign	$2V$	Extinction
	min.	max.				
natrolite	1.473	1.496	0.013	+ve	58–64	straight
mesolite	1.504	1.509	0.001	+ve	≈ 80	straight
scolecite	1.507	1.521	0.010	−ve	36–56	$\alpha \,\hat{} \, cl = 18°$
thomsonite	1.497	1.544	0.021	+ve	42–75	straight
phillipsite	1.483	1.514	0.010	+ve	60–80	$\gamma \,\hat{} \, cl = 10°\, 30'$
harmotome	1.503	1.514	0.008	+ve	≈ 80	$\alpha \,\hat{} \, cl = 65\pm$
chabazite	1.472	1.495	0.010	−ve	uniaxial	straight
heulandite	1.487	1.512	0.007	+ve	≈ 30	oblique not to cl
stilbite	1.482	1.513	0.013	−ve	30–49	oblique not to cl
laumontite	1.502	1.525	0.016	−ve	25–47	$\gamma \,\hat{} \, cl = 8–40°$

ANALCIME
C $Na[AlSi_2O_6].H_2O$

Physical properties

CS Cubic

F&H The trapezohedron {211} is common form; also massive and granular

COL Milk white, often colourless, but occasionally greyish, greenish, pink

CL Cubic {100} very poor

F Subconchoidal and uneven

L Vitreous

TR Transparent to almost opaque

T Brittle

HD 5.5

SG 2.24–2.29

Optical properties

$n = 1.479$–1.493

COL Colourless

H Usually anhedral crystals infilling interstices

CL Poor cubic cleavage

R Moderate, well below 1.54

Tests Heated in a closed tube, analcime yields water. Heated on charcoal it fuses to a clear, colourless globule. It is decomposed by HCl, with the separation of silica. It colours a flame yellow.

Occurrence Analcime may occur as a primary mineral in some alkaline basic igneous rocks, but this is questionable. It occurs as a late-stage hydrothermal mineral, crystallizing in vesicles, and occurs with the zeolites thomsonite and stilbite. Analcime occurs as an authigenic mineral in sandstones, associated with the zeolites heulandite and laumontite.

NATROLITE
C $Na_2[Al_2Si_3O_{10}].2H_2O$

Physical properties

CS Orthorhombic

F&H Usually as acicular or fibrous radiating crystals, rarely prismatic; also granular, compact or massive

COL White, rarely yellowish or reddish

CL Perfect {110} cleavages and an {010} parting may be present

L Vitreous or pearly

TR Transparent to translucent

HD 5.0

HD 2.20–2.26

See p. 439 for the main optical properties.

Tests Heated in closed tube, natrolite yields water. Heated on charcoal it fuses to a clear, colourless globule. It is decomposed by HCl, with the separation of silica. Natrolite colours a flame yellow.

Occurrence Natrolite is a common mineral in vesicles, amygdales, and other cavities in basic extrusive rocks, being deposited by hydrothermal solutions. It occurs along with analcime, stilbite, heulandite, chabazite and laumontite. Natrolite may form from the alteration of either feldspar or feldspathoid minerals.

MESOLITE
C $Na_2Ca_2[Al_2Si_3O_{10}]_3.8H_2O$

Physical properties

CS Monoclinic (pseudo-orthorhombic)

F&H Tufts of acicular crystals common, occasionally massive. Mesolite is elongated along the *b* crystallographic axis.

COL White or greyish

CL Perfect on {101} and {101}

L Vitreous, but silky when fibrous or massive

TR Transparent to translucent, but opaque when massive

HD 5.0

SG 2.26

See p. 439 for the main optical properties.

Tests Mesolite gives off water when heated in a closed tube. It gelatinizes with HCl. Heated before the blowpipe, it becomes opaque and fuses, with worm-like intumescence, to an enamel.

Occurrence Mesolite is a common deposit in amygdales and other cavities in silica-poor volcanic rocks from hydrothermal solutions. It is associated with calcite and other zeolites.

SCOLECITE
C $Ca[Al_2Si_3O_{10}].3H_2O$

Physical properties

CS Monoclinic (pseudo-tetragonal)

F&H As radiating clusters of fibrous or, rarely, prismatic crystals; also nodular or massive

COL White

CL Perfect {110} prismatic cleavages present

L Vitreous, but fibrous habits show silky lustre

TR Transparent to subtranslucent

HD 5.0

SG 2.25–2.29

See p. 439 for the main optical properties.

Tests Heated in a closed tube, scolecite gives off water. It gelatinizes with HCl. Heated before the blowpipe, it fuses, with worm-like intumescence, and often forms a frothy mass.

Occurrence It is a common mineral in cavities in basic volcanic rocks with stilbite, and other zeolites, and calcite. It appears in the zeolite facies of some metamorphosed carbonate rocks with albite, epidote, etc.

THOMSONITE

C $NaCa_2[(Al,Si)_5O_{10}]_2.6H_2O$

Physical properties

CS Orthorhombic (pseudo-tetragonal)

F&H Radiating or columnar masses of fibrous crystals

COL Snow white

CL {010} perfect and {100} distinct

L Vitreous, but fibrous habits are silky

TR Transparent to translucent; opaque when massive

HD 5.0–5.5

SG 2.10–2.39

See p. 439 for the main optical properties.

Tests Thomsonite yields water when heated in a closed tube. It fuses with intumescence to a white, blebby enamel. It is soluble in HCl with gelatinization.

Occurrence As for natrolite.

PHILLIPSITE

C $(0.5Ca,Na,K)_3[Al_3Si_5O_{16}].6H_2O$

Physical properties

CS Monoclinic

F&H Crystals as penetrating cruciform twins grouped in radiating aggregates

COL White, rarely reddish

CL Distinct prismatic cleavages {100} and {010} present

F Uneven

L Vitreous

T Brittle

HD 4.0–4.5

SG 2.20

See p. 439 for the main optical properties.

Tests Heated in a closed tube, phillipsite gives off water. Heated on charcoal, it fuses quietly to a bubbly enamel. It is decomposed by HCl with separation of silica.

Occurrence Found in cavities in mafic volcanics in association with chabazite. It is characteristic of silica-deficient igneous rocks such as phonolites and leucite-bearing rocks. Phillipsite has been reported in thermal spring deposits, and is a common constituent in the red clays of deep-sea deposits.

HARMOTOME, cross-stone

C $Ba[Al_2Si_6O_{16}].6H_2O$

Physical properties

CS Monoclinic

F&H Crystals always as cruciform penetration twins, similar to stilbite. They are either simple twins, or groups of four individuals, sometimes showing re-entrant angles (hence the name **cross-stone**). These groups may be combinations of prism faces, giving a pyramidal aspect.

COL White, shades of grey, brown or yellow

CL {010} distinct, {001} poor

F Uneven

L Vitreous

TR Subtransparent to translucent

T Brittle

HD 4.5

SG 2.40–2.50

See p. 439 for the main optical properties.

Tests Heated before the blowpipe, harmotome whitens, crumbles and fuses, without intumescence, to a white, translucent glass. It is decomposed by HCl, without gelatinization.

Occurrence Harmotome is common as a hydrothermal deposit in vesicles and fractures, in silica-poor volcanics. It is associated with Mn mineralization and occurs in veins with barite, strontianite, calcite, pyrite and galena. It may be a constituent mineral of laterites.

CHABAZITE
C $Ca[Al_2Si_4O_{12}].6H_2O$

Physical properties

CS Trigonal	**L** Vitreous
F&H Rhombohedra with simple and penetration twins common; also massive	**TR** Transparent to translucent
	HD 4.5
	SG 2.1
COL White, yellowish or reddish	See p. 439 for the main optical properties.
CL Good {1011} rhombohedral cleavage present	

Varieties   **Phacolite** is a colourless variety, occurring in lenticular crystals due to twinning.

Tests   Heated in a closed tube, chabazite gives off water. Heated before the blowpipe, it intumesces, whitens and fuses to a glass. It is decomposed by acid, with separation of silica.

Occurrence   Similar to scolecite.

HEULANDITE
C $(Ca,Na_2)[Al_2Si_7O_{18}].6H_2O$

Physical properties

CS Monoclinic	**T** Brittle
F&H Tabular crystals common, flattened on (010); rarely granular	**HD** 3.5–4.0
	SG 2.2
COL White, brick red, brown	See p. 439 for the main optical properties.
CL {010} perfect	
F Subconchoidal or uneven	
TR Transparent to subtranslucent	

Tests   Before the blowpipe, heulandite intumesces and fuses. It is decomposed by HCl, without gelatinization but with separation of silica.

Occurrence   Heulandite is most characteristic mineral of cavities in basalts and other basic volcanic rocks, associated with stilbite and chabazite. It forms by devitrification of volcanic glass and alteration of tuffs, occurring in bentonitic clays. It also occurs in cavities in metamorphic rocks and as an authigenic mineral in detrital sediments.

STILBITE, desmine
C $(Ca,Na_2,K_2)[Al_2Si_7O_{18}].7H_2O$

Physical properties

CS Monoclinic

F&H Crystals are platy on {010}, arranged as sheaf-like radiating aggregates

COL Usually white, occasionally yellow, red or brown

CL {010} perfect, {100} poor

L Vitreous; pearly on cleavage faces

TR Subtransparent to translucent

HD 3.5–4.0

SG 2.1–2.2

See p. 439 for the main optical properties.

Tests Heated in a closed tube, stilbite gives off water. Heated before the blowpipe, it fuses with a worm-like intumescence to a white enamel. It is decomposed by HCl with separation of silica.

Occurrence Similar occurrence to that of heulandite.

LAUMONTITE

C $Ca[Al_2Si_4O_{12}].4H_2O$

Physical properties

CS Monoclinic

F&H Columnar or prismatic crystals elongated along the *c* axis, or as radiating fibrous aggregates

COL white, greyish or yellowish

CL Perfect prismatic on {110} and {010}

L Vitreous if fresh; dull if altered

TR Transparent to translucent when fresh, but becomes opaque white on exposure

HD 3.0–4.0

SG 2.23–2.41

See p. 439 for the main optical properties.

Tests Heated before the blowpipe, laumontite fuses, with intumescence, to a white enamel. It is soluble in HCl, with gelatinization.

Occurrence Laumontite occurs in cavities in igneous rocks from basalts to granites, and is present in metalliferous hydrothermal veins. It forms in preference to other zeolites with increasing depth, and originates from mild metamorphic alteration of volcanic glass and feldspars appearing in tuffs and feldspathic detrital sediments.

Appendix A
Analysis by the blowpipe

A.1 The blowpipe

The blowpipe is an invaluable instrument for examining a dry mineral. It consists essentially of a tube bent at right angles, one end having a mouthpiece and the other being terminated by a finely perforated jet. The tube should bulge out between the two extremities into a cavity, where condensed moisture from one's breath may accumulate so as not to be either carried through the jet or deposited on to the **assay**, which is the name given to that portion of the mineral being tested. It is important that the aperture of the nozzle of the blowpipe should be small and circular. This can be achieved by gently tapping the nozzle on an iron surface, inserting a square needle into the aperture and, by rotation, producing a hole of the required size and shape.

The operator will probably experience some difficulty at first in keeping up a steady, continuous blast from the blowpipe, and practice will be needed to enable the instrument to be used easily. While blowing, the cheeks should be kept inflated and the air expelled by their action only, fresh air being drawn in through the nose. Trial and error are the best teachers, and practice should be continued until a steady and uninterrupted blast can be kept up for several minutes. A gas flame is very convenient for blowpipe experiments, but the flame of an oil or spirit lamp, or a candle, will do. If a lamp or candle is used, the wick should be bent in the direction in which the flame is blown. A portable blowpipe lamp is of particular use in the field.

A.2 The two types of flame

In blowpipe analysis it is necessary to be able to produce and to recognize two types of flames, the **oxidizing** flame and the **reducing** flame.

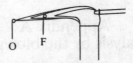

Figure A.1 The oxidizing flame, showing the position of the blowpipe and the points of oxidation (O) and fusion (F).

The oxidizing flame

An oxidizing flame is produced when the nozzle of the blowpipe is introduced into the flame so that it occupies about one-third of the breadth of the flame (see Fig. A.1). It is advisable to blow more strongly than in the production of the reducing flame. The oxidizing flame is *blue* and feebly illuminating, and complete combustion occurs since the air from the blowpipe is well mixed with the gases from the flame. There are two positions in this flame at which useful operations may be performed. These are the **point of fusion** (position F in Fig. A.1) where the hottest part of the flame occurs, and the **point of oxidation** (position O in Fig. A.1) where the assay is heated surrounded by air, and hence oxidation takes place.

The reducing flame

The reducing flame is produced when the nozzle of the blowpipe is placed some distance from the flame (Fig. A.2). The reducing flame is bright *yellow* and luminous, ragged and noisy. In this flame the stream of air from the blowpipe drives the whole flame rather feebly before it, and there is little mixing of air with the gases from the flame. The result is that these gases are not completely burnt, and hence will combine readily with the oxygen of any substance introduced into their midst. The assay must, therefore, be completely surrounded by the reducing flame, but *not* introduced too far into the flame in case a deposit of soot is formed which

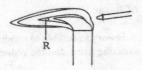

Figure A.2 The reducing flame, showing the position of the blowpipe and the point of reduction (R).

will interfere with the heating of the assay. In Figure A.2 the **point of reduction** is at R.

A.3 Supports

The portion of the substance under examination (the assay) may be supported in various ways according to the requirements of each particular case. After each experiment, all supports must be thoroughly cleaned before further use. The most common supports are charcoal blocks, platinum-tipped forceps and platinum wire.

Charcoal is a good support because of its infusibility, poor heat-conducting capacity, and its reducing action. The charcoal block is composed of carbon which readily combines with any oxygen which the assay may contain. Many metallic oxides may be reduced to their metals by heating on charcoal in the reducing flame. Sometimes, charcoal may be used as a support in oxidation, provided that its reducing action does not materially interfere with the results

In order to achieve maximum success, the assay should be placed in a small hollow scraped in the charcoal block, and there should be a large area of cool charcoal beyond the hollow on which any encrustations may form. Should the assay crackle and fly about, a fresh assay must be made by powdering the substance being examined and mixing it with water into a thick paste.

Points which should be noted include how easily the assay burns or flames, how easily it fuses and whether the fused assay is absorbed by the charcoal. The nature, colour, smell and distance from the assay of any encrustation are all important indicators about the nature of the elements present. Thus, for example, arsenic compounds give an encrustation far from the assay, whereas antimony compounds give an encrustation near the assay. White encrustations or residues, when moistened with cobalt nitrate and strongly reheated, give various colours characteristic of certain elements. Compounds containing lead or mercury or bismuth, give characteristically coloured encrustations when heated on charcoal with potassium iodide and sulphur. Section A.7 gives details of these and other tests.

Platinum-tipped forceps are useful for holding small splinters of minerals in the blowpipe flame. When substances are examined in this way, the colour of the flame is important, as also is the degree of fusibility of the mineral compared as far as is possible with the *standard scale of fusibility*, discussed in Section 2.6.

A *platinum wire* may be used to give excellent results in nearly all the operations usually carried out with the forceps. When using platinum, whether as foil, wire or metal, care should be taken to avoid supporting minerals or other substances suspected of containing iron, lead, antimony or any other metals which can form alloys with platinum.

Several elements give distinctive colours to the blowpipe flame, and this **flame test** is performed by introducing some of the finely powdered mineral, either by itself or moistened with HCl, into the flame on a platinum wire. The several important **bead tests** are carried out by fusing the mineral with a flux, in a small loop at the end of a platinum wire.

A.4   Fluxes

Certain substances are added to an assay so that fusion can be attained more rapidly than by heating the mineral on its own. Such substances are called **fluxes**, and are particularly useful when the constituents of the assay form a characteristic-coloured compound with these substances. The most important fluxes are borax, microcosmic salt and sodium carbonate.

Borax ($Na_2B_4O_5(OH)_4.8H_2O$) is a hydrous sodium borate. As a blowpipe reagent, the greater part of the attached water in the borax is first driven off by heating, and the reagent is then finely powdered. To make a borax bead, the end of a platinum wire is first formed into a loop, which is then heated to redness in the blowpipe flame and immediately dipped into the powdered borax, some of which adheres to the wire. When the loop is heated again in the blowpipe flame, the powder froths up or intumesces because of the disengagement of the water still remaining in it, and gradually fuses to a clear transparent globule, the **borax head**. The powdered substance to be examined is touched with the hot bead so that a small quantity sticks to it. The bead is then heated by a well sustained blast, and its colour and other characters noted, both when hot and cold, and in both the oxidizing and reducing flames. Some minerals should be added to the bead in very minute quantities, to enable the colour to be noted which otherwise might be masked. Borax changes substances to oxides, and the nature of the substance, to a large extent, can be identified by the colour and other properties of these oxides in the borax bead. Minerals containing arsenic and sulphur dissolve with difficulty in the borax bead, and their behaviour is different from oxides of the same metals. It is therefore advisable to roast the substance on charcoal in the oxidizing flame, before the test is carried out, so that any sulphur or arsenic is volatilized.

Microcosmic salt ($NaNH_4HPO_4.4H_2O$) is a hydrated ammonium hydrogen phosphate, which is so fluid when it is first fused that it usually drops from the platinum wire. Microcosmic salt should therefore either be heated on charcoal or platinum foil until the water and ammonia are expelled, and then taken up on a small platinum loop, or added to the loop in small quantities at a time until the complete microcosmic salt bead is formed. The substance under examination is added in the same way as with the borax bead test, and the whole fused in the blowpipe flame.

The action of microcosmic salt is to convert the oxides of metals into complex phosphates, imparting characteristic colours to the bead both when hot and when cold; these colours often differing depending upon whether the bead is produced in the oxidizing or reducing flame. Silica is insoluble in the microcosmic salt bead, so that when silicates are dealt with, a silica skeleton appears in the bead.

Sodium carbonate ($Na_2CO_3.10H_2O$) is used in the reduction of oxides or sulphides of metals to the metallic state. The mineral under examination is finely powdered and intimately mixed with sodium carbonate and charcoal. The mixture is moistened slightly, placed in a hollow on charcoal and heated in the blowpipe flame. The powdered mineral should amount to about one-third of the total mixture.

Sodium carbonate is valuable as a flux in the analysis of silicates as it then parts with carbonic acid and is converted into sodium silicate.

Manganese and chromium give characteristic colours when introduced into the sodium carbonate bead, due to the formation of sodium manganate and sodium chromate.

Sodium sulphide is formed from a sulphate by fusing the powdered mineral sulphate with sodium carbonate and charcoal, on charcoal. When placed on a silver coin and moistened, the fused mass gives a black stain of silver sulphide. Mineral sulphides give the same reaction but can be distinguished from sulphates by other tests.

A.5 Tube tests

Reactions using **closed** and **open tubes** are of great importance in blowpipe analysis. The closed tube consists of narrow, soft tubing cut into 50–80 mm lengths and sealed at one end. In the closed tube, the assay is heated virtually out of contact with the oxygen of the air. In the open tube, which consists of hard glass tubing of 100–120 mm lengths and open at both ends, heating takes place in a stream of hot air and oxidation results.

A small quantity of the powdered assay is introduced into the closed tube and heated. In many cases a deposit called the **sublimate** is formed on the cooler parts of the tube, and the colour and nature of this sublimate may give an indication of one or more of the elements present in the assay. Water driven out of the assay collects as droplets towards the mouth of the tube. The assay may be converted by heat into the oxides of the metals present, and some of these oxides have characteristic colours and properties. Thus, for example, brown limonite (hydrated iron oxide; $FeO(OH).nH_2O$) is converted into black magnetic oxide (Fe_3O_4) by the expulsion of water, which collects on the cooler parts of the tube.

With the open tube, the assay is placed towards one end and the tube inclined. The assay is thus heated in a current of air and is oxidized; characteristic *smells* or *sublimates* are formed.

A.6 Reactions

The detection of several of the acid radicles present in minerals depends on the use of reagents, such as the usual acids, powdered magnesium, granulated tin, etc. For instance, carbonates give off carbon dioxide (CO_2) on being treated with HCl, and some silicates gelatinize on being heated with the same acid. These, and other reactions, are given in the following tables.

A.7 Tables of blowpipe analyses

Flame test

The substance, either alone or moistened with HCl, HNO_3 or H_2SO_4, is heated on a clean platinum wire, and colours the outer part of the blowpipe flame.

calcium	brick red
strontium	crimson
lithium	deep crimson
sodium	yellow
potassium	violet (masked by sodium, use blue glass filter to view flame)
barium	yellow–green
copper	emerald green with HNO_3: sky blue with HCl
thallium	bright green
boron	momentary yellow–green with H_2SO_4

Indefinite blue flames are given by lead, arsenic and antimony; and indefinite green flames by zinc, phosphorus and molybdenum. These elements are more satisfactorily detected by other tests.

Borax bead test

Element	Oxidizing flame	Reducing flame
iron	yellow hot, colourless cold	bottle green
copper	blue	opaque red
chromium	yellowish-green	emerald green
manganese	reddish-violet	colourless
cobalt	deep blue	deep blue
nickel	reddish-brown	opaque grey
uranium	yellow	pale green

Microcosmic salt bead test

iron	colourless to brownish-red	reddish
copper	blue	opaque red
chromium	red when hot, green cold	green
manganese	violet	colourless
cobalt	blue	blue
nickel	yellow	reddish-yellow
uranium	yellow when hot, yellow–green cold	yellow–green hot, bright green cold
tungsten	colourless	blue–green
molybdenum	bright green	green
titanium	colourless	yellow hot, violet cold
silica	remains undissolved in microcosmic bead	
chlorine	saturate microcosmic salt bead with copper oxide; if a powdered chloride is added, a rich blue flame surrounds the bead	

Sodium carbonate bead test

Element	Oxidizing flame
manganese	opaque blue–green
chromium	opaque yellow–green

Reactions on charcoal

Oxidation

(a) Substance heated alone in oxidizing flame on charcoal:

Element	Encrustation or smell
arsenic	white, far from assay; smell garlic
antimony	white, near assay
zinc	yellow when hot, white when cold
lead	dark yellow when hot, yellow when cold
bismuth	dark orange when hot, paler when cold
sulphur	smell of sulphur dioxide
tin	yellow when hot, paler or colourless when cold
molybdenum	yellow when hot, yellow or colourless when cold; in reducing flame, blue

(b) White encrustations and residues, obtained from (a) above, moistened with cobalt nitrate and strongly reheated:

Element, etc.	Colour
zinc	encrustation grass green
tin	encrustation blue–green
antimony	encrustation dirty green
magnesium	residue pink
aluminium	residue blue and unfused
fusible silicates, phosphates and borates	residue blue fused and glassy-looking

(c) Substance heated in oxidizing flame with potassium iodide and sulphur:

Element	Encrustation
lead	brilliant yellow
bismuth	scarlet; yellow near assay
mercury	greenish-yellow; and greenish-yellow fumes

Reduction

(a) Substance mixed with powdered charcoal and sodium carbonate and heated in oxidizing flame:

Element	Bead or residue obtained
lead	soft malleable metallic bead; easily fused; marks paper
tin	tin white bead, soft and malleable; not marking paper
silver	silver white malleable bead
gold	yellow bead, soft and malleable
bismuth	silver white bead, brittle
copper	red spongy mass
iron	residue strongly magnetic
cobalt	residue feebly magnetic
nickel	residue feebly magnetic

(b) *Special reduction tests for titanium and tungsten.* Substance fused with powdered charcoal and sodium carbonate; the residue boiled with hydrochloric acid and few grains of granulated tin:

Metal	Colour of solution
titanium	violet
tungsten	Prussian blue

Substance fused with powdered charcoal and sodium carbonate; the residue dissolved in concentrated sulphuric acid with an equal volume of water added; the solution is cooled, water added, and then hydrogen peroxide added:

Metal	Colour of solution
titanium	amber

Closed tube test

Assay heated in closed tube, either alone, or with sodium carbonate and powdered charcoal, or with magnesium:

Element, etc.	Observation
sulphur	orange sublimate
arsenic	black sublimate; smell of garlic
mercury, with sulphur	black sublimate, red on rubbing
arsenic, with sulphur	reddish-yellow sublimate, deep red while liquid
antimony, with sulphur	brownish-red sublimate, black while hot
water	colourless drops
mercury	heat with sodium carbonate and charcoal, globules of mercury as sublimate
arsenic	heat with sodium carbonate and charcoal; black mirror of arsenic, soluble in sodium hypochlorite
phosphates	heat with magnesium and add water; characteristic smell of phosphoretted hydrogen

Open tube test

Assay heated in open tube:

Element	Observation
sulphur	sulphurous fumes of sulphur dioxide
arsenic	white sublimate, crystalline, volatile, far from assay; smell of garlic
antimony	white sublimate near assay
tellurium	whitish sublimate, fusible to colourless drops

Reactions for acid radicle

Acid radicle	Test
carbonate	With hydrochloric acid, carbon dioxide evolved, turning lime-water milky
sulphides (some)	With hydrochloric acid, sulphuretted hydrogen evolved. Also indicated by closed tube, open tube and charcoal tests, q.v.
fluoride	With strong sulphuric acid, greasy bubbles of hydrofluoric acid evolved, causing deposition of a white film on silica on a drop of water held at the mouth of the tube
chloride	With sulphuric acid and manganese dioxide, greenish chlorine evolved. Also detected by microcosmic salt bead saturated with copper oxide, q.v.
bromide	With sulphuric acid and manganese dioxide, brown bromine evolved
iodide	With sulphuric acid and manganese dioxide, violet iodine evolved
nitrate	With sulphuric acid, brown nitrous fumes evolved
silicates (some)	With hydrochloric acid, gelatinize. Silica skeleton in microcosmic salt bead
sulphate	Heat substance on charcoal with sodium carbonate and powdered charcoal; place residue on silver coin and moisten. Black stain indicates sulphate (or sulphide)
phosphate	Heat with magnesium in closed tube, add water; phosphoretted hydrogen evolved. Also detected by giving a fused blue mass when heated on charcoal, moistened with cobalt nitrate and strongly reheated
telluride	Heat powdered mineral with a little strong sulphuric acid; reddish-violet solution; colour disappears on adding water to the cold solution and a grey precipitate is deposited

Summary of tests for metals

aluminium	Heated on charcoal, moistened with cobalt nitrate, strongly reheated, blue unfused residue
antimony	Roasted on charcoal, white encrustation near assay. Heated in open tube, white sublimate near assay. Heated in closed tube, red–brown sublimate; black when hot
arsenic	Roasted on charcoal, white encrustation far from assay; garlic smell. Heated in open tube, white volatile sublimate. Heated in closed tube with sodium carbonate, black arsenic mirror, soluble in sodium hypochlorite
barium	Flame test, yellow–green
bismuth	Reduction on charcoal, brittle bead. Roasted with potassium iodide and sulphur, yellow encrustation near assay; outer parts scarlet
calcium	Flame test, brick red
cadmium	Heated on charcoal with sodium carbonate, reddish-brown sublimate
chromium	Borax bead, green; microcosmic salt bead, green; sodium carbonate bead, yellow–green, opaque
cobalt	Borax bead, deep blue; microcosmic salt bead, deep blue
copper	Flame test, emerald green with nitric acid, sky blue with hydrochloric acid. Borax bead, blue in oxidizing flame; opaque red in reducing flame. Reduction on charcoal, red metallic copper
gold	Reduction on charcoal, soft malleable gold bead
iron	Borax bead, yellow hot, colourless cold, in oxidizing flame; bottle green in reducing flame. Reduction on charcoal, magnetic residue
lead	Reduction on charcoal, malleable metallic bead, marking paper. Roasted with potassium iodide and sulphur, brilliant yellow encrustation
lithium	Flame test, deep crimson; deeper than strontium flame
magnesium	Heated on charcoal, moistened with cobalt nitrate, strongly reheated, pink residue
manganese	Borax bead, reddish-violet in oxidizing flame; colourless in reducing flame. Microcosmic bead, violet in oxidizing flame; colourless in reducing flame. Sodium carbonate bead, blue–green opaque
mercury	Heated on charcoal with potassium iodide and sulphur, greenish-yellow encrustation and greenish-yellow fumes. Heated in closed tube with sodium carbonate and charcoal, globules of mercury as sublimate

molybdenum	Microcosmic salt bead, bright green in oxidizing flame; dirty green hot, fine rich green cold, in reducing flame. Roasted on charcoal, yellow hot, yellow or colourless cold; in reducing flame, blue
nickel	Borax bead, reddish-brown in oxidizing flame; opaque grey in reducing flame
potassium	Flame test, violet, view through blue glass filter
silver	Reduction on charcoal, silver bead
sodium	Flame test, yellow
strontium	Flame test, crimson
tellurium	Heated in open tube, whitish sublimate, fusible to colourless drops. Heated with strong sulphuric acid, reddish-violet solution
thallium	Flame test, bright green
tin	Reduction on charcoal, tin bead
titanium	Microcosmic salt bead, yellow hot, violet cold, in reducing flame. Reduction with tin, violet solution. Hydrogen peroxide test, amber solution
tungsten	Microcosmic salt bead, blue–green in reducing flame. Reduction with tin, blue solution
uranium	Microcosmic salt bead, yellow hot, yellow–green cold, in oxidizing flame; yellow–green hot, bright green cold, in reducing flame
zinc	Roasted on charcoal, encrustation yellow when hot; white when cold. Heated on charcoal, moistened with cobalt nitrate, and strongly reheated, grass green encrustation

Appendix B
Hydrocarbons

B.1 Introduction

Many substances are included in this appendix, differing from each other in mode of occurrence, physical properties and chemical composition, but with each consisting mainly of carbon, and also containing oxygen and hydrogen. The substances are considered in two groups: **coals** and **bitumens**.

B.2 Coals

The name 'coal' is applied to a number of different substances largely made up of carbon, hydrogen and oxygen, all of which more or less represent the altered remains of land vegetation transformed by slow chemical changes (principally the elimination of oxygen and hydrogen from the original woody tissue), into a material containing a higher percentage of carbon. Two hypotheses can explain the origin of coal: (1) *growth in place* considers coal to result from the decay of vegetable matter *in situ*, and this explains the origin of pure well-bedded extensive coals, such as bituminous coal; while (2) *drift* has been advanced to explain the formation of impure current-bedded coals, such as some types of cannel coal, which are thought to be the result of drifted vegetable manner becoming buried in a delta or estuary. According to the geological history it has undergone, coal contains varying amounts of carbon, oxygen, hydrogen and nitrogen, with the least altered varieties such as lignite containing large amounts of the gaseous elements, and the most altered varieties such as anthracite containing as much as 95% carbon. Several varieties of coal can be distinguished, and typical chemical analyses of the main varieties are given in Table B.1. Whatever the origin of coal is, its final character has been largely influenced by the processes of organic decay.

Coal-bearing seams may be of differing geological ages. Most coals are of Carboniferous age, and the coal-bearing strata of the UK, Pennsylvania, West Germany, India, etc. are of this age. In Britain, the Carboniferous rocks have been gently folded and, although the coal-bearing beds have been removed from the crest of the folds by subsequent erosion, they have been preserved in the troughs or coal basins. Examples of such basins are

Table B.1 Composition of types of coals (expressed as percentages).

	C	O	H	N
wood	49.65	43.20	6.23	0.92
peat	55.44	35.56	6.28	1.72
lignite	72.95	20.50	5.24	1.31
bituminous coal	84.24	8.69	5.55	1.52
anthracite	93.50	2.72	2.81	0.97

afforded by the Lancashire coalfield and the Yorkshire coalfield, which occupy downfolded troughs on either side of the Pennines.

Other extensive coals also occur, however, which are of different geological ages from the Carboniferous, although these deposits are usually less valuable both in quality and thickness. A small Jurassic coal seam was recently worked at Brora in the far north of Scotland, and Cretaceous coals occur in the USA and Europe. In some countries Tertiary coals have also been worked.

Coals are classified or *ranked* in various ways, including:

(a) The *fuel ratio*, which is the ratio of fixed carbon to volatiles.
(2) The *calorific value*, which is the amount of heat produced by the burning of a standard unit of coal. This used to be measured in British thermal units, but now is measured in *joules* (1 Btu = 1055.16 joules.
(3) *Carbon content, content of volatiles* and *water content*.
(4) The *nature of coke* produced by the coal.

From the above, Table B.2 can be compiled, and is shown overleaf.

The main varieties of coal are described below.

PEAT

Peat results from the accumulation of vegetable matter such as mosses and other bog plant, and forms extensive deposits in Eire, the USSR, the USA and elsewhere. Its organic nature is evident throughout the entire deposit, although the bottom layers may become compressed through time into a compact, homogenous substance with an increased carbon content.

LIGNITE, BROWN COAL

Lignite marks a further stage in the alteration of the vegetable matter which, although compact and possessing a brilliant lustre, still contains impressions and remains of fragments of vegetation, leaves, etc. Lignites

Table B.2 Coal classification based on calorific value, carbon content, water content and percentage of volatiles present.

Coal type	Calorific value (MJ)[a]	Carbon (%)	Water (%)	Volatiles (%)
lignite	7.4–11.6	45–65	>20	
lignitic or semi-bituminous	10.5–13.7	60.75	6–20	
cannel	12.7–16.9			30–40
low-carbon bituminous	12.7–14.8	70–80	< 6	<35
bituminous	14.8–16.9	75–90	—	12–26
anthracitic and high-carbon bituminous	16.0–16.9	80–90	—	12–15
semi-anthracite	15.8–16.4	90–93	—	7–12
anthracite	15.3–15.9	93–95	—	3– 5

[a] MJ = megajoules; i.e. 10^6 joules.

contain a large amount of water (more than 20%, Table B.2), and may change to powder on drying. Lignite beds are found at several horizons in more recent geological formations, as in East Germany (see Chapter 7 under Carbon; East Germany produces 300 Mt per year of lignite or soft coal), the USSR, West Germany, Czechoslovakia and Yugoslavia. The name **brown coal** is often restricted to a coal in which the evidence of vegetable matter is not so obvious as in lignite. **Jet** is a resinous, hard, black variety of lignite, capable of taking a high polish, and therefore suitable for ornaments. It is found at Whitby in Yorkshire and elsewhere.

CANNEL COAL

Cannel is a variety of coal which ignites in a candle flame, and burns with a smoky flame. It is one of the bituminous coals, but differs from the usual types in its composition, lustre, fracture and colour. It is dense, has no lustre, has a conchoidal fracture, is dull grey or black and contains a large amount of gas. Microscopic examination reveals that cannel is typically composed of spore and pollen remains, with an abundance of those of oil-bearing algae. On distillation, cannel produces a large amount of volatiles (30–40%), and is valuable for the amount of different oils it produces. **Torbanite**, or **boghead coal**, is a variety of cannel arising from the deposition of vegetable matter in lakes. It is found at Torban and Boghead, and other localities in the Central Valley of Scotland, and forms lenticular deposits in New South Wales. Its exploitation in Scotland gave rise to a celebrated lawsuit which involved an accurate description of the term 'coal'.

BITUMINOUS COALS
SG 1.14–1.40

Bituminous coals vary considerably in character, but they all burn with a smoky flame, and during combustion soften and swell in a manner resembling the fusion of pitch or bitumen. This, however, is merely the first stage in the process of their destructive distillation, and there is no bitumen present. They all have a bright, pitchy lustre, and different varieties are distinguished by their manner of burning such as, for example, coking and non-coking coal. Most household coal is bituminous coal, which usually shows banding parallel to its bedding. The bedding planes are marked by a soft, powdery, charcoal-like material called **fusain**; other bands parallel to the fusain layers include **durain**, which is hard and dull, **clarain**, which has a brighter lustre, and **vitrain**, which occurs as bright, glassy-looking streaks. All these layers should be identified in the hand specimen.

ANTHRACITE
SG 1.0–1.8   **HD** 0.5–2.5

Anthracite is black or brownish black in colour, and sometimes iridescent. It has a black streak and does not soil the fingers. It has a brilliant lustre and breaks with a conchoidal or uneven fracture.

It is less easily set alight than other coals, and burns with little flame and virtually no smoke, which explains its importance as a fuel in 'smokeless zones', and during its combustion it gives out much heat. Passages from ordinary coal into anthracite have been recorded, and anthracite usually occurs where coal-bearing strata have been subjected to increased temperatures and pressures. There are exceptions, and some anthracites may be formed by the alteration of the vegetable matter before its entombment. Anthracite occurs in the coalfields of South Wales, Scotland and Pennsylvania.

B.3 Bitumens

Bitumens are essentially hydrocarbons belonging to both the paraffin series, C_nH_{2n+2}, and the napthalene series, C_nH_{2n}. Different proportions of these give different bitumens, although each bitumen may contain smaller amounts of allied hydrocarbon series. The bitumens include liquid, light yellow oils, with an SG of 0.771, solid asphalts, and waxy materials such as ozokerite. The various bitumens are now described.

Table B.3 The geological ages of some of the world's oilfields.

Age	Oilfields
Tertiary	Middle East (Saudi Arabia, Iraq, Iran, Gulf States, etc.) California, Rumania, Venezuela, Mexico, Burma, southern USSR oilfields
Mesozoic	North Sea, Texas, Wyoming, Galicia
Upper Palaeozoic	Texas, Oklahoma, Kansas, Pennsylvania, Illinois, Canada
Lower Palaeozoic	Indiana

CRUDE PETROLEUM, NAPTHA, MINERAL OIL

Petroleums include brownish or blackish liquids, often with a greenish tinge, generally lighter than water, and with a powerful smell. Fractional distillation produces various economic oils such as petroleum spirit, benzene, etc. The light products are used in the internal combustion engine, the intermediate products for lighting, and the heavy products for lubrication and fuel oils. Crude oil is the basis of the petrochemical industry.

Crude oil is a mineral of organic origin found in *host* rocks (sandstones, etc.) which have been folded or faulted so that the oil has accumulated, and then been trapped within the host rock by overlying impervious rock layers, which prevent the oil migrating. The presence of crude oil may be shown by either oil seepage at the surface, or the presence of bitumen or pitch deposits caused by the evaporation and oxidation of volatile hydrocarbons. Table B.3 gives the ages of the main oilfields of the world.

ASPHALT, asphaltum, mineral pitch

Asphalt is a mixture of different hydrocarbons, black or brownish in colour, usually soft, but sometimes hard with a conchoidal fracture. In suitable solvents such as carbon bisulphide, asphalt may be dissolved either into various hydrocarbons, or into a non-bituminous organic substance containing any inorganic matter present in the crude material.

Asphalt formed from the oxidation of crude oil occurs in quantity in the famous pitch lakes of Trinidad, and also in Alberta, Canada, Venezuela and Cuba. Asphalt has been found in various localities in England, as at Castleton (Derbyshire), Pitchford near Shrewsbury, and Stanton Harold in Leicestershire where it is found encrusting crystals of galena and chalcopyrite; but all these occurrences have no commercial value whatsoever. In some occurrences porous sedimentary rocks such as sandstones become impregnated with the asphalt, and from these the asphalt can be extracted.

Asphalt is a particularly important material since most road surfaces are composed of aggregate, which may be either crushed rock or sand and gravel of a particular grading, mixed with hot asphalt or *bitumen* (as it is known commercially), and this *coated material* is then spread over the bottom layers of a road, with a top layer, called the *wearing course*, finally being laid down.

ELATERITE, elastic bitumen, mineral caoutchouc

Elaterite is a soft, elastic, solid bitumen, not unlike rubber in its physical properties. It has been reported from Castleton (Derbyshire), Neufchatel and elsewhere.

ALBERTITE, gilsonite, grahamite, uintaite, wurzillite

All of these substances are varieties of solid bitumen which differ slightly in their physical and chemical properties. They usually occur infilling fissures, as in New Brunswick, and have been derived from oil-bearing rocks.

OZOKERITE

Ozokerite resembles beeswax in appearance. It is dark yellow or brownish in colour, often with a greenish opalescence. It is found associated with crude oil in Utah, Moldavia and Galicia, where it was mined. In Galicia, ozokerite has been squeezed into fractures, where it forms vein-like bodies. Material oozing upwards from depth refills these fractures as the ozokerite is removed. When purified it forms **ceresine**, which is used in the manufacture of candles.

HATCHETTINE

This is a colourless or yellowish, soft, waxy substance, resembling ozokerite. It has been found in cracks in ironstone nodules at Merthyr Tydfil in Wales.

AMBER

SG 1.1   **HD** 2.0–2.5   **RI** 1.54

Amber is a fossil resin much used for beads and pendants, and in ornaments, although this latter use is found only in objects of great antiquity. Amber varies in colour from deep orange–yellow to pale yellow, and to sometimes white. It is often cloudy, and contains leaves, fossil insects, etc. When heated, amber leaves a black residue which is used in the manufacture of the best varnishes. Amber occurs as irregular nodules in recent sediments, deposited under estuarine (shallow-water) conditions, and is worked commercially on the southern coast of the Baltic.

COPALITE, Highgate resin

Copalite is a pale yellow or brownish, waxy substance, found in small films or fragments in the London Clay, at Highgate Hill, London. It burns easily with a smoky flame, and leaves little ash.

GUM COPAL

Gum copal is resin found buried in modern sands, as in New Zealand. It is of inferior quality to amber.

Bibliography

Ahrens, L. H. 1952. The use of ionisation potentials, Part 1. *Geochimica et Cosmochimica Acta* **2**, 155–69.

Best, M. G. 1982. *Igneous and metamorphic geology*. New York: W. H. Freeman.

Bravais, A. 1848. In *Ostwald's Klassiker der exakten Wissenschaften* **90** (1897).

Bloss, F. D. 1971. *Crystallography and crystal chemistry*. New York: Holt, Rinehart & Winston.

Deer, W. A., R. A. Howie & J. Zussman 1966. *An introduction to the rock-forming minerals*. London: Longman.

Deer, W. A., R. A. Howie & J. Zussman 1978 *et seq. Rock-forming minerals* (various volumes). London: Longman.

Dixon, C. J. 1979. *Atlas of economic mineral deposits*. London: Chapman & Hall.

Evans, A. M. 1987. *An introduction to ore geology*, 2nd edn. Oxford: Blackwell.

Gillen, C. 1982. *Metamorphic geology*. London: Allen & Unwin.

Goldschmidt, V. M. 1937. The principles of distribution of chemical elements in minerals and rocks. *Journal of the Chemical Society of London* **1937**, 655–73.

Henderson, P. 1982. *Inorganic chemistry*. Oxford: Pergamon Press.

Lindgren, W. 1933. *Mineral deposits*, 2nd edn. New York: McGraw-Hill.

Mason, B. 1966. *Principles of geochemistry*, 3rd edn. New York: John Wiley.

Mining Journal 1980 *et seq. Mining annual review.*

Miyashiro, A. 1973. *Metamorphism and metamorphic belts*. London: Allen & Unwin.

Palache, C., H. Berman & C. Frondel 1944. *Dana's system of mineralogy*. New York: John Wiley.

Pauling, L. 1960. *The nature of the chemical bond*, 3rd edn. Ithaca, New York: Cornell University Press.

Phillips, F. C. 1963. *An introduction to crystallography*, 3rd edn. London: Longman.

Phillips, W. R. & D. T. Griffin 1981. *Optical mineralogy: the non-opaque minerals*. New York: W. H. Freeman.

Ringwood, A. E. 1955. Principles governing trace element distribution during magmatic crystallization, I. The influence of electronegativity. *Geochimica et Cosmochimica Acta* **7**, 189–202.

Stanton, R. L. 1972. *Ore petrology*. New York: McGraw-Hill.

Vernon, R. H. 1976. *Metamorphic processes*. London: Allen & Unwin.

Index